Plant Electrophysiology

Alexander G. Volkov
Editor

Plant Electrophysiology

Signaling and Responses

Springer

Editor
Alexander G. Volkov
Department of Chemistry
Oakwood University
Adventist Blvd. 7000
Huntsville, AL 35896
USA

ISBN 978-3-642-29109-8   ISBN 978-3-642-29110-4 (eBook)
DOI 10.1007/978-3-642-29110-4
Springer Heidelberg New York Dordrecht London

Library of Congress Control Number: 2012937217

© Springer-Verlag Berlin Heidelberg 2012
This work is subject to copyright. All rights are reserved by the Publisher, whether the whole or part of the material is concerned, specifically the rights of translation, reprinting, reuse of illustrations, recitation, broadcasting, reproduction on microfilms or in any other physical way, and transmission or information storage and retrieval, electronic adaptation, computer software, or by similar or dissimilar methodology now known or hereafter developed. Exempted from this legal reservation are brief excerpts in connection with reviews or scholarly analysis or material supplied specifically for the purpose of being entered and executed on a computer system, for exclusive use by the purchaser of the work. Duplication of this publication or parts thereof is permitted only under the provisions of the Copyright Law of the Publisher's location, in its current version, and permission for use must always be obtained from Springer. Permissions for use may be obtained through RightsLink at the Copyright Clearance Center. Violations are liable to prosecution under the respective Copyright Law.
The use of general descriptive names, registered names, trademarks, service marks, etc. in this publication does not imply, even in the absence of a specific statement, that such names are exempt from the relevant protective laws and regulations and therefore free for general use.
While the advice and information in this book are believed to be true and accurate at the date of publication, neither the authors nor the editors nor the publisher can accept any legal responsibility for any errors or omissions that may be made. The publisher makes no warranty, express or implied, with respect to the material contained herein.

Printed on acid-free paper

Springer is part of Springer Science+Business Media (www.springer.com)

Preface

Plant electrophysiology is the study of the electrochemical phenomena associated with biological cells and tissues in plants. It involves measurements of electrical potentials and currents on a wide variety of scales from single ion channels to whole plant tissues. Electrical properties of plant cells mostly derive from the electrochemical properties of their membranes. Electrophysiological study of plants includes measurements of the electrical activity of the phloem, xylem, plasmodesmata, stomata, and particularly the electrical signal's propagation along the plasma membrane. Action potentials are characteristic responses of excitation that can be induced by stimuli such as: applied pressure, chemical substances, thermal stimuli, electrical or magnetic stimuli, and mechanical stimuli.

There are two major divisions of electrophysiology: intracellular recording and extracellular recording.

The electrical phenomena in plants have attracted researchers since the eighteenth century and have been discussed in a variety of books (Baluška et al. 2006; Bertholon 1783; Bose 1907, 1913, 1918, 1926, 1928; Lemström 1902; Ksenzhek and Volkov 1998, 2006; Volta 1816). The identification and characterization of bioelectrochemical mechanisms for electrical signal transduction in plants would mark a significant step forward in understanding this underexplored area of plant physiology. Although plant mechanical and chemical sensing and corresponding responses are well known, membrane electrical potential changes in plant cells and the possible involvement of electrophysiology in transduction mediation of these sense-response patterns represents a new dimension of plant tissue and whole organism integrative communication. Plants continually gather information about their environment. Environmental changes elicit various biological responses. The cells, tissues, and organs of plants possess the ability to become excited under the influence of certain environmental factors. Plants synchronize their normal biological functions with their responses to the environment. The synchronization of internal functions, based on external events, is linked with the phenomenon of excitability in plant cells. The conduction of bioelectrochemical excitation is a fundamental property of living organisms.

Electrical impulses may arise as a result of stimulation. Once initiated, these impulses can propagate to adjacent excitable cells. The change in transmembrane potential can create a wave of depolarization which can affect the adjoining resting membrane. Action potentials in higher plants are the information carriers in intracellular and intercellular communication during environmental changes.

The conduction of bioelectrochemical excitation is a rapid method of long distance signal transmission between plant tissues and organs. Plants promptly respond to changes in luminous intensity, osmotic pressure, temperature, cutting, mechanical stimulation, water availability, wounding, and chemical compounds such as herbicides, plant growth stimulants, salts, and water potential. Once initiated, electrical impulses can propagate to adjacent excitable cells. The bioelectrochemical system in plants not only regulates stress responses, but photosynthetic processes as well. The generation of electrical gradients is a fundamental aspect of signal transduction.

The first volume entitled "Plant Electrophysiology—Methods and Cell Electrophysiology" consists of a historical introduction to plant electrophysiology and two parts. The first part introduces the different methods in plant electrophysiology. The chapters present methods of measuring the membrane potentials, ion fluxes, trans-membrane ion gradients, ion-selective microelectrode measurements, patch-clamp technique, multi-electrode array, electrochemical impedance spectroscopy, data acquisition, and electrostimulation methods. The second part deals with plant cell electrophysiology. It includes chapters on pH banding in Characean cells, effects of membrane excitation and cytoplasmic streaming on photosynthesis in Chara, functional characterization of plant ion channels, and mechanism of passive permeation of ions and molecules through plant membranes.

The second volume entitled "Plant Electrophysiology—Signaling and Responses" presents experimental results and theoretical interpretation of whole plant electrophysiology. The first three chapters describe electrophysiology of the Venus flytrap, including mechanisms of the trap closing and opening, morphing structures, and the effects of electrical signal transduction on photosynthesis and respiration. The Venus flytrap is a marvelous plant that has intrigued scientists since the times of Charles Darwin. This carnivorous plant is capable of very fast movements to catch insects. The mechanism of this movement has been debated for a long time. The Chap. 4 describes the electrophysiology of the Telegraph plant. The role of ion channels in plant nyctinastic movement is discussed in Chap. 5. Electrophysiology of plant–insect interactions can be found in Chap. 6. Plants can sense mechanical, electrical, and electromagnetic stimuli, gravity, temperature, direction of light, insect attack, chemicals and pollutants, pathogens, water balance, etc. Chapter 7 shows how plants sense different environmental stresses and stimuli and how phytoactuators response to them. This field has both theoretical and practical significance because these phytosensors and phytoactuators employ new principles of stimuli reception and signal transduction and play a very important role in the life of plants. Chapters 8 and 9 analyze generation and transmission of electrical signals in plants. Chapter 10 explores bioelectrochemical aspects of the plant-lunisolar gravitational relationship. Authors of Chap. 11

describe the higher plant as a hydraulic-electrochemical signal transducer. Chapter 12 discusses properties of auxin-secreting plant synapses. The coordination of cellular physiology, organ development, life cycle phases and symbiotic interaction, as well as the triggering of a response to changes is the environment in plants depends on the exchange of molecules that function as messengers. Chapter 13 presents an overview of the coupling between ligands binding to a receptor protein and subsequent ion flux changes. Chapter 14 summarizes data on physiological techniques and basic concepts for investigation of Ca^{2+}-permeable cation channels in plant root cells.

All chapters are comprehensively referenced throughout.

Green plants are a unique canvas for studying signal transduction. Plant electrophysiology is the foundation of discovering and improving biosensors for monitoring the environment; detecting effects of pollutants, pesticides, and defoliants; monitoring climate changes; plant-insect interactions; agriculture; and directing and fast controlling of conditions influencing the harvest.

We thank the authors for the time they spent on this project and for teaching us about their work. I would like to thank our Acquisition Editor, Dr. Cristina Eckey, and our Production Editor, Dr. Ursula Gramm, for their friendly and courteous assistance.

Prof. Alexander George Volkov Ph.D.

References

Baluška F, Mancuso S, Volkmann D (2006) Communication in plants. Neuronal aspects of plant life. Springer, Berlin.

Bertholon M (1783) De l'electricite des vegetaux: ouvrage dans lequel on traite de l'electricite de l'atmosphere sur les plantes, de ses effets sur leconomie des vegetaux, de leurs vertus medico. P.F. Didot Jeune, Paris

Bose JC (1907) Comparative electro-physiology, a physico-physiological study. Longmans, Green & Co., London

Bose JC (1913) Researches on Irritability of Plants. Longmans, London

Bose JC (1918) Life Movements in Plants. B.R. Publishing Corp., Delhi

Bose JC (1926) The Nervous mechanism of plants. Longmans, Green and Co., London

Bose JC (1928) The Motor Mechanism of Plants. Longmans Green, London

Ksenzhek OS, Volkov AG (1998) Plant energetics. Academic Press, San Diego

Lemström S (1902) Elektrokultur. Springer, Berlin

Stern K (1924) Elektrophysiologie der Pflanzen. Springer, Berlin

Volkov AG (ed) (2006) Plant electrophysiology. Springer, Berlin

Volta A (1816) Collez ione dell' opera del cavaliere Conte Alessandro Volta, vol 1. Nella stamperia di G. Piatti, Firence

Contents

1   **Morphing Structures in the Venus Flytrap** 1
     Vladislav S. Markin and Alexander G. Volkov

2   **The Effect of Electrical Signals on Photosynthesis
     and Respiration** 33
     Andrej Pavlovič

3   **Mathematical Modeling, Dynamics Analysis and Control
     of Carnivorous Plants**................................... 63
     Ruoting Yang, Scott C. Lenaghan, Yongfeng Li, Stephen Oi
     and Mingjun Zhang

4   **The Telegraph Plant:** *Codariocalyx motorius* **(Formerly
     Also** *Desmodium gyrans***)**................................ 85
     Anders Johnsson, Vijay K. Sharma and Wolfgang Engelmann

5   **Regulatory Mechanism of Plant Nyctinastic Movement:
     An Ion Channel-Related Plant Behavior** 125
     Yasuhiro Ishimaru, Shin Hamamoto, Nobuyuki Uozumi
     and Minoru Ueda

6   **Signal Transduction in Plant–Insect Interactions:
     From Membrane Potential Variations to Metabolomics** 143
     Simon Atsbaha Zebelo and Massimo E. Maffei

7   **Phytosensors and Phytoactuators** 173
     Alexander G. Volkov and Vladislav S. Markin

| 8 | Generation, Transmission, and Physiological Effects of Electrical Signals in Plants | 207 |

Jörg Fromm and Silke Lautner

| 9 | The Role of Plasmodesmata in the Electrotonic Transmission of Action Potentials | 233 |

Roger M. Spanswick

| 10 | Moon and Cosmos: Plant Growth and Plant Bioelectricity | 249 |

Peter W. Barlow

| 11 | Biosystems Analysis of Plant Development Concerning Photoperiodic Flower Induction by Hydro-Electrochemical Signal Transduction | 281 |

Edgar Wagner, Lars Lehner, Justyna Veit, Johannes Normann and Jolana T. P. Albrechtová

| 12 | Actin, Myosin VIII and ABP1 as Central Organizers of Auxin-Secreting Synapses | 303 |

František Baluška

| 13 | Ion Currents Associated with Membrane Receptors | 323 |

J. Theo M. Elzenga

| 14 | Characterisation of Root Plasma Membrane Ca^{2+}-Permeable Cation Channels: Techniques and Basic Concepts | 339 |

Vadim Demidchik

Index 371

Contributors

Jolana T. P. Albrechtová Institute of Biology II, University of Freiburg, Schänzlestr. 1, 79104 Freiburg, Germany

František Baluška IZMB, University of Bonn, Kirschallee 1, 53115 Bonn, Germany

Peter W. Barlow School of Biological Sciences, University of Bristol, Woodland Road, Bristol BS8 1UG, UK

Vadim Demidchik Department of Physiology and Biochemistry of Plants, Biological Faculty, Belarusian State University, 4 Independence Ave., 220030 Minsk, Belarus

J. Theo M. Elzenga Plant Electrophysiology, University of Groningen, Nijenborgh 7, 9747 AG Groningen, The Netherlands

Wolfgang Engelmann Botanisches Institut, Universität Tübingen, Auf der Morgenstelle 1, 72076 Tübingen, Germany

Jörg Fromm Institute for Wood Biology, Universität Hamburgh, Leuschnerstrasse 91, 21031 Hamburg, Germany

Shin Hamamoto Faculty of Engineering, Tohoku University, 6-3 Aramaki-aza-Aoba, Aoba-Ku, Sendai 980-8578, Japan

Yasuhiro Ishimaru Faculty of Science, Tohoku University, 6-3 Aramaki-aza-Aoba, Aoba-Ku, Sendai 980-8578, Japan

Anders Johnsson Department of Physics, Norwegian University of Science and Technology, 7041 Trondheim, Norway

Silke Lautner Institute for Wood Biology, Universität Hamburgh, Leuschnerstrasse 91, 21031 Hamburg, Germany

Lars Lehner Institute of Biology II, University of Freiburg, Schänzlestr. 1, 79104, Freiburg, Germany

Scott C. Lenaghan Department of Mechanical, Aerospace and Biomedical Engineering, University of Tennessee, Knoxville, TN 37996-2210, USA

Yongfeng Li Division of Space Life Science, Universities Space Research Association, Houston, TX 77058, USA

Massimo Maffei Plant Physiology Unit, Department of Plant Biology, Innovation Centre, University of Turin, Via Quarello 11/A, 10135 Turin, Italy

Stefano Mancuso Department of Plant, Soil and Environment, University of Firenze, Viale delle Idee 30, 50019 Sesto Fiorentino, Italy

Vladislav S. Markin Department of Neurology, University of Texas Southwestern Medical Center, Dallas, TX 75390-8833, USA

Johannes Normann Institute of Biology II, University of Freiburg, Schänzlestr. 1, 79104, Freiburg, Germany

Stephen Oi Department of Mechanical, Aerospace and Biomedical Engineering, University of Tennessee, Knoxville, TN , 37996-2210, USA

Andrej Pavlovič Department of Plant Physiology, Faculty of Natural Sciences, Comenius University in Bratislava, Mlynská dolina B-2, 842 15, Bratislava, Slovakia

Vijay K. Sharma Chronobiology Laboratory, Evolutionary and Organismal Biology Unit, Jawaharlal Nehru Centre for Advanced Scientific Research, Jakkur, PO Box. 6436, Bangalore, Karnataka 560064, India

Roger Spanswick Department of Biological and Environmental Engineering, Cornell University, 316 Riley-Robb Hall, Ithaca, NY 14853-5701, USA

Minoru Ueda Faculty of Science, Tohoku University, 6-3 Aramaki-aza-Aoba, Aoba-Ku, Sendai 980-8578, Japan

Nobuyuki Uozumi Faculty of Engineering, Tohoku University, 6-3 Aramaki-aza-Aoba, Aoba-Ku, Sendai 980-8578, Japan

Justyna Veit Institute of Biology II, University of Freiburg, Schänzlestr. 1, 79104, Freiburg, Germany

Alexander G. Volkov Department of Chemistry and Biochemistry, Oakwood University, 7000 Adventist Blvd., Huntsville, AL 35896, USA

Edgar Wagner Institute of Biology II, University of Freiburg, Schänzlestr. 1, 79104, Freiburg, Germany

Ruoting Yang Institute for Collaborative Biotechnologies, University of California, Santa Barbara, CA 93106-5080, USA

Simon Atsbaha Zebelo Plant Physiology Unit, Department of Plant Biology, Innovation Centre, University of Turin, Via Quarello 11/A, 10135 Turin, Italy

Mingjun Zhang Department of Mechanical, Aerospace and Biomedical Engineering, University of Tennessee, Knoxville, TN , 37996-2210, USA

Chapter 1
Morphing Structures in the Venus Flytrap

Vladislav S. Markin and Alexander G. Volkov

Abstract Venus flytrap is a marvelous plant that intrigued scientists since times of Charles Darwin. This carnivorous plant is capable of very fast movements to catch insects. Mechanism of this movement was debated for a long time. Here, the most recent Hydroelastic Curvature Model is presented. In this model the upper leaf of the Venus flytrap is visualized as a thin, weakly curved elastic shell with principal natural curvatures that depend on the hydrostatic state of the two surface layers of cell, where different hydrostatic pressures are maintained. Unequal expansion of individual layers A and B results in bending of the leaf, and it was described in terms of bending elasticity. The external triggers, either mechanical or electrical, result in the opening of pores connecting these layers; water then rushes from the upper layer to the lower layer, and the bilayer couple quickly changes its curvature from convex to concave and the trap closes. Equations describing this movement were derived and verified with experimental data. The whole hunting cycle from catching the fly through tightening, through digestion, and through reopening the trap was described.

1.1 Introduction

All biological organisms continuously change their shapes both in the animal kingdom and in plant kingdom. These changes include the internal properties of plants. Among them there are interesting examples that are able to morph extremely

---

V. S. Markin (✉)
Department of Neurology, University of Texas Southwestern
Medical Center at Dallas, Dallas, TX 75390-8833, USA
e-mail: markina@swbell.net

A. G. Volkov
Department of Chemistry, Oakwood University, Huntsville,
AL 35896, USA

Fig. 1.1 Venus flytrap in open and closed states

fast. They not only adjust to the changing environment but they also receive signals from the external world, process those signals, and react accordingly. The world "morphing" is defined as efficient, multipoint adaptability and may include macro, micro, structural, and/or fluidic approaches (McGowan et al. 2002).

Some carnivorous plants are able to attack their preys. The most famous of these is the Venus flytrap (*Dionaea muscipula* Ellis). This is a sensitive plant whose leaves have miniature antennae or sensing hairs that are able to receive, process, and transfer information about an insect's stimuli (Fig. 1.1). Touching trigger hairs, protruding from the upper leaf epidermis of the Venus flytrap, activates mechanosensitive ion channels, and generates receptor potentials (Jacobson 1974; Volkov et al. 2008a), which can induce action potentials (Burdon-Sanderson J. 1873; Volkov et al. 2007; Sibaoka 1969; Hodick and Sievers 1988, 1989; Stuhlman and Darden 1950). It was found that two action potentials are required to trigger the trap closing (Brown 1916).

The history of studying the Venus flytrap spans more than a century. Although the sequence of actions is clearly described in the existing literature, the exact mechanism of the trap closure is still poorly understood. Indeed, quite a bit is known about how the flytrap closes: stimulating the trigger hair twice within 40s unleashes two action potentials triggering curvature changes, which helps the plant rapidly close its upper leaf. When trigger hairs in the open trap receive mechanical stimuli, a receptor potential is generated (Benolken and Jacobson 1970; DiPalma et al. 1966). Two mechanical stimuli are required for closing the trap in vivo (Darvin 1875; Lloid 1942). However, at high temperatures (36–40°C) only one stimulus is required for trap closure (Lloyd 1942). Receptor potentials generate action potentials (Jacobson 1974; Volkov et al. 2008a; Burdon-Sanderson J. 1873; Volkov et al. 2007a; Jacobson 1965), which can propagate in the plasmodesmata of the plant to the midrib (Volkov et al. 2007). Uncouplers and blockers of fast anion and potassium channels can inhibit action potential propagation in the Venus flytrap (Volkov et al. 2008c;

Volkov et al. 2007; Hodick and Sievers 1988; Krol et al. 2006). The trap accumulates the electrical charge delivered by an action potential. Once a threshold value of the charge is accumulated, ATP hydrolysis (Jaffe 1973) and fast proton transport starts (Rea 1983, 1984; Williams and Bennet 1982), and aquaporin opening is initiated (Volkov et al. 2008a, 2011). Fast proton transport induces transport of water and a change in turgor (Hodick and Sievers 1989).

A number of contradictory models were proposed (Bobji 2005; Brown 1916; Darwin 1875; Forterre et al. 2005; Hill and Findley 1981; Hodick and Sievers 1989; Jacobson 1974; Nelson and Cox 2005; Williams and Bennet 1982; Yang et al. 2010), and still there is no agreement between the researchers (Hodick and Sievers 1989). Recently, the focus of interest returned to the original ideas proposed by Darwin in the nineteenth century. In his seminal work, Darwin (1875) demonstrated that the basic catching movement of the Venus flytrap involves the transformation of the leaf curvature from convex to concave resulting in the closing of the trap. Darwin wrote: "We know that the lobes, whilst closing, become slightly incurved throughout their whole breadth. This movement appears to be due to the contraction of the superficial layers of cells over the whole upper surface. In order to observe their contraction, a narrow strip was cut out of one lobe at right angles to the midrib, so that the surface of the opposite lobe could be seen in this part when the leaf was shut. After the leaf had recovered from the operation and had re-expanded, three minute black dots were made on the surface opposite to the slit or window, in a line at right angles to the midrib. The distance between the dots was found to be 40/1000 of an inch, so that the two extreme dots were 80/1000 of an inch apart. One of the filaments was now touched and the leaf closed. On again measuring the distances between the dots, the two next to the midrib were nearer together by 1–2/1000 of an inch, and the two further dots by 3–4/1000 of an inch, than they were before; so that the two extreme dots now stood about 5/1000 of an inch (0.127 mm) nearer together than before. If we suppose the whole upper surface of the lobe, which was 400/1000 of an inch in breadth, to have contracted in the same proportion, the total contraction will have amounted to about 25/1000 or 1/40 of an inch (0.635 mm)."

Darwin established that the upper leaf of the Venus flytrap includes two distinct layers of cells at upper and lower surfaces that behave quite differently in the process of trap closure. The finding of these two independent layers was later confirmed by many authors and their role was related to the turgor pressure (Fagerberg and Allain 1991; Fagerberg and Howe 1996; Mozingo et al. 1970). It is well known that some functions in plants and fungi can only be driven by exploiting hydrodynamic flow, such as stomata guard cell opening and closing, leaf pulvini motor organ, mechanical traps of carnivorous plants, and fungal appressorial penetration (Beilby et al. 2006; Shimmen 2006; Zonia and Munnik 2007).

It is common knowledge that the leaves of the Venus flytrap actively employ turgor pressure and hydrodynamic flow for fast movement and catching insects. In these processes the upper and lower surfaces of the leaf behave quite differently. The loss of turgor by parenchyma, lying beneath the upper epidermis, accompanied by the active expansion of the tissues of the lower layers of parenchyma near the lower epidermis closes the trap (Brown 1916; Brown and Sharp 1910; Darwin 1875;

Fig. 1.2 Hydroelastic curvature model

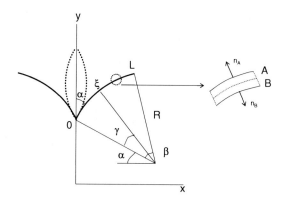

De Candolle 1876; Lloyd 1942; Munk 1876). The cells on the inner face of the trap jettison their cargo of water, shrink, and allow the trap lobe to fold over. The cells of the lower epidermis expand rapidly, folding the trap lobes over (Brown 1916).

Recently, Forterre et al. (2005) reproduced the Darwin (1875) work at the modern technical level with high-speed video imaging and noninvasive microscopy techniques. Forterre et al. (2005) documented in minute details the change of the geometry of the leaf in two dimensions and brought up the idea that elastic energy might play an important role in the closing of the trap.

Recently it was found that the trap of the Venus flytrap can be also closed by electrical stimulation of a midrib (Volkov et al. 2007; Volkov et al. 2008a, b, c, 2009a, b).

1.2 Anatomy and Mechanics of the Trap

It is important to understand the mechanics of the trap closure. One could compare the leaf of this plant to the open book with a fly sitting on the page; the fly can be caught by swift shutting of the book. However, this comparison would be very wrong. In the "book model" there is a pivot at the midrib of the leaf and two flat parts of the book would rotate around this pivot and crush the poor fly. Actual closing of the trap occurs in a different way. The midrib is not a pivot.

The cross-section of the leaf is presented in Fig. 1.2. In the open state (cocked state) the lobes of the leaf have the convex shape (when looking from above). Angle α is the initial angle between the lobe and the vertical line at the midrib. The total angle between two lobes at the midrib is 2α. This angle does not change (at least does not change noticeably) in the process of trap closing (Fagerberg and Allain 1991). The lobes do not rotate around the midrib, but only change their curvature. As a result the distant parts of the leaf move in the space and approach each other—the trap closes. Every point ξ of the lobe moves with different velocity $v(\xi, t)$.

In this text we shall designate the width of the lobe by L. The initial radius of curvature is R, ξ is an arbitrary point along the leaf with corresponding angle γ. We measured a number of typical leafs of Venus flytraps to find the following averaged parameters: $\alpha = 34° = 0.593$ rad, length $L = 2$ cm. These two parameters remain constant. The angle at the center of curvature is β. It changes in the process of closing; its initial value is $\beta_1 = 52°$, initial radius of curvature is $R_1 = 2.2$ cm, or curvature $C_1 = 0.454$ cm^{-1}. After closing angle β changes to $\beta_2 = -2\alpha = -1.186$ rad, $C_2 = -0.593$ cm^{-1}.

1.3 The Hydroelastic Curvature Model of Venus Flytrap

As we previously mentioned, there is a number of models attempting to explain the fast morphing in this plant. In very detailed work, Forterre et al. (2005) reproduced the Darwin (1875) work with high-speed video imaging and documented, in minute details, the change of the geometry of the leaf in two dimensions and brought up the idea that elastic energy might play an important role in the closing of the trap. They described the kinematics of closing of the Venus flytrap as a relaxation of the leaf to the new equilibrium state after triggering. They wrote: "Upon stimulation, the plant 'actively' changes one of its principal natural curvatures, the microscopic mechanism for which remains poorly understood." So, the mechanism of active change of one of principal curvatures of the leaf remained beyond the scope of their work.

This issue was addressed by Markin et al. (2008) who tried to elucidate what causes the change of spontaneous curvature of the leaf. The accumulated data suggest that elastic energy does play an important role, but driving force behind this event involves another process that determines the transformation from an open to a closed state. Markin et al. (2008) developed the *Hydroelastic Curvature Model* that includes bending elasticity, turgor pressure, and water jets. The closure of the Venus flytrap represents the nonmuscular movement based on hydraulics and mechanical elasticity. The nastic movements in various plants involve a large internal pressure (turgor) actively regulated by plants.

In the *Hydroelastic Curvature Model* (Markin et al. 2008) the leaf of Venus flytrap is visualized as a thin, weakly curved elastic shell with principal natural curvatures that depend on the hydrostatic state of the two surface layers of cell A and B (Fig. 1.2), where different hydrostatic pressures P_A and P_B are maintained.

Two layers of cells, mechanically connected to each other, behave like a very popular in membrane mechanics bilayer couple where the inplane expansion or contraction of any of them causes the change of curvature of the whole leaf. The bilayer couple hypothesis was first introduced by Sheetz and Singer (1974). They noticed that the proteins and the phospholipids of membranes are asymmetrically distributed in the two halves of the bilayer, which is most substantial for the erythrocyte membrane.

The two halves of the closed membrane bilayer may respond differently to various perturbations while remaining coupled to one another. One half of the bilayer may expand in the plane of the membrane relative to the other half of the bilayer, while the two layers remain in contact with one another. This leads to various functional consequences, including shape changes of the intact cell. This concept is called the bilayer couple hypothesis because of the analogy to the response of a bimetallic couple to changes in temperature. It remains very popular and applied to explanation of numerous phenomena, such as red blood cell transformations (see for example Lim et al. (2002) and references within) and the gating of mechanosensitive channels (Qi et al. 2005).

The bilayer couple properties were also extensively studied in connection with bilayer fusion, fission, endo and exocytosis (Markin and Albanesi 2002; Volkov et al. 1998). This technique was applied for the design and analysis of the hydroelastic curvature model of the Venus flytrap. The model is based on the assumption that the driving force of closing is the elastic curvature energy stored and locked in the leaves due to pressure differential between the outer and inner layers of the leaf (Fig. 1.2).

Unequal expansion of individual layers A and B results in bending of the leaf, and it was described in terms of bending elasticity. Unequal expansion means that the torque M appears in the leaf. The energy of the bent layer A is described by the equation.

$$E_A = \frac{1}{2}\kappa_0(C_{AM} - C_{A0})^2 + \kappa_G C_{AG} \quad (1.1)$$

Here C_{AM} is the total curvature of the layer A, C_{AG} is the Gaussian curvature, C_{A0} is the spontaneous or intrinsic curvature of the layer, and κ designates the elasticity. Usually, spontaneous curvature of layers is considered a constant b_A, depending on the composition of the layer, and describes the intrinsic tendency of the layer to bend. There is an additional source of bending—different pressure in two adjacent layers. One can easily visualize the number of mechanical models in which spontaneous curvature is proportional to the pressure in which curvature is $C_{A0} = a_A P_A + b_A$. The same equations are valid for layer B.

The geometrical mean and Gaussian curvatures are defined as $C_{AM} = 1/R_1 + 1/R_2$ and $C_{AG} = 1/(R_1 R_2)$, where R_1 and R_2 are the main radii of curvature of the layer. The shape of the leaf was approximated by a spherical surface; then $C_{AM} = 2/R$ and $C_{AG} = C_{AM}^2/4$. The leaf is thin and hence the two layers have the mean curvatures that are equal in absolute value but have opposite signs: $C_{AM} = -C_{BM} = C_M$. The sign of curvature was defined with respect to the normal directed outside of the layer (Fig. 1.2). Total elastic energy of the lobe was presented as

$$E_L = \frac{1}{2}\kappa_0\left[(C_M - aP_A - b_A)^2 + (-C_M - aP_B - b_B)^2\right] + \frac{1}{2}\kappa_G C_M^2 \quad (1.2)$$

Here, the coefficients a_A and a_B are assumed to be equal to a.

At the given pressures P_A and P_B, the equilibrium value of the mean curvature can be found from the minimum value of elastic energy (1.2):

1 Morphing Structures in the Venus flytrap

$$C_M = \frac{1}{2 + \kappa_0/\kappa_G}[a(P_A - P_B) + b_L] \quad (1.3)$$

Here b_L designates the difference between two intrinsic curvatures, $b_L = b_A - b_B$. This equilibrium shape is maintained if the pressure difference does not change.

In the open state, the pressure in the upper layer is higher than in the lower layer, maintaining the convex shape of the leaf. The fact, that the hydrostatic pressure in different parts of the plant can vary, is very well known. It is also known (Tamiya et al., 1988) that stimulation of a *Mimosa* plant causes very fast redistribution of water. Tamiya et al. (1988) found that after stimulation, water in the lower half of the main pulvinus is transferred to the upper half of the main pulvinus. Movement of the water in conjunction with *Mimosa* movement was visualized by a noninvasive NMR imaging procedure (Detmers et al. 2006). This fast water redistribution is obviously driven by the pressure difference between different parts of the plant, and exchange occurs through open pores. Unfortunately, the anatomy and the nature of these pores are not currently known. So, for the mechanical analysis their existence was simply accepted.

At the resting state water pores between the two hydraulic layers are closed. The external trigger, either mechanical or electrical, results in the opening of these connecting pores; water rushes from the upper to the lower layer, the bilayer couple quickly changes its curvature from convex to concave and the trap closes.

If the trigger reaches threshold value at the moment t_s and the characteristic time of the opening kinetics is τ_a then the open probability of the pores (after $t \geq t_s$) will be given by $n_{op}(t) = 1 - Exp[-(t-t_s)/\tau_a]$. The rate of fluid transfer can be presented as $J = n_{op}L_H(P_A - P_B)$, where L_H is the hydraulic coefficient of pore permeability. If the pressure in the layer is proportional to the amount of fluid confined in it, the pressure will change with a rate proportional to the fluid transfer between the layers: $dP_A/dt = -k_r J = -dP_B/dt$. This means that the sum of the two pressures remains constant: $P_A + P_B = const = P_{total}$. Then the variation of pressure can be described by the equation

$$\frac{dP_A}{dt} = -k_r n_{op} L(P_A - P_B) = -\frac{n_{op}}{\tau_r}\left(P_A - \frac{1}{2}P_{total}\right) \quad (1.4)$$

Here the characteristic time of fluid transfer, $\tau_r = 1/(2k_r L_H)$, is introduced. A similar equation for the mean curvature can be easily obtained from Eqs. 1.3 and 1.4:

$$\frac{dC_M}{dt} = -\frac{n_{op}}{\tau_r}\left(C_M - \frac{b_L}{2 + \kappa_G/\kappa_0}\right) \quad (1.5)$$

Initial curvature C_1 in the open state can be introduced arbitrarily, while the final curvature C_2 in the closed state is found from Eq. 1.5 automatically: $C_2 = \frac{b_L}{2+\kappa_G/\kappa_0}$.

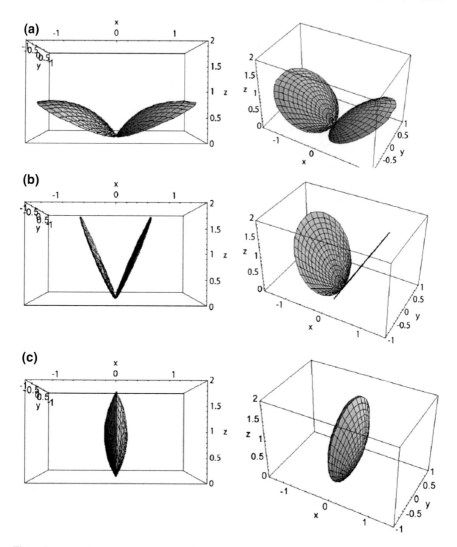

Fig. 1.3 Computer modeling of venus flytrap closing

When solving this equation, one has to have in mind that the open probability n_{op} is the function of time found above. If initial mean curvature at the moment $t = t_s$ is $C_M = C_1$, the solution of Eq. 1.5 at $t \geq t_s$ is

$$C_M(t) = (C_1 - C_2)\ \exp\left\{\frac{\tau_a}{\tau_r}\left[1 - \exp\left(-\frac{t-t_s}{\tau_a}\right)\right] - \frac{(t-t_s)}{\tau_r}\right\} + C_2 \qquad (1.6)$$

The Gaussian curvature was calculated in a similar way. Based on these equations the computer modeling gave the sequence of shapes in the process of trap closing. This sequence A, B, C is presented in Fig. 1.3. The left column gives

1 Morphing Structures in the Venus flytrap

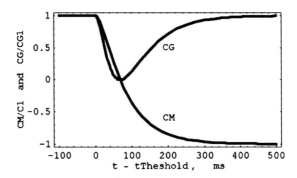

Fig. 1.4 Variation of curvature during Venus flytrap closing

the front view, and the right column the right and above. In panel A the trap is open, the lobes have convex shape. In the process of closing the curvature is changing and panel B shows intermediate state with flat lobes. The final closed state is presented in panel C with lobes having concave shape. Notice that the angle between the lobes at the midrib does not change.

Variation of mean and Gaussian curvature of the lobes is illustrated in Fig. 1.4 with parameters $\tau_a = 20$ ms, $\tau_r = 70$ ms, and $C_2 = -C_1$. As was shown before these curvatures in the given approximation are $C_{AM} = 2/R$ and $C_{AG} = C_{AM}^2/4$. In Fig. 1.4 they are normalized by their initial curvatures, so that initially both of them are equal to 1. It means that the lobes are convex. When closing begins ($t = t_{\text{thershold}}$) both curvatures start to decrease and at some point reach zero. This corresponds to flat lobes in panel B of Fig. 1.3. After that the mean curvature continues to decrease and goes into negative range until it reaches value of -1. This signifies that the lobes became concave. In contrast to that the Gaussian curvature does not become negative; it is the product of two principal curvatures both of which change sign at flat position so that the Gaussian curvature remains positive and returns to value of $+1$.

When experimentally observing the closing of the Venus flytrap, one can register the change of leaf curvature, but it is easier to measure the change of the distance, X, between the edges of the leaves of the Venus flytrap. Let us designate the initial distance as X_1, and the final distance as X_2. We shall use the normalized distance defined as $x = X/X_1$. It was shown that both distance and mean curvature of the leaf are described by the same function of time.

When the trigger signal opens the pores between the hydraulic layers at the moment $t = 0$, the fluid rushes from one layer to another. The leaf relaxes to its equilibrium state corresponding to the closed configuration. The distance between the edges of the trap was found to vary with time as

$$x(t) = (1 - x_2) \exp\left\{\frac{\tau_a}{\tau_r}\left[1 - \exp\left(-\frac{t - t_s}{\tau_a}\right)\right] - \frac{(t - t_s)}{\tau_r}\right\} + x_2 \qquad (1.7)$$

This function was experimentally verified by studying the closure of the Venus flytrap.

Fig. 1.5 Kinetics of the trap closing stimulated by a cotton thread

1.4 Comparison with Experiment

The Venus flytrap can be closed by mechanical stimulation of trigger hairs using a cotton thread or wooden stick to gently touch one or two of the six trigger hairs inside the upper leaf of the Venus flytrap. The cotton thread was removed before the leaves closed. Consecutive photos of the trap are presented in Fig. 1.5. It could also be closed by small piece of gelatin. Plants were fed a 6 × 6 × 2 mm cube of 4% (w/v) gelatin. This induces closing by stimulating 2 of the 6 trigger hairs of the Venus flytrap. The photos are presented in Fig. 1.6.

The Venus flytrap could also be closed by an electrical pulse between the midrib and a lobe of the upper leaf without mechanical stimulation. The closing was achieved by electrical stimulation with a positive electrode connected to the midrib and a negative electrode located in one of the lobes. It is interesting that inverted polarity pulse was not able to close the plant, and the closed trap would not open by electrical stimulus lasting up to 100 s.

A single electrical pulse exceeding a threshold (mean 13.63 µC, median 14.00 µC, std. dev. 1.51 µC, n = 41) causes closure of a trap and induces an electrical signal propagating between the lobes and the midrib. When charges were

1 Morphing Structures in the Venus flytrap

Fig. 1.6 Kinetics of the trap closing stimulated by a piece of gelatin

smaller, the trap did not close. Repeated application of small charges demonstrates a summation of stimuli. Two or more injections of electrical charges within a period of less then 50 s closed the trap as soon as a total of 14 µC charge is applied. Traps closing by electrical stimulus obeys the all-or-none law: there is no reaction for stimulus under the threshold and the speed of closing does not depend on stimulus strength above threshold.

Experimental points in Fig. 1.7 shows the kinetics of closing the upper leaf induced by mechanical or electrical stimuli. Closing consists of three distinctive phases (Fig. 1.7). Immediately after stimulation, there is a mechanically silent period with no observable movement of the plant. This is followed by a period when the lobes begin to accelerate. The third period of fast movement is actual trapping when the leaves quickly relax to the new equilibrium state.

The processes of closing by mechanical or electrical stimuli qualitatively are very similar though parameters of these processes are somewhat different. These parameters were found from curve fitting. They are presented in Table 1.1 (rows A and B). The closing develops at just a fraction of a second. The first mechanically silent phase lasts between 68 and 110 ms and the opening of water channels takes between 10 and 20 ms. These two stages are about two times faster with mechanical stimulation than with electrical one. However, this is not the case for relaxation stage: it is two times slower with mechanical stimulation. Therefore, the fastest stage both with mechanical and electrical stimulation is the opening of

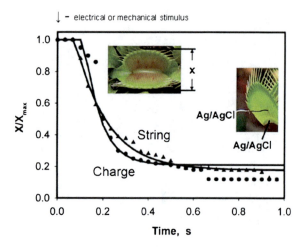

Fig. 1.7 Kinetics of a trap closing stimulated by a cotton thread (*triangle*) or by 14 μC charge (*circle*); x is the distance between the trap margins. Solid lines are plotted according to Eq. 1.7 with parameters from Table 1.1. All these results were reproduced at least ten times. Reproducibility of the initial mechanically silent period with no observable movement of the trap is ±33 ms

water channels. The table shows that the limiting stage of the process is the fluid transfer in the leaf, though it is also very quick due to the small distance between the layers. In both experiment the characteristic time τ_a is always less than τ_r. This means that pore opening is relatively fast and it is not a limiting stage. Final relaxation of the trap to the closed state is much slower.

1.5 Interrogating Consecutive Stages of Trap Closing

The hydroelastic curvature model described three consecutive stages of trap closing: mechanically silent stage of impulse transduction with characteristic time τ_s, opening of water channels with characteristic time τ_a, and relaxation of the elastic shell to the equilibrium closed state with characteristic time τ_r. To verify basic assumptions of this model Volkov et al. (2008c) interrogated these stages using specific inhibitors of different mechanical and biochemical processes involved in the closing process. They presented detailed experiments for comparative study of the effects of inhibitors of ion channels, aquaporins, and uncouplers on kinetics of the trap closing, stimulated by mechanical or electrical triggering of the trap. This gave the opportunity to justify the basic assumption of the hydroelastic curvature model and to determine the variation of kinetic parameters of the Venus flytrap closure.

The first mechanically silent stage of the trap closing involves transduction of electrical signals and hence it is related to ion channel gating. Therefore it should be sensitive to agents interfering with ion channels. To check this hypothesis, the ion channel blockers Ba^{2+}, Zn^{2+}, and tetraethylammonium chloride (TEACl) were used. The altered kinetics of trap closing is presented in Fig. 1.8, panel b. For convenience of comparison, panel a presents control experiment before modification of the plant. Both Ba^{2+} and Zn^{2+} had similar effects on the Venus flytrap:

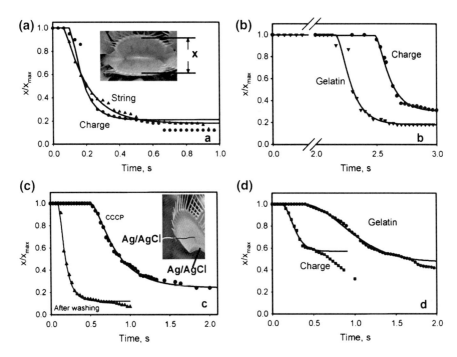

Fig. 1.8 a Kinetics of a trap closing stimulated by a cotton thread (*triangle*) or by 14 μC charge, *x* is the distance between the trap margins; (*circle*). **b** Kinetics of a trap closing stimulated by a gelatin (*triangle*) or by 28 μC charge (*circle*). 50 mL of 5 mM BaCl$_2$ was added to soil 55 h before experiments. **c** Kinetics of a trap closing after 70 μC electrical stimulation. 50 mL of 10 μM CCCP was added to soil 4.5 h before experiments (*circle*). Soil around Venus flytrap was washed during 2 days with 200 mL distilled water per day to decrease CCCP concentration (*triangle*). **d** Kinetics of a trap closing after stimulation of trigger hairs by a small piece of gelatin (*circle*) or by 28 μC electrical stimulation (*square*). 50 mL of 10 mM TEACl was added to the soil 55 h before experiments. *Solid* lines are theoretical dependencies estimated from Eq. 1.7 with parameters from Table 1.1. Reproducibility of the initial mechanically silent period with no observable movement of the trap is ±33 ms

they significantly extended mechanically silent stage—processing of electrical signals (Fig. 1.8b and Table 1.1). They changed the duration of then first stage from tens of milliseconds to a few seconds—more than an order of magnitude. The effect was more pronounced for electrical stimulation than for mechanical stimulation that intuitively seems quite expected. These blockers of ion channels did not interfere with other two stages: the speed of closing when it started remained similar to nontreated plants both for electrical and mechanical stimulation.

Another agent TEACl is known as a blocker of potassium channels in plants (Volkov 2006a, b). It was found that 10 mM aqueous solution of TEACl decreased the speed of the trap closure induced both by mechanical and electrical stimuli (Fig. 1.8d and Table 1.1). The effect is more pronounced for mechanical stimulation.

The next group of active substances studied in this work included uncouplers of oxidative phosphorylation. They are soluble in both water and lipid phases,

Table 1.1 Estimated kinetic parameters with and without inhibitors

Experiment		t_s (ms)	τ_a (ms)	τ_r (ms)	$\tau_a + \tau_r$ (ms)	x_2
A	Mechanical stimulation	68	10	140	150	0.178
B	Electrical stimulation	110	20	70	90	0.21
C	Electrical stimulation with BaCl$_2$ added to soil	2,480	50	100	150	0.30
D	Mechanical stimulation with BaCl$_2$ added to soil	2,150	50	100	150	0.18
E	Electrical stimulation with CCCP added to soil	480	80	320	400	0.24
F	Electrical stimulation after CCCP washed out	90	20	80	100	0.12
G	Electrical stimulation with TEACl added to soil	120	220	60	280	0.57
H	Mechanical stimulation with TEACl added to soil	370	1900	120	2020	0.475

permeate the lipid phase of a membrane by diffusion and transfer protons across the membrane, thus eliminating the proton electrochemical gradient and/or a membrane potential (Volkov et al. 1998). Hodick and Sievers (1988) reported an excitability inhibition of *Dionaea* leaf mesophyll cells using uncoupler 2,4-dinitrophenol. In our experiments uncoupler CCCP caused the delay of the trap closing (Fig. 1.8c and Table 1.1, electrical stimulation) in addition to significant decrease of the speed of closing as a result of membrane depolarization or dissipation of a proton gradient during ATP hydrolysis. This effect is reversible when concentration of CCCP was decreased by soil washing with distilled water, if an uncoupler was added to soil less than 5 h before. After soil washing with distilled water, the closing time of Venus flytrap treated by CCCP returned back to 0.3 s, but a higher electrical charge is needed for trap closure (Fig. 1.8c). After 48 h incubation of CCCP in the soil, the inhibitory effect of CCCP on the trap closure became irreversible and could not be washed out by distilled water. We found similar effects of significant increase of time closing of the trap in the presence of uncouplers FCCP, pentachlorophenol, and 2,4-dinitrophenol.

Millimolar solutions of BaCl$_2$ and TEACl may affect physiology of the plant since the Venus flytrap is notoriously sensitive to some ions (Hodick and Sievers 1988; Lloyd 1942). Control plants were exposed to similar concentrations of KCl (Fig. 1.9a) and CaCl$_2$ (Fig. 1.9b) added to soil and no inhibitory effect of these salts on the trap closure was found. Usually, concentration of salts in water from lakes and ponds is much higher and varies from 100 to 400 mg/L (Drever 1997). Water from lakes, ponds, and rivers is the traditional source of water for the Venus flytrap in natural habitat and in vitro.

The rate of cellular movement is determined by the water flux induced by a very rapid change in osmotic pressure, monitoring by a fast and transient opening of aquaporins. HgCl$_2$, TEACl, and Zn^{2+} inhibit water channel activity (Detmers et al. 2006; Maurel 1997; Savage and Stroud 2007). According to literature, 1 mM HgCl$_2$ is an efficient blocker of aquaporins (Maurel and Chrispeels 2001; Tyerman et al. 2002). Figure 1.10 shows that the inhibitor of aquaporins HgCl$_2$ hinders the trap closing after mechanical stimulation independently on the type of extraction.

Figure 1.11 shows kinetics of a trap closing after extraction of TEACl from 2 drops of the TEACl solution placed on the midrib. Mechanical stimulation of three

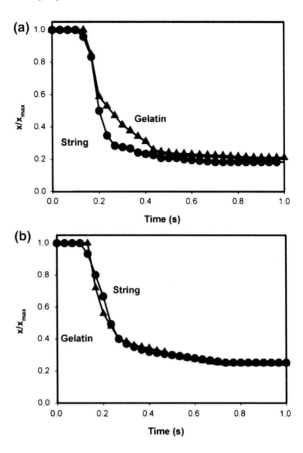

Fig. 1.9 Kinetics of a trap closing stimulated by a cotton thread (*circle*) or by gelatin (*triangle*), x is the distance between the trap margins. **a** 50 mL of 10 mM KCl was added to soil 7 h before experiments. **b** 50 mL of 5 mM CaCl$_2$ was added to soil 7 h before experiments. Reproducibility of the initial mechanically silent period with no observable movement of the trap is ±33 ms

mechanosensitive hairs did not induce the closing of the trap (Fig. 1.11a, line 1). If these mechanosensors stimulated again after short period of time (10–30 s), the trap slowly closes (Fig. 1.11a, line 2). Figure 1.11b shows similar dependencies after electrical stimulation by 14 µC (line 1) and 28 µC (line 2). To close the trap after TEACl treatment, a double electrical charge is required. TEACl is known as a blocker of aquaporins (Demeters et al. 2006) and K$^+$-channels (Volkov 2006b) in plants.

Figure 1.12 shows kinetics of the trap closing stimulated by a cotton thread after phytoextraction of BaCl$_2$ (Fig. 1.12a) or ZnCl$_2$ (Fig. 1.12b) from aqueous solution placed on the midrib. BaCl$_2$ induces 3 s delay before the trap closing and ZnCl$_2$ decreases dramatically the speed of trap closing.

Figure 1.13 shows inhibitory effects induced by two drops of uncoupler CCCP at the midrib. Other uncouplers FCCP, 2, 4-dinitrophenol, and pentachlorophenol decrease speed and increase time of the trap closing similar to inhibitory effects of CCCP.

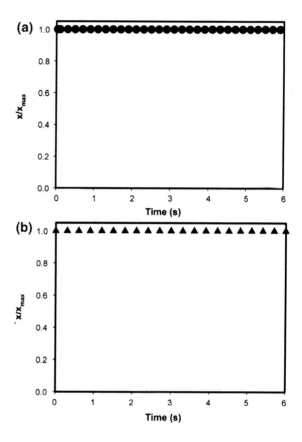

Fig. 1.10 Kinetics of a trap closing stimulated by a cotton thread: **a** 50 mL of 1 mM HgCl$_2$ was added to soil 7 h before experiments. Concentration of HgCl$_2$ in soil was 1 mM. **b** Two 10 μL drops of 1 mM HgCl$_2$ were placed on the midrib 24 h before a mechanical stimulation by a cotton thread of three mechanosensitive hairs. These results were reproduced 16 times on different Venus flytrap plants

So it was found that extraction of ZnCl$_2$, BaCl$_2$, HgCl$_2$, TEACl, and CCCP by the Venus flytrap from soil and from drops of inhibitor solutions placed on the midrib gives the similar results. All these results are summarized in Table 1.1.

The closing of the Venus flytrap develops very quickly; it takes just a fraction of a second. The curve fitting shows that the limiting stage of the process is the fluid transfer in the leaf, though it is also very quick due to the small distance between the layers. In six experiments (A-F, Table 1.1) characteristic time τ_a is always less than τ_r. This means that water channel opening is relatively fast and it is not a limiting stage. Looking for a way to interrogate this stage, we turned to TEACl, which is known as a blocker of K$^+$ channels and aquaporins in plants. Table 1.1 (rows G and H) shows that in the presence of the aquaporin blocker, τ_r is less than τ_a, and the limiting step of the whole process is the water transport between two layers in the presence of TEACl.

When trigger hairs in the open trap receive mechanical stimuli, a receptor potential is generated (Benolken and Jacobson 1970; DiPalma et al. 1966). Two mechanical stimuli are required for closing the trap in vivo (Darwin 1875; Lloid 1942). Receptor potentials generate action potentials (Burdon-Sanderson 1873; Jacobson 1965, 1974; Volkov et al. 2008a), which can propagate in the

Fig. 1.11 Kinetics of a trap closing stimulated by a cotton thread (**a**) or by electrical charge (**b**). Two 10 μL drops of 10 mM TEACl were placed on the midrib 24 h before a mechanical or electrical stimuli applications. **a** Mechanical stimulation by a cotton thread of three mechanosensitive hairs (1) and repeating of this stimulation in 16 s (2). **b** electrical stimulation by 14 μC (1) or 28 μC (2) between a midrib (+) and a lobe (−). These results were reproduced 25 times on different Venus flytrap plants

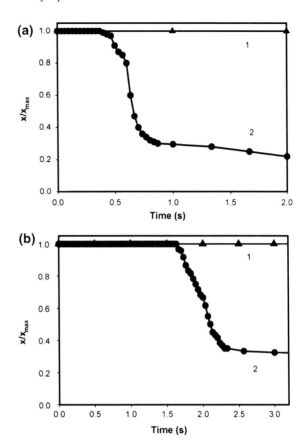

plasmodesmata of the plant to the midrib (Volkov et al. 2007, 2008a). Uncouplers and blockers of fast anion and potassium channels can inhibit action potential propagation in the Venus flytrap (Hodick and Sievers 1988; Krol et al. 2006; Volkov et al. 2008c). Once a threshold value of the charge is accumulated, ATP hydrolysis (Jaffe 1973) and fast proton transport start (Rea 1983), and aquaporin opening is initiated. In the presence of aquaporin blocker $HgCl_2$ the trap does not close during sufficiently long period of time. Fast proton transport induces transport of water and a change in turgor (Markin et al. 2008; Volkov et al. 2007, 2008c).

1.6 Electrical Memory in Venus Flytrap

The Venus flytrap has a short-term electrical memory (Volkov et al. 2008a, b, c). It was shown by new charge injection method. As mentioned before, the application of an electrical stimulus between the midrib (positive potential) and the lobe

Fig. 1.12 Kinetics of a trap closing stimulated by a cotton thread: Two 10 μL drops of 5 mM BaCl$_2$ **a** or ZnCl$_2$ **b** were placed on the midrib 24 h before a mechanical stimulation by a cotton thread of three mechanosensitive hairs. These results were reproduced 30 times on different venus flytrap plants

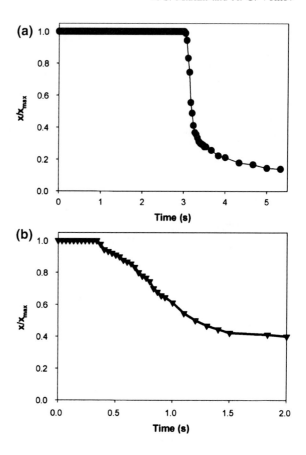

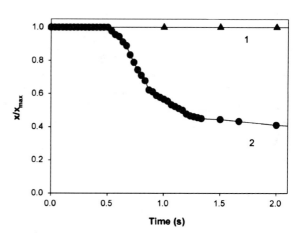

Fig. 1.13 Kinetics of a trap closing stimulated by a cotton thread: Two 10 μL drops of 10 μM CCCP were placed on the midrib 24 h before a mechanical stimulation by a cotton thread of three mechanosensitive hairs (1) and repeating of this stimulation in 16 s (2). These results were reproduced 19 times on different venus flytrap plants

1 Morphing Structures in the Venus flytrap

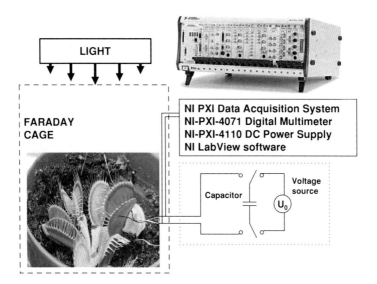

Fig. 1.14 The charged capacitor method

(negative potential) causes the Venus flytrap to close the trap without any mechanical stimulation. The average stimulation pulse voltage sufficient for rapid closure of the Venus flytrap was 1.5 V (standard deviation is 0.01 V, n = 50) for 1 s. The inverted polarity pulse with negative voltage applied to the midrib did not close the plant (Volkov et al. 2007). Applying impulses in the same voltage range with inverted polarity did not open the trap, even with pulses of up to 100 s (Volkov et al. 2007; Volkov et al. 2008a). Energy for trap closure is generated by ATP hydrolysis (Jaffe 1973). ATP is used for a fast transport of protons. The amount of ATP drops from 950 µM per midrib before mechanical stimulation to 650 µM per midrib after stimulation and closure.

The action potential delivers the electrical signal to the midrib, which can activate the trap closing. We studied the amount of transmitted electrical charge from the charged capacitors between the lobe and the midrib of the Venus flytrap (Fig. 1.14). Application of a single electrical charge initially stored on the capacitor when it is connected to the electrodes inserted in the trap of *D. muscipula* (mean 13.63 µC, median 14.00 µC, std. dev. 1.51 µC, n = 41) caused trap closure and induces an electrical signal propagating between the lobes and the midrib (Markin et al. 2008; Volkov et al. 2008a, 2009a).

The electrical signal in the lobes was not an action potential, because its amplitude depended on the applied voltage from the charged capacitor. Charge induced closing of a trap plant can be repeated 2–3 times on the same Venus flytrap plant after complete reopening of the trap.

The Venus flytrap has the ability to accumulate small charges. When the threshold value is reached, the trap closes (Markin et al. 2008; Volkov et al. 2008a, 2009a).

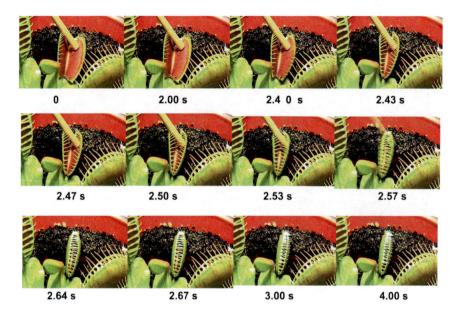

Fig. 1.15 Closing of the *Dionaea muscipula* trap by pressing of only one trigger hair at 21°C. These results were reproduced 14 times on different Venus flytrap plants. Reproducibility of the movement of the trap was ±33 ms

A summation of stimuli was demonstrated through the repetitive application of smaller charges (Volkov et al. 2009a). Previous work by Brown and Sharp (1910) indicated that strong electrical shock between lower and upper leaves can cause the Venus flytrap to close, but the amplitude and polarity of applied voltage, charge, and electrical current were not reported. The trap did not close when we applied the same electrostimulation between the upper and lower leaves as we applied between the midrib and the lobe, even when the injected charge was increased from 14 to 750 µC charges (Volkov et al. 2008a). It is probable that the electroshock induced by Brown and Sharp (1910) had a very high voltage or electrical current.

Mechanical touching of one sensitive hair twice or two different hairs with an interval up to 40 s induces the trap closing. One sustained mechanical stimulus applied to only one trigger hair can close the trap of *D. muscipula* in a few seconds (Fig. 1.15). Prolonged pressing of the trigger hair generates two electrical signals similar to action potentials reported in reference (Volkov et al. 2007) within an interval of about 2 s, which stimulate the trap closing.

The Venus flytrap can capture insects by any of these three ways of mechanical stimulation of trigger hairs. Insects only have to touch a few trigger hairs for just 1 or 2 s. In biology, this effect is referred to as molecular or sensory memory. It was shown that sustained membrane depolarization induced a "molecular" memory phenomenon and has profound implication on the biophysical properties of voltage-gated ion channels (Nayak and Sikdar 2007).

1 Morphing Structures in the Venus flytrap

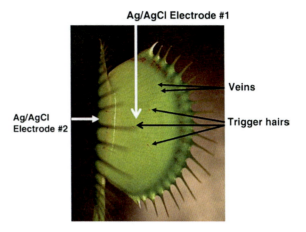

Fig. 1.16 Position of electrodes in the Venus flytrap

For the analysis of short-term electrical memory in the Venus flytrap two electrodes were inserted in the midrib and one of the lobes (Fig. 1.16). Transmission of an electrical charge between electrodes in a lobe and the midrib causes closure of the trap. If two or more injections of electrical charges are applied within a period of less than 40 s, the Venus flytrap upper leaf closes as soon as a 14 μC charge is passed. The capacitor slowly discharges with time, so although a 14 μC charge was applied, not all of this charge was accumulated during 20–40 s to assist in closing the trap. This experiment gave the opportunity of finding the exact electrical charge utilized by the Venus flytrap to facilitate trap closing. The voltage between electrodes #1 and #2 inserted in the midrib and one of the lobes was measured. Charge was estimated as the voltage U between two Ag/AgCl electrodes multiplied by capacitance C of the capacitor: $Q = UC$. The closing charge of the trap was estimated as the difference between the initial charge and the final charge of the capacitor, when the trap began to close. In the experiments, when a few small charges were applied to close the trap, the closing charge was estimated as the sum of all charges transmitted from the capacitor. This shows that repeated application of smaller charges demonstrates a summation of stimuli (Figs. 1.17 and 1.18).

The capacitor discharge on small traps with a midrib length of 1.0 cm and large traps with a midrib length of 3.5 cm at room temperature is presented in Fig. 1.17. As soon as 8.05 μC charge (mean 8.05 μC, median 9.00 μC, std. dev. 0.06 μC, n = 34) for small trap or 9.01 μC charge (mean 9.01 μC, median 8.00 μC, std. dev. 0.27 μC, n = 34) for large trap was transmitted between a lobe and midrib from the capacitor, the trap began to close at room temperature. At temperatures of 28–36°C a smaller electrical charge of 4.1 μC (mean 4.1 μC, median 4.1 μC, std. dev. 0.24 μC, n = 34) is required to close the trap of the *D. muscipula* (Fig. 1.18). There should be an essential transformation in the circuitry of at temperature between 25 and 28°C and due to this reason only one mechanical stimulus or a smaller electrical charge was required to close the trap.

Fig. 1.17 Closing of the *Dionaea muscipula* trap at 21°C by electrical charge Q injected between a lobe and a midrib: **a** small trap, midrib length 1.0 cm; Q = 8.05 µC, **b** large trap midrib length 3.5 cm; Q = 9.1 µC

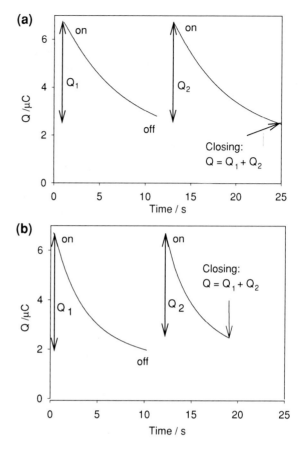

Fig. 1.18 Closing of the *Dionaea muscipula* trap at 30°C by electrical charge Q injected between a lobe and a midrib: *Large* trap midrib length 3.5 cm. Q = 4.1 µC

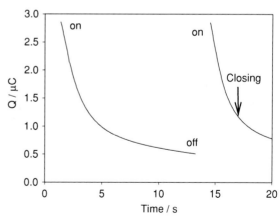

The general classification of memory in plants is based on the duration of memory retention, and identifies three distinct types of memory: sensory memory, short-term memory and long-term memory. Sensory memory corresponds approximately to the initial 0.2–3.0 s after an item is perceived. Some information in sensory memory can be transferred then to short-term memory. Short-term memory allows recalling something from several second to as long as a minute without rehearsal. The storage in sensory memory and short-term memory generally has a strictly limited capacity and duration, which means information is available for a certain period of time, but is not retained indefinitely. Long-term memory can store much larger quantities of information for potentially unlimited duration up to the whole life span of the plant. The Venus flytrap can accumulate small subthreshold charges, and when the threshold value is reached, the trap closes. The cumulative character of electrical stimuli points to the existence of short-term electrical memory in the Venus flytrap.

1.7 Complete Hunting Cycle of the Venus Flytrap

The trap closing. Touching trigger hairs protruding from the upper epidermal layer of the Venus flytrap's leaves activate mechanosensitive ion channels. As a result, receptor potentials are generated which in turn induce a propagating action potential throughout the upper leaf of the Venus flytrap (Benolken and Jacobson 1970; Jacobson 1965; Volkov et al. 2007, 2008a, 2011). A receptor potential always precedes an action potential and couples the mechanical stimulation step to the action potential step of the preying sequence (Jacobson 1965). A possible pathway of action potential propagation to the midrib includes vascular bundles and plasmodesmata in the upper leaf (Buchen et al. 1983; Ksenzhek and Volkov 1998). Pavlovič et al. (2010, 2011) found that the irritation of trigger hairs and subsequent generation of action potentials resulted in a decrease in the effective photochemical quantum yield of photosystem II and the rate of net photosynthesis.

When the Venus flytrap catches the insect it does not "crash" the prey, but instead hugs it by building the cage around it. This is achieved by bending the lobes. The curvature of the lobes changes during closing of the trap from convex to concave configuration. The trap changes from a convex to a concave shape in 100 ms. There is a very small tightening of lobes during the first 5 min. Cilia on the rims of the lobes bend over and lock the edges. The trap can stay in such a position for a few hours before opening if the prey is too small for digesting.

On closure, the cilia protruding from the edge of each lobe form an interlocking wall that is impenetrable to all except the smallest prey. The trap uses the double-trigger mechanism and shuts when the prey touches its trigger hairs twice in succession within a 25 s window of time. Partial closure allows the cilia to overlap, however, the lobes are still slightly ajar. This partial closure occurs in a fraction of a second, and several minutes may be required for the lobes to fully come together. When a prey is caught, the lobes seal tightly and thus remain for

Fig. 1.19 Photos of different stages of the trap closing, looking and digestion by a piece of gelatin: **a** open state, **b** locking after closing, (C)–(H) constricting of lobes and digestion during 5 days

5–7 days, allowing digestion to take place (Jaffe 1973). The stalk and basal cells containing lipid globules and the common wall between these two cells are traversed by numerous plasmodesmata (Williams and Mozingo 1971). Electron micrographs of the trigger hairs reveal three regions where the cells differ in size, shape, and cytoplasmic content. The basal walls of the indentation cells contain many plasmodesmata. Plasmodesmata found in anticlinal and podium cells pass through constricted zones in the cell wall; there are numerous plasmodesmata in the peripheral podium cells (Mozingo et al. 1970).

The majority of publications concerning the Venus flytrap have addressed the process of the trap closing, but few have discussed the biochemistry of digestion and the secretory cycle of *D. muscipula* Ellis (Fagerberg and Howe 1996).

The trap tightening. After closing occurs the next stage is the tightening the trap and digesting of the prey. The distance between the centers of lobes decreases during tightening of the trap and digestion of the prey. After the lobes close, the trap should be locked when cilia, finger-like protrusions, bend around the edges and overlap them. This is the locked state. Next, the lobes flatten, constrict the prey, and seal the margins as a watertight chamber and digestion begins (Fig. 1.19). During digestion, cilia from one lobe overlap the edges perpendicular to lobes as cilia from another lobe stand up parallel to this lobe (Fig. 1.19). This effect might be caused by a small difference in turgor between constricted lobes. Darwin (1875) found that mechanical stimulation of the trigger hairs is not essential for digestion. The trap continues to narrow and secrete digestive compounds whether the prey is alive or substituted by pieces of gelatin or meat. Burdon-Sanderson and Page (1876) observed a greater force exerted by opposing lobes on one another at this stage. Affolter and Olivo (1975) monitored electrical signaling from the Venus flytrap after live prey was captured. Indeed, stimulation of trigger hairs after trap closure results in additional constricting of the trap.

Lichtner and Williams (1977) found that in 50% of traps, mechanical stimulation after closing leads to an additional narrowing phase and secretion of viscous

1 Morphing Structures in the Venus flytrap

Fig. 1.20 Photo of the upper leaf of the venus flytrap. veins between the midrib and cilia are shown

digestive substances with pH of 1.5–3.5 coating the interior of traps. Scala et al. (1969) observed phosphatase, proteinase, nuclease, and amylase in the digestive secretion. Maximum secretion of the Venus flytrap enzymes occurs within the first 5 days. A number of different enzymes are responsible for digestion and transport of amino acids. The adaxial surfaces of the trap lobes possess an H^+ extrusion mechanism stimulated by prolonged exposure to secretion elicitors during digestion (Rea 1983). The acidity of the secretion may play an important role in the facilitation of the carrier-mediated uptake of amino acids from the trap cavity (Rea 1984).

In the processes of water and ion transport during the trap closing, sealing, digestion and opening, an important role is played by veins shown in Fig. 1.20.

Figure 1.21 shows the long scale kinetics of the trap closing after mechanostimulation of a trigger hairs without a prey (Fig. 1.21a), electrostimulation by 15 µC charge using two Ag/AgCl electrodes located in a midrib (+) and in one of the lobes (−) (Fig. 1.21b), and stimulation by a piece of a gelatin as a prey (Fig. 1.21c). Traps closed in less than 1 s after mechanical or electrical stimulation (Fig. 1.21a, b), but after the tightening of the lobes, which takes place during 4–5 min, traps can stay in this state for hours. We measured distance d between the middle of the lobes outside the trap during the process of trap closing. Figure 1.21c shows that if the trap is closed by a piece of gelatin, there is additional slow kinetics of lobe constricting during a few days of gelatin digestion.

If after closing the trap by electrical discharge we submit an additional charge from the capacitor (the arrow, Fig. 1.22); the fast and significant constricting of the lobes takes place during 15 min. This tightening occurs about two hundred times faster than in the presence of a gelatin, probably because there is no mechanical resistance from the prey inside the trap (Figs. 1.21c and 1.22).

The trap opening. After complete digestion and absorption of nutrients the trap is able to reopen. If the trap was closed by electrostimulation of *Dionaea muscipula* Ellis (15 µC, 1.5 V) using two Ag/AgCl electrodes located in a midrib (+) and in one of the lobes (−) (Fig. 1.23a), so that there is no digestion stage, the trap

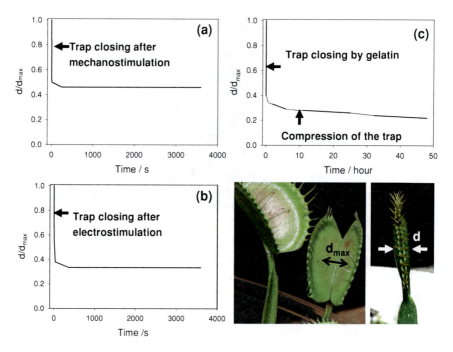

Fig. 1.21 Long term kinetics of the trap closing after mechanostimulation of a trigger hairs (**a**), electrostimulation by 15 μC charge using two Ag/AgCl electrodes located in a midrib (+) and in one of the lobes (−) (**b**), and stimulation by a piece of a gelatin (**c**)

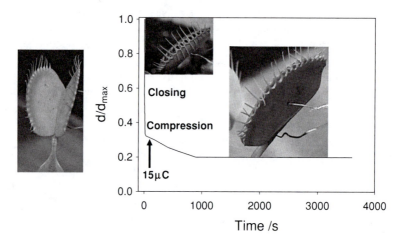

Fig. 1.22 Kinetics of the trap closing and constricting after additional electrostimulation by 15 μC charge using two Ag/AgCl electrodes located in a midrib (+) and in one of the lobes (−)

1 Morphing Structures in the Venus flytrap

Fig. 1.23 The trap opening after electrostimulation of *Dionaea muscipula* Ellis (15 µC, 1.5 V) using two Ag/AgCl electrodes located in a midrib (+) and in one of the lobes (−)

opened in 24 h (Fig. 1.23b, c), but transformation from the concave to convex shape (Fig. 1.23d) needs an additional 20 h. We tried to open the closed trap by applying an electrical stimulus of opposite polarity. However, changing the polarity of electrodes and increasing the charge up to 300 µC did not result in the opening of the trap. If the trap caught the prey, the digestion begins and lasts for a week. After 1 week of digestion the trap starts to open; in another day the lobes will be reopen with the original convex shape.

The exact mechanism of the trap opening is still unknown though we might guess that at this time active pumping of water between two hydraulic layers occurs.

Inhibition of trap opening. We showed earlier that ion and water channel blockers such as $HgCl_2$, TEACl, $ZnCl_2$, $BaCl_2$, as well as uncouplers CCCP, FCCP, 2, 4-dinitrophenol, and pentachlorophenol decrease speed and increase time of the trap closing. Blockers of ion channels and uncouplers inhibit electrical signal transduction in the Venus flytrap. Glycol-bis(2-aminoethylether)-N,N,N',N'-tetra-acetic acid, antrhracene-9-carboxylic acid (A-9-C), neomycin, ruthenium red, lanthanum ions, ethylene glycol-bis(β-aminoethyl ether)N,N,N',N'-tetraacetic acid, and NaN_3, all inhibit the excitability of the Venus flytrap, which indicates that ion channels are responsible for propagation of action potentials. Uncouplers, which are soluble in both water and lipid phases, eliminate the proton concentration gradient and/or a membrane potential. Uncouplers create proton conductivity and block cotransport of amino acids. Control plants were exposed to 10 mM of KCl or $CaCl_2$ and no inhibitory effects of these salts on the trap closure were found.

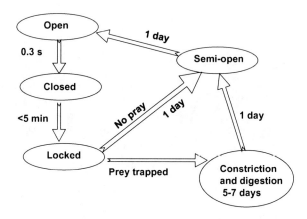

Fig. 1.24 The schematic representation of event during the trap closing and opening

We applied 20 μL of one of the inhibitors to the midrib of the Venus flytrap before the trap was closed by addition of a small piece of gelatin into the trap. Strong inhibition of the trap opening was found in the Venus flytraps treated by 5 mM $BaCl_2$, 10 μM CCCP or FCCP during 14 days after traps closing. Inhibitors of aquaporins 1 mM $HgCl_2$, 10 mM TEACl and $ZnCl_2$ also showed some delaying in the trap opening. 10 mM Anthracene-9-carboxylic acid (A-9-C), inhibitor of anion channels, showed inhibition of the trap constricting. The Venus flytrap absorbs chemical compounds through leaf-surface glands by energy dependent processes (Lloyd 1942). There is the H^+-cotransport of amino acids by the digestive glands of the Venus flytrap (Rea 1983, 1984). Uncouplers can work as a shunt of H^+-pump during the digestion process or transport of amino acids.

Stages of complete hunting cycle. The trap closing stage does not exhaust the whole process of catching and digesting insects by the Venus flytrap. We found that the distance between the rims of lobes after closing is not equal to zero, but remains in range of 15–20% of the initial distance between these edges. After closing, the trap should be locked when cilia, finger-like protrusions, bend around the edges and tighten the gap. We call this third phase the locked state. After that the lobes flatten, depress the prey and the digestion process begins. The trap starts to open after 5–7 days of digestion and after a day it will be open with the concave shape of lobes. Additionally, another day is required for changing of the trap from a concave to a convex shape and the trap will be completely open. Therefore, the total hunting cycle of the Venus flytrap consists of five stages: 1. Open State → 2. Closed State → 3. Locked State → 4. Constriction and Digestion → 5. Semiopen State → 1. Open State (Volkov et al. 2011).

Figure 1.24 shows the sequence of events during the trap closing and opening. The completely open trap with convex shape can be closed by a prey, electrical, or mechanical stimulation in less than 1 s. If the trap is not completely open and has a concave shape, the trap closing will be slow. Locking and small tightening takes a few minutes. If the prey was not captured, the trap will be subsequently open and another day will be required for transition from the concave to convex shape. If prey is captured by the Venus flytrap, constricting and digesting takes at least 5 days and depends on the size of a prey. Partial opening of the trap takes place during 24 h and a trap will be completely open the next day after a transition in shape.

Acknowledgment This work was supported by the grant from the U.S. Army Research Office.

References

Affolter JM, Olivo RF (1975) Action potentials in Venus's-flytraps: long-term observations following the capture of prey. Am Midl Nat 93:443–445
Beilby MJ, Bisson MA, Shepherd VA (2006) Electrophysiology of turgor regulation in charophyte cells. In: Volkov AG (ed) Plant electrophysiology—theory and methods. Springer, Berlin, pp 375–406
Benolken RM, Jacobson SL (1970) Response properties of a sensory hair excised from Venus's flytrap. J Gen Physiol 56:64–82
Bobji MS (2005) Springing the trap. J Biosci 30:143–146
Brown WH (1916) The mechanism of movement and the duration of the effect of stimulation in the leaves of *Dionaea*. Amer J Bot 3:68–90
Brown WH, Sharp LW (1910) The closing response in Dionaea. Bot Gaz 49(1910):290–302
Buchen B, Hensel D, Sievers A (1983) Polarity in mechanoreceptor cells of trigger hairs of *Dionaea muscipula* Ellis. Planta 158:458–468
Burdon-Sanderson J (1873) Note on the electrical phenomena, which accompany stimulation of the leaf of *Dionaea muscipula* Ellis. Phil Proc R Soc Lond 21:495–496
Burdon-Sanderson J, Page FJM (1876) On the mechanical effects and on the electrical disturbance consequent on excitation of the leaf of *Dionaea muscipula*. Philos Proc R Soc Lond 25:411–434
Darwin C (1875) Insectivorous plants. Murray, London
De Candolle CP (1876) Sur la structure et les mouvements des feuilles du Dionaea muscipula. Arch Sci Phys Nat 55:400–431
Detmers FJM, De Groot BL, Mueller EM, Hinton A, Konings IBM, Sze M, Flitsch SL, Grubmueller H, Deen PMT (2006) Quaternary ammonium compounds as water channel blockers: specificity, potency, and site of action. J Biol Chem 281:14207–14214
DiPalma JR, McMichael R, DiPalma M (1966) Touch receptor of Venus flytrap, Dionaea muscipula. Science 152:539–540
Drever JI (1997) The geochemistry of natural waters: surface and groundwater environments. Prentice Hall, Englewood Cliffs
Fagerberg WR, Allain D (1991) A quantitative study of tissue dynamics during closure in the traps of Venus's flytrap *Dionaea muscipula* Ellis. Amer J Bot 78:647–657
Fagerberg WR, Howe DG (1996) A quantitative study of tissue dynamics in Venus's flytrap *Dionaea muscipula* (Droseraceae) II. Trap reopening. Amer J Bot 83:836–842
Forterre Y, Skothelm JM, Dumals J, Mahadevan L (2005) How the Venus flytrap snaps. Nature 433:421–425
Hill BS, Findlay GP (1981) The power of movement in plants: the role of osmotic machines. Q Rev Biophys 14:173–222
Hodick D, Sievers A (1988) The action potential of *Dionaea muscipula* Ellis. Planta 174:8–18
Hodick D, Sievers A (1989) The influence of Ca^{2+} on the action potential in mesophyll cells of *Dionaea muscipula* Ellis. Protoplasma 133:83–84
Jacobson SL (1965) Receptor response in Venus's flytrap. J Gen Physiol 49:117–129
Jacobson SL (1974) The effect of ionic environment on the response of the sensory hair of Venus's flytrap. Can J Bot 52:1293–1302

Jaffe MJ (1973) The role of ATP in mechanically stimulated rapid closure of the Venus's flytrap. Plant Physiol 51:17–18

Krol E, Dziubinska H, Stolarz M, Trebacz K (2006) Effects of ion channel inhibitors on cold- and electrically-induced action potentials in *Dionaea muscipula*. Biol Plantarum 50:411–416

Ksenzhek OS, Volkov AG (1998) Plant energetics. Academic Press, San Diego

Lichtner FT, Williams SE (1977) Prey capture and factors controlling trap narrowing in *Dionaea* (Doseraceae). Am J Bot 64:881–886

Lim HWG, Wortis M, Mukhopadhyay R (2002) Stomatocyte-discocyte-echinocyte sequence of the human red blood cell: evidence for the bilayer-couple hypothesis from membrane mechanics. Proc Natl Acad Sci USA 99:16766–16769

Lloyd FE (1942) The carnivorous plants. Ronald, New York

Markin VS, Albanesi JP (2002) Membrane fusion: stalk model revisited. Biophys J 82:693–712

Markin VS, Volkov AG, Jovanov E (2008) Active movements in plants: mechanism of trap closure by *Dionaea muscipula* Ellis. Plant Signal Behav 3:778–783

Maurel C (1997) Aquaporins and water permeability of plant membranes. Annu Rev Plant Physiol Plant Mol Biol 48:399–429

Maurel C, Chrispeels MJ (2001) Aquaporins. A molecular entry into plant water relations. Plant Physiol 125:135–138

McGowan AMR, Washburn AE, Horta LG, Bryant RG, Cox DE, Siochi EJ, Padula SL, Holloway NM (2002) Recent results from NASA's morphing project, smart structures and materials. In: Proceedings of SPIE—International Society for Optical Engineering (USA), San Diego, CA, vol 4698, doi:10.1117/12.475056

Mozingo HN, Klein P, Zeevi Y, Lewis ER (1970) Venus's flytrap observations by scanning electron microscopy. Amer J Bot 57:593–598

Munk H (1876) Die electrischen und Bewegungserscheinungen am Blatte der Dionaeae muscipula. Arch Anat Physiol Wiss Med pp 30–203

Nayak TK, Sikdar SK (2007) Time-dependent molecular memory in single voltage-gated sodium channel. J Membr Biol 219:19–36

Nelson DL, Cox MM (2005) Lehninger principles of biochemistry, 4th edn. Freeman, New York, pp 58–59

Pavlovič A, Demko V, Hudak J (2010) Trap closure and prey retention in Venus flytrap (*Dionaea muscipula*) temporarily reduces photosynthesis and stimulates respiration. Ann Bot 105:37–44

Pavlovič A, Slovakova L, Pandolfi C, Mancuso S (2011) On the mechanism underlying photosynthetic limitation upon trigger hair irritation in the carnivorous plant Venus flytrap (*Dionaea muscipula* Ellis). J Exp Bot 62:1991–2000

Qi Z, Chi S, Su X, Naruse K, Sokabe M (2005) Activation of a mechanosensitive BK channel by membrane stress created with amphipaths. Mol Membr Biol 22:519–527

Rea PA (1983) The dynamics of H^+ efflux from the trap lobes of *Dionaea muscipula* Ellis (Venus's flytrap). Plant, Cell Environ 6:125–134

Rea PA (1984) Evidence for the H^+-co-transport of D-alanine by the digestive glands of *Dionaea muscipula* Ellis. Plant, Cell Environ 7:363–366

Savage DF, Stroud RM (2007) Structural basis of aquaporin inhibition by mercury. J Mol Biol 368:607–617

Scala J, Iott K, Schwab DW, Semersky FE (1969) Digestive secretion of *Dionaea muscipula* (Venus's-Flytrap). Plant Physiol 44:367–371

Sheetz MP, Singer SJ (1974) Biological membranes as bilayer couples. A molecular mechanism of drug-erythrocyte interactions. Proc Natl Acad Sci USA 71:4457–4461

Shimmen T (2006) Electrophysiology in mechanosensing and wounding response. In: Volkov AG (ed) Plant electrophysiology—theory and methods. Springer, Berlin, pp 319–339

Sibaoka T (1969) Physiology of rapid movements in higher plants. Annu Rev Plant Physiol 20:165–184

Stuhlman O, Darden E (1950) The action potential obtained from Venus's flytrap. Science 111:491–492

Tamiya T, Miyazaki T, Ishikawa H, Iriguchi N, Maki T, Matsumoto JJ, Tsuchiya T (1988) Movement of water in conjunction with plant movement visualized by NMR imaging. J Biochem 104:5–8

Tyerman SD, Niemietz CM, Bramley H (2002) Plant aquaporins: multifunctional water and solute channels with expanding roles. Plant, Cell Environ 25:173–194

Volkov AG (ed) (2006a) Plant electrophysiology. Springer, Berlin

Volkov AG (2006b) Electrophysiology and phototropism. In: Balushka F, Manusco S, Volkman D (eds) Communication in plants. Neuronal aspects of plant life. Springer, Berlin, pp 351–367

Volkov AG, Deamer DW, Tanelian DL, Markin VS (1998) Liquid interfaces in chemistry and biology. Wiley, New York

Volkov AG, Adesina T, Jovanov E (2007) Closing of Venus flytrap by electrical stimulation of motor cells. Plant Signal Behav 2:139–145

Volkov AG, Adesina T, Markin VS, Jovanov E (2008a) Kinetics and mechanism of *Dionaea muscipula* trap closing. Plant Physiol 146:694–702

Volkov AG, Carrell H, Adesina T, Markin VS, Jovanov E (2008b) Plant electrical memory. Plant Signal Behav 3:490–492

Volkov AG, Coopwood KJ, Markin VS (2008c) Inhibition of the *Dionaea muscipula* Ellis trap closure by ion and water channels blockers and uncouplers. Plant Sci 175:642–649

Volkov AG, Carrell H, Baldwin A, Markin VS (2009a) Electrical memory in Venus flytrap. Bioelectrochcmistry 75:142–147

Volkov AG, Carrell H, Markin VS (2009b) Biologically closed electrical circuits in Venus flytrap. Plant Physiol 149:1661–1667

Volkov AG, Pinnock MR, Lowe DC, Gay MS, Markin VS (2011) Complete hunting cycle of *Dionaea muscipula*: Consecutive steps and their electrical properties. J Plant Physiol 168:109–120

Williams ME, Mozingo HN (1971) The fine structure of the trigger hair in Venus's flytrap. Amer J Botany 58:532–539

Williams SE, Bennet AB (1982) Leaf closure in the Venus flytrap: an acid growth response. Science 218:1120–1121

Yang R, Lenaghan SC, Zhang M, Xia L (2010) A mathematical model on the closing and opening mechanism for Venus flytrap. Plant Signal Behav 5:968–978

Zonia L, Munnik T (2007) Life under pressure: hydrostatic pressure in cell growth and function. Trends Plant Sci 12:90–97

Chapter 2
The Effect of Electrical Signals on Photosynthesis and Respiration

Andrej Pavlovič

Abstract Electrical signals are initial response of plant to the external stimuli. This type of signal may trigger different physiological responses. The most famous is the rapid leaf movement in carnivorous and sensitive plants. However, a lot of less visible changes in plant physiology may occur. This chapter focuses on the effect of action (APs) and variation potentials (VPs) on photosynthesis and respiration. First, experimental methods and setup for measurements of photosynthesis and respiration in response to electrical signals are described. Then detailed information about effect of AP and VP on CO_2 metabolism in different plant species are summarized. Both light and dark reactions of photosynthesis, as well as rate of respiration, are affected by electrical signals, but the effect is often adverse (from inhibition to stimulation of photosynthesis). In addition, the stomatal conductance (g_s), an important component of gas exchange, is also differently affected by electrical signals. Summarizing the data from numerous authors, the hypothesis about mechanism underlying photosynthetic limitation and stimulation of respiration is proposed.

Abbreviations

1–qP	excitation pressure at photosystem II
ABA	abscisic acid
AP	action potential
DCMU	3-(3′,4′-dichlorophenyl)-1,1-dimethylurea
DW	dry weight
E	rate of transpiration
F_0	minimal fluorescence level in dark-adapted leaves

A. Pavlovič (✉)
Faculty of Natural Sciences, Department of Plant Physiology,
Comenius University in Bratislava, Mlynská dolina B-2,
SK-842 15 Bratislava, Slovakia
e-mail: pavlovic@fns.uniba.sk

F_0'	minimal fluorescence level "in the light"
F_m	maximal fluorescence level from dark adapted leaves
F_m'	maximal fluorescence level in light-adapted leaves
F_t	steady-state fluorescence in the light
F_v	variable fluorescence
F_v/F_m	maximum quantum yield of photosystem II
g_s	stomatal conductance
IAA	indole-3-acetic acid
JA	jasmonic acid
NPQ	non-photochemical quenching
OEC	oxygen evolving complex
OPDA	12-oxo-phytodienoic acid
PAR	photosynthetic active radiation
Pheo	pheophytin
P_G	rate of gross photosynthesis
P_N	rate of net photosynthesis
PSI	photosystem I
PSII	photosystem II
PQ	plastoquinone pool
Q_A	plastoquinone A
Q_B	plastoquinone B
qE	energy-dependent quenching
qI	photoinhibitory quenching
qT	state-transition quenching
qP	photochemical quenching
R_D	rate of respiration
VP	variation potential
Φ_{PSII}	effective photochemical quantum yield of photosystem II

2.1 Introduction

Life on Earth depends on energy derived from the sun. Photosynthesis is the only process of biological importance that can harvest this energy and provides building carbon blocks that plants depend on. On the other hand, respiration releases the energy stored in carbon compounds for cellular use. Although respiration is common for all eukaryotic organisms, the photosynthesis is confined only to the plant kingdom and also to some prokaryotic bacteria. Many environmental factors can affect rate of photosynthesis (P_N) and respiration (R_D). Concentration of O_2, CO_2, temperature, water and nutrient supply are the most often designated (Taiz and Zeiger 2002). However, not very often mentioned, electrical signals generated by plants have also impact on both processes. And it is not only the case of sensitive (*Mimosa pudica*) and carnivorous plants (*Dionaea muscipula*,

Aldrovanda vesiculosa, *Drosera* spp.) but of all green plants around us. Besides photosynthesis and respiration, many others physiological responses to electrical signals have been reported, among them are, for example: decrease of elongation growth (Shiina and Tazawa 1986), changes in transpiration and stomata opening/closing (Koziolek et al. 2003; Hlaváčková et al. 2006; Kaiser and Grams 2006; Grams et al. 2009), phloem unloading and translocation (Fromm and Eshrich 1988a; Fromm 1991; Fromm and Bauer 1994), changes in transcription and translation (Davies and Schuster 1981; Graham et al. 1986; Wildon et al. 1992, Peña-Cortéz et al. 1995; Stankovič and Davies 1998), decrease of circumnutation rate in *Helianthus annuus* (Stolarz et al. 2010), induction of jasmonic acid (JA) and abscisic acid (ABA) synthesis (Fisahn et al. 2004; Hlaváčková et al. 2006) and rapid trap closure and secretion of digestive fluid in carnivorous plants (Affolter and Olivo 1975; Lichtner and Williams 1977). You may find detailed description of these responses in other chapters of this book. This chapter summarizes data on the effect of action (APs) and variation potentials (VPs) generated by different stimuli on photosynthesis and respiration in plants.

2.2 Methodology and Experimental Setup

The best way how to study the effect of electrical signals on photosynthesis is simultaneous measurements of electrical potential, gas exchange, and chlorophyll fluorescence. Such experimental setup together with modified inverted microscope for observation of stomata aperture has been used rarely (Kaiser and Grams 2006). If it is not possible to do simultaneous measurements, then the measurements of gas exchange and chlorophyll fluorescence can be done separately (e.g. Koziolek et al. 2003; Lautner et al. 2005; Hlaváčková et al. 2006). Because the techniques for measurements of electrical signals in plants are well summarized in Chaps. 7 (Jovanov and Volkov) and 8 (Fromm and Lautner) of this book, I will focus on measurement of gas exchange and chlorophyll fluorescence in response to electrical signals.

2.2.1 Gas Exchange Measurements

The most commonly used method of measuring the gas exchange of leaf is to enclose it in a cuvette, pass a known flow rate of air over the leaf, and measure the exchange in concentration of CO_2 and H_2O in the air. Heteroatomic molecules such as CO_2, H_2O, NO, and NH_3 absorb infrared radiation in specific infrared wavebands. Gas molecules with two identical atoms (e.g. N_2, O_2, H_2) do not absorb infrared radiation. A large number of infrared gas analyzers are available commercially for plant science application (Ciras-2, PP-Systems; LI-6400, Li-Cor Biosciences; LC*pro*, ADC Bioscientific). The most of analyzers act as

Fig. 2.1 Experimental setup. The trap of Venus flytrap (*D. muscipula*) is enclosed in hermetically closed gas exchange cuvette with attached fluorocamera. The thin wire in the needle from syringe is hermetically sealed and used for trigger hair stimulation

absorptiometers measuring at ~4.26 μm (CO_2) and ~2.60 μm (H_2O). This means, that besides the rate of photosynthesis (P_N) and respiration (R_D), you can also measure the transpiration rate (E) and stomatal conductance (g_s). Absorption at any wavelength follows the Beer–Lambert Law so that if gas containing CO_2 and H_2O is flushed through a tube with an infrared light source at one end and an infrared light detector at the other end, the signal from the detector will decline with increasing CO_2 and H_2O concentration (Hunt 2003). The best equipment for such measurements is true differential system with four independent infrared gas analyzers, two for CO_2 (one for analysis air from the cuvette with leaf and one for reference air) and two for H_2O (also one for analysis air from the cuvette and one for reference air) eliminating the problems associated with "gas switching." The difference between analysis and reference air is used for calculation of P_N, R_D, g_s and E, in modern equipment this is automatically done by computer software (care must be taken, whether the leaf covers the entire cuvette area, otherwise the recalculation per area, or per dry weight is necessary). Equipments for high level research has usually independent automatic CO_2 and H_2O control and self calibration checking facility as well as control of temperature, light and humidity with powerful, flexible software allowing simple, individual measurements to more complex, automated and preprogrammable measurements. Wide range of cuvettes and chambers are also commercially available. All modern automatic cuvettes are constructed from materials that have minimal absorption of CO_2 and H_2O, and have closed cell gaskets that give perfect hermetically leaf seal. Automatic cuvettes contain sensors to measure leaf and air temperature (these are important for calculation of parameters, e.g. g_s) and photosynthetically active radiation (PAR). As a source of PAR, LED light sources produce very little heat and become increasingly popular. They usually produce light in narrow wave length, the most common are red and blue LEDs. It is very important, especially in studying the effect of electrical signals on photosynthesis, to take into account the time taken for gas to pass from the cuvette to the infrared gas analyzer, because the changes in CO_2 and H_2O are recorded with constant delay at a given flow rate of air (Fig. 2.1).

2.2.2 Chlorophyll Fluorescence Measurements

Gas-exchange measurements are powerful tool for studying respiration, the dark reaction of photosynthesis (CO_2 assimilation) and water transpiration, the methods of chlorophyll fluorescence are useful for quantification of light reaction of photosynthesis occurring at the level of photosystems. Ground state molecules of chlorophylls in a leaf can absorb the light energy. Absorption of light excites the chlorophyll to a higher energy state, where it can be stable for a maximum of several nanoseconds (10^{-9} s) and soon return to its ground state. The energy difference between the ground and excited state can be used to drive photosynthesis (photochemistry), excess energy can be dissipated as heat or it can be reemitted as light (chlorophyll fluorescence). These three processes are in competition. Increase in the efficiency of one will results in a decrease in the yield of the other two. Under physiological condition, fluorescence signal is assumed to originate mainly from photosystem II (Maxwell and Johnson 2000).

Considering the competition of fluorescence with photochemistry, two extreme situations are possible. All reaction center can be open (i.e. electron acceptor plastoquinone A, Q_A is oxidized) or closed (i.e. Q_A is reduced), with fluorescence yield being minimal (F_0) and maximal (F_m), respectively. Prior to F_0 measurement, it is important that the leaf sample was sufficiently dark-adapted (20–30 min). Then minimal fluorescence (F_0) can be determined at very low photosynthetic active radiation (PAR) below 0.1 µmol m^{-2} s^{-1} PAR. It is a fluorescence signal that comes from excited chlorophylls of light harvesting antenna of PS II before the excitation reaches the reaction center. Maximal fluorescence (F_m) can be determined by strong saturation pulse of light (several thousand µmol m^{-2} s^{-1} PAR) with duration around 1 s. This rise of fluorescence signal has been explained as a consequence of reduction of plastoquinone and in particular Q_A. Reduced Q_A decrease the capacity for PSII photochemistry almost to zero by inhibition of charge separation (P680$^+$ Pheo$^-$) and an increase in a charge recombination (backward electron transport from Pheo$^-$ to P680$^+$) by electrostatic repulsion of the charge on Q_A^-. This means that under physiological condition, the rate of energy conversion at PSII reaction center is acceptor side limited [sufficient electrons are available from H_2O splitting oxygen evolving complex (OEC)] and, hence, fluorescence yield is controlled by the quencher Q_A. From F_0 and F_m variable fluorescence (F_v) can be calculated as $F_m - F_0$ and maximum quantum efficiency of PSII photochemistry F_v/F_m as ($F_m - F_0$)/F_m, which is proportional to quantum yield of O_2 evolution. Dark-adapted values of F_v/F_m are used as a sensitive indicator of plant photosynthetic performance, with optimal values of around 0.83 (Maxwell and Johnson 2000).

After the initial rise of chlorophyll fluorescence signal, the fluorescence level typically starts to fall during few minutes and steady-state level is achieved (F_t). This phenomenon termed fluorescence quenching (Kautsky curve) is explained in two ways. Photochemical quenching (qP) is process by which the electrons move away from PSII due to light-induced activation of enzymes involved in carbon metabolism. Nonphotochemical quenching (NPQ) is increase in the efficiency with which the absorbed energy is converted to heat. Both qP and NPQ help to minimize production of triplet ^{3}Chl*, which can transfer energy to O_2 to generate singlet oxygen ($^1O_2^*$), an extremely damaging reactive oxygen species. Quenching analysis can distinguish between qP and NPQ. After initial assessment of F_0 and F_m in dark-adapted samples actinic light is turn on, and at appropriate intervals, further saturation flashes are applied. It is recommended to use actinic light at low intensity (<100 µmol m^{-2} s^{-1} PAR) for studying the electrical signals on chlorophyll fluorescence, because at higher PAR, electrical signals produce no additional fluorescence quenching as was found by Krupenina and Bulychev (2007). Thus, maximal fluorescence in light-adapted state (F_m') as well as steady-state fluorescence (F_t) is determined. The value of F_m' is usually lower than F_m in dark-adapted sample, because NPQ is competing for the excited states and the difference between F_m' and F_m reflects fluorescence quenching due to heat dissipation. Finally, the minimal fluorescence in the light-adapted state (F_0') can be measured by turning off the actinic light or by applying a far-red light. From these data several parameters can be calculated. The most useful parameter is effective photochemical quantum yield of PSII (Φ_{PSII}, but many alternative terms exist in literature, for overview see Roháček 2002), which is calculated as $\Phi_{PSII} = (F_m' - F_t)/F_m'$. This parameter measures the proportion of light absorbed by chlorophylls associated with PSII that is used in photochemistry and strong linear relationship exists between this parameter and P_N. Another parameter, qP = $(F_m' - F_t)/(F_m' - F_0')$ measures the proportion of reaction centers that are open (i.e. the proportion of plastoquinone molecules that are oxidized). On the contrary, 1 − qP gives an indication of reaction centers that are closed and is sometimes termed as excitation pressure at PSII. NPQ parameter can be calculated as $(F_m - F_m')/F_m'$ and measures the thermal dissipation process at PSII. NPQ can be divided into at least three different components according to their relaxation kinetics. The major component of NPQ is pH-dependent or energy state quenching (qE) activated by light. Absorption of light that exceeds a plant capacity for CO_2 fixation results in a build up of thylakoid ΔpH that is generated by electron transport. Low pH activates violaxanthin de-epoxidase which transforms violaxanthin to zeaxanthin and harmlessly dissipates excess light energy as heat; however, zeaxanthin-independent quenching in PSII core and lutein may also contribute to heat dissipation (Finazzi et al. 2004; Johnson et al. 2009). A second component, state transition quenching (qT) is caused by uncoupling of light harvesting complexes from PSII. The third component of NPQ is related to photoinhibition (qI). The best way, how to distinguish among these three components, is relaxation kinetic measurements. Whereas, qE relaxes within seconds to minutes in the dark, qI shows very slow relaxation kinetics in the range of hours (for details see Müller et al. 2001). Sequence of a typical chlorophyll fluorescence trace is shown in Fig. 2.2. If the plant response to electrical signals is leaf folding (e.g. *Mimosa* and *Dionaea*), it is necessary to fix the leaf in proper position to avoid changes in light interception. Such changes may be source of errors in calculation of Φ_{PSII} and other parameters and the changes in Φ_{PSII} are then not caused by electrical signal

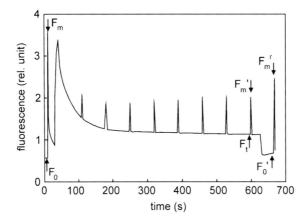

Fig. 2.2 Sequence of a typical fluorescence trace. Minimum fluorescence (F_0) is measured by weak light (<0.1 μmol m^{-2} s^{-1} PAR). Then saturation flash is applied (4,000 μmol m^{-2} s^{-1} PAR, 800 ms duration) and maximum fluorescence (F_m) is measured. Actinic light is turn on (in this case 300 μmol m^{-2} s^{-1} PAR) and fluorescence is quenched to a new steady-state level in the light (F_t). Simultaneously, saturation pulses of the same intensity and duration are given at regular interval. This allows the maximum fluorescence in the light (F_m') to be measured. Turning off the actinic light allows the minimum fluorescence "in the light" (F_0') to be estimated. Recovery of maximum fluorescence after switching off the light (F_m^r) is measured by saturation pulse

itself, but by changes of leaf position and light interception. For more details about chlorophyll fluorescence measurements see Maxwell and Johnson (2000), Müller et al. (2001), Blankenship (2002), Roháček (2002) and Papageorgiou and Govindjee (2004).

On onset of illumination in dark-adapted leaf, fluorescence yield rises in two steps (J, I) from minimal level (F_0) to a peak level (F_m). These changes in fluorescence yield have been called fluorescence induction or fluorescence transient (O-J-I-P) and can be used as a quick monitor of the electron acceptor side reaction (Fig. 2.3). It has been proposed that O step (40–50 μs, the same as F_0) is the fluorescence signal coming from excited chlorophylls of light harvesting antenna before the excitations reach reaction center of PSII, J step (2–3 ms) reflects light-driven accumulation of Q_A^-, and I (30–50 ms) and P step (200–500 ms) reflect light-driven accumulation of Q_B^- and Q_B^{2-} respectively; however, several other explanations have been proposed, for example, electron transport reaction beyond PSII also effect the shape of induction curve (for review see: Lazár 2006, 2009; Vredenberg 2011). When a sample is treated with DCMU (3-(3′,4′-dichlorophenyl)-1,1-dimethylurea; an inhibitor of electron transport between Q_A and Q_B) fluorescence induction is characterized by a steep fluorescence increase, reaching maximal saturation level at the position of J step confirming that it is mainly Q_A^- that accumulates in the position of the J step (Strasser and Govindjee 1992). In the absence of DCMU, Q_A may not be fully reduced in the J step, because electrons move from Q_A^- further towards PSI, and fluorescence reaches maximum at the

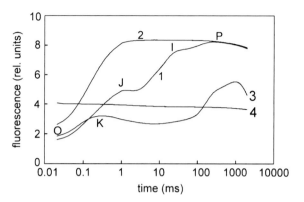

Fig. 2.3 Chlorophyll fluorescence rise on onset of illumination in dark-adapted control leaf (*1*), leaf incubated in 100 mM DCMU (*2*), heat-treated leaf (*3*) and chlorophyll solution in 80% acetone (no variable fluorescence is detected, *4*). Five distinct phases is possible to recognize (O-K-J-I-P). The transient is plotted on logarithmic time scale

position of P step (Fig. 2.3). Apart from light-driven accumulation of reduced PQ pool, it has been proposed that electrical voltage across thylakoid membrane (thylakoid membrane potential) may affect I-P phase (Vredenberg and Bulychev 2002; Vredenberg 2004). This assumption is important because APs have significant effect on I-P rise of chlorophyll fluorescence as has been documented recently (Pavlovič et al. 2011). When the sample is treated with high temperature, a new step at 300–400 µs denoted as K-step appears reflecting an inhibition of OEC or donor side of PSII (Srivastava et al. 1997). Chlorophyll in acetone solution does not show any variable fluorescence (Fig. 2.3). For details of chlorophyll a fluorescence transient see Schreiber (2004), Strasser et al. (2004) or Lazár (2006).

The experimental protocols developed for non-imaging fluorometry described above can be successfully adapted also to imaging fluorometers. Due to technical limitation the fast chlorophyll fluorescence transient (O-J-I-P) can not be measured by imaging fluorometers. Chlorophyll fluorescence imaging has the capacity to resolve photosynthetic performance over the leaf surface. Its application is growing rapidly including also plant electrophysiology. In contrast to non-imaging fluorometry, which measures an average fluorescence signals over the sample area, imaging fluorometry may reveal spatial heterogeneity in photosynthetic performance in response to electrical signals. Such spatial heterogeneity was documented, for example, in carnivorous Venus flytrap (*D. muscipula*) in response to trigger hair irritation (Pavlovič et al. 2010, 2011) as well as in maize (Grams et al. 2009) and *Mimosa* leaf in response to heat treatment (Koziolek et al. 2003; Kaiser and Grams 2006). Simultaneous measurement of chlorophyll fluorescence imaging and gas exchange is the best way how to monitor the photosynthetic response to electrical signals (Fig. 2.1). Fortunately, such systems are now commercially available (e.g. Photon System Instruments and PP-Systems). For details about chlorophyll fluorescence imaging see Nedbal and Whitmarsh (2004).

2.2.3 Polarographic O_2 Measurements

Not very often used in terrestrial but the indispensable method in aquatic plants or algae is the polarographic measurements of O_2 evolution which come from photolysis of water (Hill reaction). Regarding studies on electrical signaling this method was successfully used by Alexander Bulychev group in *Chara corallina* (Bulychev and Kamzolkina 2006a, b). An O_2 electrode (Clarke type) consists of a platinum cathode and a silver anode linked by an electrolyte. The current flowing between anode and cathode is directly proportional to the O_2 concentration. For more details see Hunt (2003).

2.3 Effect of APs on Photosynthesis

APs are rapidly propagated electrical messages that are well known in animal kingdom. The first message about APs in plants comes from Burdon–Sanderson (1873) who conducted experiments with carnivorous plant Venus flytrap (*D. muscipula*) on Charles Darwin request. The APs are usually induced by non-damaging stimuli (e.g. cold, mechanical and electrical stimulation, irrigation etc.) and have three major features: (1) AP transmits at constant velocity and amplitude, (2) AP follows the all-or-none law, (3) after AP is generated, the cell membrane enters into refractory periods (Yan et al. 2009). While the ionic mechanism of APs in animals depends on inward-flowing Na^+ (depolarization) and outward-flowing K^+ ions (repolarization), excitation of plant cells depends on Ca^{2+}, Cl^-, and K^+ ions (Fromm and Lautner 2007). After perception of external stimulus, Ca^{2+} flow into the cell. The elevation of Ca^{2+} concentration in the cytoplasm activates the anion channel and Cl^- efflux depolarizes the plasma membrane. Then outward rectifying K^+ channel would repolarize the membrane (Yan et al. 2009). In higher plants, the APs can travel over short as well as long distances through plasmodesmata and phloem, respectively (Sibaoka 1962, 1991; Fromm and Eschrich 1988b; Fromm and Lautner 2007).

It is known that electric signals arising at the plasma membrane are transmitted to the level of thylakoid membrane and may affect the photosynthetic reaction, but the mechanism is still not fully understood (Bulychev and Kamzolkina 2006a, b). Below you will find some typical examples for different group of plants and algae.

2.3.1 Chara cells

Internodal cells of characean algae represent a convenient model for studying plant cell excitability (Beilby 2007). The internodes of Characeae are suitable for exploring the interaction between electric excitation, photosynthesis, and ion transport for the presence of single layer of chloroplasts in the peripheral part of cytoplasm. *Chara* cells exposed to illumination form domains with H^+ extrusion and H^+ sink that account for pH banding (Lucas 1975). The term pH banding designates the parts of cell producing alkaline and acid regions along the cell length in the outer medium. This spatial arrangement has significant effect on photosynthesis. Plieth et al. (1994) and Bulychev et al. (2001) found by using chlorophyll fluorescence imaging and microfluorometry, respectively, that electron transport was higher in the acid than in the alkaline bands. Polarographic measurements also revealed longitudinal profile of O_2 evolution from Hill reaction. Plieth et al. (1994) suggested better supply of CO_2 in the acid region, because equilibrium between CO_2 and bicarbonate is shifted towards CO_2, in contrast to alkaline region, where is shifted toward bicarbonate, which poorly permeates through the membranes. In general, changing CO_2 concentration modulates the activity of the Calvin cycle and magnitude of electron transport from PSII and thus chlorophyll fluorescence (Schreiber 2004).

APs have significant effect on pH banding. After electric stimulation and subsequent generation of AP, the pH decreases in alkaline regions resulting in the smoothening of the pH profile. The pH decrease in alkaline zone is accompanied by a reduction of F_m' and Φ_{PSII} in contrast to the acid zone, where such changes were not observed (Bulychev et al. 2004; Bulychev and Kamzolkina 2006a, b). Whereas, AP develops on a time scale from milliseconds to seconds, subsequent pH and Φ_{PSII} changes proceed within minutes. Krupenina and Bulychev (2007) called it as a long-lived state effect of AP on photosynthesis. Electrical excitation caused parallel decrease of Φ_{PSII} and pH what disrupts the usual inverse relation between Φ_{PSII} and pH found by Plieth et al. (1994), thus CO_2 availability in ambient medium is unlikely to play any role in the AP-induced decrease of Φ_{PSII} (Bulychev et al. 2001).

AP has no effect on F_m but have on F_m' indicating that excess energy absorbed by chlorophylls was dissipated by NPQ as heat. The effect of AP on Φ_{PSII} and F_m' was most pronounced at light-limiting condition. The increase in light intensity elevated NPQ and diminished the AP-induced F_m' changes. Rapid dark relaxation of F_m' and experiments with protonophorous uncoupler monensin and nigericine confirmed that NPQ is related to the formation of thylakoid ΔpH and qE and not to photoinhibition (Bulychev and Kamzolkina 2006a, b; Krupenina and Bulychev 2007). Application of ionophore A23187 which increase Ca^{2+} concentration in cytoplasm has similar effect as APs. Ca^{2+} plays a central role in a membrane excitation of plant cells and may be involved also in suppression of photosynthesis by indirect or direct way. Increase of cytosolic Ca^{2+} during AP may activate Ca-dependent protein kinases that modulate ion channels and pumps activity including transmembrane H^+ flow. This may affect cytoplasmic pH and subsequently the function of chloroplasts by indirect way. However, it can not be excluded a direct role of Ca^{2+}. Chloroplast envelope and thylakoid membrane contain electrogenic Ca^{2+} pumps and Ca^{2+}/H^+ antiporters which play role in light-induced depletion of cytosolic and stroma free Ca^{2+}, respectively (Kreimer et al. 1985; Ettinger et al. 1999). In chloroplast Ca^{2+} acts as cofactor in the OEC and may also influence the activity of Calvin cycle enzymes. Suppression of Calvin cycle decreases ATP consumption. This would inhibit ATP synthesis, raise ΔpH across the thylakoid membrane and activates NPQ (Bulychev et al. 2004; Krupenina and Bulychev 2007).

2.3.2 Carnivorous Plant Venus Flytrap (D. muscipula)

The most famous carnivorous plant, the Venus flytrap (*D. muscipula*), catches prey by rapid leaf movement. The plant produces a rosette of leaves, each divided into two parts: the lower is called the lamina and the upper is called the trap. The trap consists of two lobes, which close together after irritation of trigger hairs, which protrude from digestive zone of the trap. Two touches of trigger hairs activate the trap which snaps in a fraction of second at room temperature and AP play an indispensable role in this movement (Juniper et al. 1989). AP is generated at the base of the stimulated trigger hair after mechanical, electrical, or chemical stimulation and spreads in all directions over the whole surface of both lobes (Williams and Pickard 1980; Hodick and Sievers 1988; Sibaoka 1991; Volkov et al. 2008a, b). Volkov et al. (2007) found that the AP had duration 1.5 ms and velocity of 10 m s^{-1}. After the rapid closure secures the prey, struggling of the entrapped prey in the closed trap against trigger hairs results in generation of further APs, which cease to occur when the prey stop moving (Affolter and Olivo 1975).

APs spread over the trap surface and were not recorded in adjacent lamina (Volkov et al. 2007). This is in accordance with the observations that rapid changes of photosynthesis are confined to the digestive zone of trap and were not found in photosynthetic lamina (Pavlovič et al. 2010; Pavlovič and Mancuso 2011). Trigger hair irritation and rapid trap closure resulted in rapid inhibition of P_N and Φ_{PSII}, whereas g_s was not affected. It can be argued that such changes may be associated with changes of trap geometry; however, the effect is similar and even stronger after repeated trigger hair irritation in closed trap. The changes in P_N and Φ_{PSII} are rapid and occur usually within 2–4 s after irritation what is in accordance with rapid propagation of AP through the trap lobe (Pavlovič et al. 2010). A detailed analysis of chlorophyll a fluorescence kinetics revealed the possible mechanism underlying photosynthetic limitation. The decrease of Φ_{PSII} is caused at first by increase in F_t, whereas F_m' was not affected immediately after irritation (Figs. 2.4d, 2.5). This indicates that plastoquinone pool became more reduced (lower qP) what results in a traffic jam of electrons and increase of excitation pressure at the PSII reaction center, promoting photoinhibition. Later F_t is quenched by NPQ as indicates large drop of F_m' thus preventing PSII against photoinhibition. Rapid relaxation of F_m' confirmed that the large drop of F_m' is associated with qE and not to qI. Uncoupling the light and dark enzymatic reaction of photosynthesis by unavaibility of CO_2 had no inhibition effect of AP on Φ_{PSII}, indicating that changes in Φ_{PSII} are mainly caused by inhibition of dark reaction of photosynthesis by feedback mechanism. Repeating the experiments in the dark,

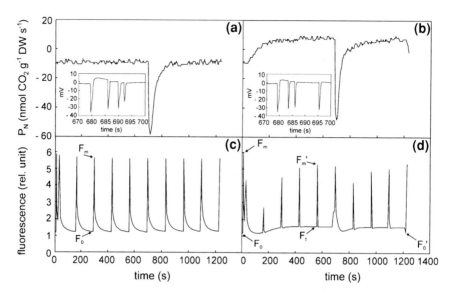

Fig. 2.4 Response of Venus flytrap (*D. muscipula*) to trigger hair irritation and subsequent generation of APs (figure *insets*) on gas exchange (**a, b**) and chlorophyll fluorescence (**c, d**). The trigger hairs were irritated between 670 and 700 s in the dark (**a–c**) and at light intensity 100 μmol m^{-2} s^{-1} PAR (**b, d**), F$_0$—minimal fluorescence in the dark adapted trap, F$_m$—maximum fluorescence in the dark after saturation pulse (4,000 μmol m^{-2} s^{-1} PAR), F$_m'$—maximum fluorescence in the light after saturation pulse (4,000 μmol m^{-2} s^{-1} PAR), F$_t$—steady-state fluorescence at light intensity 100 μmol m^{-2} s^{-1} PAR, F$_0'$—minimum fluorescence after light treatment measured in the dark at <0.1 μmol m^{-2} s^{-1} PAR. **a, c**, and **b, d** measurements were recorded simultaneously in the dark and light, respectively. The APs were recorded by extracellular measurements as described in Pavlovič et al. (2011)

where the enzymes of Calvin cycle are inactivated, revealed that APs have only very small but significant effect on decrease of F$_0$ and increase of F$_m$ and thus increase of maximal quantum yield of PSII (F$_v$/F$_m$, Pavlovič et al. 2011). However, this assumption is valid only if we consider that changes in fluorescence yield are determined only by the redox-state of PSII reaction center (reduction of Q$_A$) according to Strasser et al. (2004) what seems is not *apriori* true. A three state trapping model proposed by Vredenberg (2000, 2004) suggests that the saturation of photochemistry does not necessarily result in saturation of changes in fluorescence yield. Analysis of O-J-I-P curve after AP revealed that the decrease in the O-J-I rise and increase in the I-P rise found by Pavlovič et al. (2011) is in accordance with electrochemical stimulation of the fluorescence yield supplementary to qP (Vredenberg and Bulychev 2002, 2003; Vredenberg 2004; Vredenberg et al. 2009). Electrical field in the vicinity of the reaction center can influence the chlorophyll fluorescence (Meiburg et al. 1983; Dau and Sauer 1991, 1992; Bulychev and Vredenberg 1999; Vredenberg and Bulychev 2002). Apart from Q$_A$ oxidation, electrical field exerts its effect on charge separation and recombination reaction by decrease in Gibbs free-energy difference between the excited states in reaction center II and the charge separated state (P$_{680}^+$ Pheo$^-$, Dau and Sauer 1991, 1992).

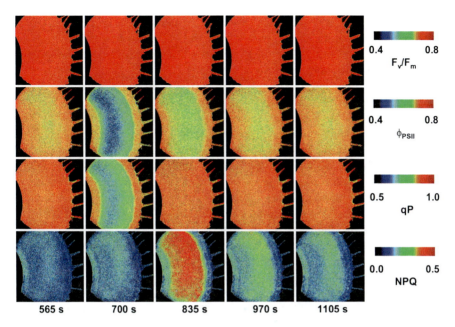

Fig. 2.5 Chlorophyll fluorescence parameters showed in 2-D images from experiment presented in Fig. 2.4 in response to trigger hair irritation (between 670 and 700 s) in Venus flytrap (*D. muscipula*). F_v/F_m—maximum quantum yield of PSII, Φ_{PSII}—effective photochemical quantum yield of PSII, *qP*—photochemical quenching, *NPQ*—nonphotochemical quenching

However, the shape of O-J-I-P curve, mainly I-P phase, is also influenced by reactions beyond PSII (cyclic electron flow around PSI, Schansker et al. 2005; Lazár 2009) so the effect of APs on PSI and cyclic electron flow can not be excluded. Indeed, Vredenberg and Bulychev (2010) and Vredenberg (2011) supposed that photoelectrical control is exerted by the PSI-powered proton pump associated with cyclic electron transport. These data indicate that besides the effect of APs on dark reaction of photosynthesis there is also a significant effect on light reactions (electron transport) but this will need further examination.

These results may have more general consequences in measurements of chlorophyll fluorescence. The pulse of light, which is commonly used during chlorophyll fluorescence measurement, may trigger AP at plasma membrane or even changes in thylakoid membrane potentials (Trebacz and Sievers 1998; Bulychev and Vredenberg 1999; From and Fei 1998; Pikulenko and Bulychev 2005). If changes of membrane potentials have effect on fluorescence yield then variable fluorescence in dark as well as light-adapted state are not exclusively related to quantum photochemical yield of PSII and qP, but incorporate also photoelectrochemical and photoelectrical component (Vredenberg et al. 2009; Vredenberg 2011). In other words, the F_m measured by saturation light pulse is different from the original hypothetic dark-adapted state, because pulse of light triggers also changes in membrane potential in the vicinity of reaction center, which affect reaction in PSII. However, at high light intensity (routinely used for estimation of F_v/F_m), the I-P phase contributes little to the variable fluorescence and its effect on quantum yield is not substantial.

The above mentioned response to APs is direct and short term. However, it is tempting to assume that long term and AP-indirect inhibition of photosynthesis may occur in response to prey capture in Venus flytrap. Recently, Escalante–Pérez et al. (2011) found, that 30 min and 1 week after prey capture, the level of 12-oxophytodienoic acid (OPDA, precursor of JA) significantly increased. The increased level of OPDA is systemic and makes other traps more sensitive to mechanical irritation. Taken into account that JA inhibits photosynthesis (Herde et al. 1997), prey capture, retention and digestion may further inhibits photosynthetic assimilation even after the stop of prey moving. This will need further examination.

We can expect similar response of AP on photosynthesis also in related carnivorous plant *A. vesiculosa*; however, this plant species has not been investigated in this respect. Extensive electrical activity has also been recorded in the tentacle of carnivorous genus *Drosera*; however, the APs do not propagate to the photosynthetically active leaf lamina and have no effect on photosynthesis (Williams and Pickard 1972a, b, 1974, 1980; Williams and Spanswick 1976; Pavlovič 2010).

2.3.3 Mimosa Pudica

When the tip of leaf pinna in *M. pudica* is stimulated by cooling, touching, or electrically (Volkov et al. 2010) an AP is evoked and transmitted basipetally with a speed 20–30 mm s^{-1}, what results in leaflet folding by a sudden loss of turgor pressure in the motor cells of the pulvinus (Fig. 2.6a, Fromm and Eschrich 1988c, Fromm and Lautner 2007). In contrast to *D. muscipula*, where all the major tissues of trap lobes (sensory cells of trigger hairs, upper, and lower epidermis, mesophyll cells) are excitable with similar resting potentials (150–160 mV, Hodick and Sievers 1988), excitable cells in *Mimosa* are found in vascular bundles and pulvinus (Sibaoka 1962; Samejima and Sibaoka 1983; Fromm and Eschrich 1988b). In *Mimosa,* the phloem and pulvinus is surrounded by sclerenchyma and collenchyma sheath to restrict electrical signaling to phloem and lateral propagation (Fleurad-Lessard et al. 1997; Fromm 2006; Fromm and Lautner 2007). Hoddinott (1977) found that following the mechanical stimulation and leaf folding, P_N declines by 40%. However, this declines is not caused by AP itself, but by leaf folding and reduced leaf area able to receive incident illumination. My unpublished results are in accordance with the observation that mesophyll cells of leaflets of *Mimosa* are probably not excitable in contrast to *Dionaea* and colling and touching of pinna has no inhibition effect on Φ_{PSII} in fixed paired leaflets (in contrast to VP evoked by heating, Figs. 2.6, 2.7, 2.8).

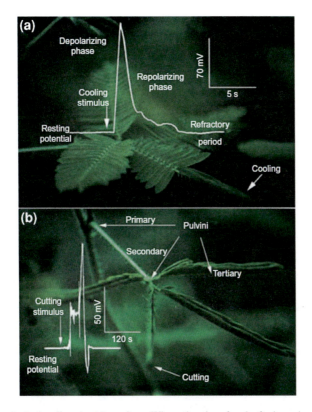

Fig. 2.6 Electrical signaling in *M. pudica*. When the tip of a leaf pinna is stimulated by spontaneous cooling with ice water or mechanically by touch, an AP is evoked and transmitted basipetally within the rachis with the speed of 20–30 mm s^{-1}. The tertiary pulvini at the base of the leaflets respond to the AP, causing ion and water fluxes that lead to leaf movements. This type of signal stops at the base of the pinna, and no further transmission occurs (**a**). When the leaf is stimulated by cutting or heating, a basipetally moving VP is generated in the rachis, irregular in shape, and long in duration. Its speed is slower (5–6 mm s^{-1}) than that of the AP; however, it is able to pass through secondary pulvini at the base of the pinna and causes leaflet folding of neighbouring pinna, and also bending of the primary pulvinus at the base of the petiolus (from Fromm and Lautner 2007, courtesy of prof. Jörg Fromm and John Wiley and Sons) (**b**)

2.3.4 Other Plant Species

APs are not confined only to the carnivorous and sensitive plants but occur in all green plants around us. Whereas in *Chara* and *Dionaea* the APs generated by electrical or mechanical stimulation, respectively, had inhibition effect on photosynthesis, several other examples have been documented with no or even stimulatory effect on photosynthesis (Fromm and Fei 1998; Lautner et al. 2005; Grams et al. 2007). Such contrasting response is difficult to explain, but may be based on amplitude or on restriction of AP propagation mainly to phloem, without reaching

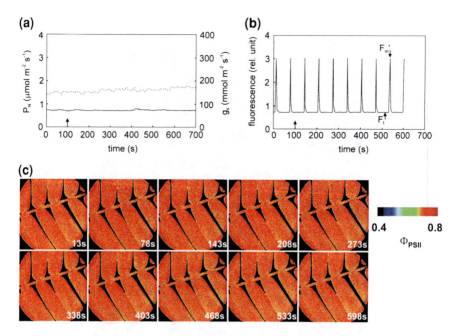

Fig. 2.7 Photosynthetic response of *M. pudica* to cooling. The tip of pinna was cooled by ice water at time 100 s while the basal part of the same pinna was enclosed in gas exchange cuvette. Rate of net photosynthesis (P_N, *dashed line*), stomatal conductance (g_s, *solid line*) (**a**), trace of chlorophyll fluorescence (**b**), and spatio-temporal changes of effective photochemical quantum yield of PSII (Φ_{PSII}) calculated from fluorescence trace (**c**). *Arrow* on x-axis indicates the time of cooling. F_m'—maximum fluorescence in the light after saturation pulse (4,000 µmol m^{-2} s^{-1} PAR), F_t—steady-state fluorescence at light intensity 100 µmol m^{-2} s^{-1} PAR

the mesophyll cells (no response), or stimulation of stomata opening in drought stress plants with subsequent increase of intercellular CO_2 concentration (c_i) for photosynthesis and thus stimulation of photosynthesis.

Stimulation of leaf or stem with ice water in poplar tree resulted in generation of typical APs with amplitude around 25 mV in phloem (measured using aphid stylet); however, no photosynthetic response in leaves was found (Lautner et al. 2005). After watering the roots of maize in drying soil, an AP with average amplitude of 50 mV was evoked and measured on the leaf surface and phloem. At the same time the P_N and g_s began to rise. Application of dye solution and microscopic observations revealed that changes in gas exchange were not triggered by water ascent but by AP (Fromm and Fei 1998). After watering the plant, VPs are also often generated as a result of pressure increase in the xylem (Stahlberg and Cosgrove 1997; Mancuso 1999). To eliminate the effect of hydraulic signals, Grams et al. (2007) applied compensatory pressure to the root system of maize and found similar effect of AP on P_N and g_s as Fromm and Fei (1998). It is interesting that Φ_{PSII} did not change significantly. Similarly, application of auxin (IAA) and cytokinins triggered AP and 3 min later the P_N and g_s increased in willow (*Salix*

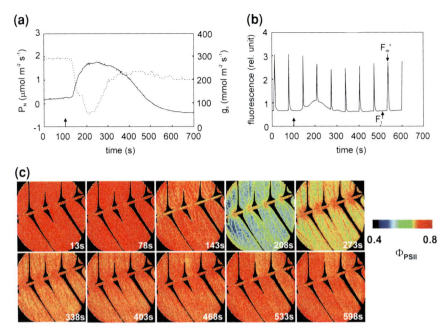

Fig. 2.8 Photosynthetic response of *M. pudica* to heating by flame. The tip of pinna was heated by flame at time 100 s while the basal part of the same pinna was enclosed in gas exchange cuvette. Rate of net photosynthesis (P_N, *dashed line*), stomatal conductance (g_s, *solid line*) (**a**), trace of chlorophyll fluorescence (**b**), and spatio-temporal changes of effective photochemical quantum yield of PSII (Φ_{PSII}) calculated from fluorescence trace (**c**). *Arrow* on x-axis indicates the time of heating. F_m'—maximum fluorescence in the light after saturation pulse (4,000 µmol m^{-2} s^{-1} PAR), F_t—steady-state fluorescence at light intensity 100 µmol m^{-2} s^{-1} PAR

viminalis, Fromm and Eschrich 1983). Authors believe that the response is too fast to be explained by moving a signaling substance to the leaves via transpiration stream.

The response described above is in contrast to the response found in *D. muscipula*, where AP generation has no effect on g_s and inhibits P_N and Φ_{PSII} (Pavlovič et al. 2010; 2011). To explain the adverse effect of APs on photosynthesis in different plant species need further examination.

2.4 Effect of VPs on Photosynthesis

VPs are generated in response to damaging stimuli (wounding, organ excision, flaming etc.) Many studies have suggested a chemical means of propagation in which wounding releases a chemical substance Ricca's factor, which is moved by the xylem flow causing electrical changes along the path (Ricca 1916). Published evidences indicate that hydraulic event produced by wounding is the signal responsible for systemic induction of the VP (Malone and Stankovič 1991). The hydraulic surge precedes the electrical changes, thus VP seems to be a local response to the passage of an hydraulic wave (Stahlberg and Cosgrove 1997; Mancuso 1999). In contrast to APs, magnitude and shape of VPs vary with the intensity of the stimulus and decrease with increasing distance from the injured site, have prolonged repolarization and are able to pass through dead plant tissue (Mancuso 1999; Fromm and Lautner 2007; Yan et al. 2009). The ionic mechanism differs from APs and involves a transient inactivation of proton pumps in the plasma membrane (Stahlberg and Cosgrove 1996). Sometimes APs and VPs may be confused and in some cases their affiliation is difficult, not to mention that are often generated together (Mancuso 1999; Grams et al. 2009).

In contrast to APs, where different response to photosynthesis has been recorded, in the case of VPs only inhibition of photosynthesis has been documented. Below you will find same typical effects of VP on photosynthesis.

2.4.1 Mimosa Pudica

The best documented effect of VPs on photosynthesis is in the sensitive plants *M. pudica*. In contrast to the AP, which is generated by touching or cooling and is stopped at the base of pinna, VP generated by cutting or heating passes through the secondary pulvinus at the base of a pinna and moves into three neighboring pinna and causes a bending movement (Fig. 2.6b). Further, the VP propagates basipetally through petiolus causes a bending movement of the primary pulvinus and pulvini from distant leaves (Fromm and Lautner 2007). Heat stimulation by the flame of one pinna resulted in generation of VP, which spread with velocity of 4.0–8.0 mm s^{-1} to neighboring fixed pinna where decrease of Φ_{PSII} was recorded starting 90–170 s after heat stimulation of the first pinna. This inhibition propagated acropetally throughout the pinna. The decrease of Φ_{PSII} has spatio-temporal distribution, as chlorophyll fluorescence imaging indicates. In the intervein area, where the photosynthetically most active mesophyll is located, the strongest response was detected, but delayed in comparison to leaflet vein, what indicates that the VP spread via the veins. After 450 s Φ_{PSII} reached a similar level as before the stimulus. Concurrently, the P_N sharply dropped to about compensation point and g_s first rapidly increased and then declined to a lower values, indicating that stomata did not contribute to rapid suppression of P_N. When the leaflets of investigated pinna were allowed to close, the first rapid increase of g_s was not recorded (Koziolek et al. 2003). The microscopic observations and stomata aperture measurements revealed a fast stomata opening movement leading to a doubling of aperture within 1–2 min after heat stimulation and followed by a closing movements, which was completed within 4–5 min after heat treatment (Fig. 2.8; Kaiser and Grams 2006). The authors concluded that the fast opening movements is not osmotically driven but caused by a sudden loss of epidermal turgor. The most likely explanation is that epidermal cells, similar to the pulvinar motor cells, respond to depolarization and decrease a turgor with subsequent hydropassive stomata opening. On the other hand, the second phase—stomata closure can be caused either by a decrease in guard cell turgor or a recovery of epidermal turgor. Besides direct effect of electrical signals on guard cells, it is also possible that stomata closure is triggered indirectly, by increase c_i during inhibition of P_N. To exclude the possibility that chemical signals are responsible for such rapid response in *Mimosa*, Koziolek et al. (2003) determined the speed of a chemical signals by exposing the part of pinna to $^{14}CO_2$. Then the translocation rate of $^{14}CO_2$ was observed. $^{14}CO_2$ signal in neighboring pinna was detected after 12 h, confirming that chemical signals are too slow to account for the photosynthetic response.

2.4.2 Other Plant Species

Flame stimulation of poplar leaf evoked a VP that was recorded in stimulated as well as distant leaf. Signals moving in basipetal direction change toward negative direction (hyperpolarization), which contrasts with the acropetally traveling signals (depolarization). Application of K^+ channel by TEA^+ resulted in inhibition of basipetally transmitted signals, indicating that K^+ efflux caused the hyperpolarization of the plasma membrane. Both, hyperpolarization as well as depolarization had significant effect on photosynthesis. Flaming of the leaf caused substantial decrease of Φ_{PSII} after 80 s at a distance 3 cm from stimulated leaf tip as well as in distant leaf after 240 s. In stimulated leaf, Φ_{PSII} decreased in veins and intervein regions simultaneously, in distant leaf the response was delayed by about 60 s in intervein regions in comparison to veins, indicating that long-distance signaling occurs through the veins. Rate of net photosynthesis decreased; however, g_s was not affected in stimulated and distant leaf. Calcium deficiency, as well as chilling the stem with ice water, disrupted VP transition and photosynthetic response, indicating the direct role of electrical signaling in regulation of photosynthesis (Lautner et al. 2005).

Hlaváčková et al. (2006) measured gas exchange and chlorophyll fluorescence induction curves on the fifth leaf after local burning of the sixth leaf in tobacco. The VPs were evoked with amplitude 10–25 mV which preceded the response in photosynthetic parameters in distant leaf. Five minutes following the burning of the sixth leaf, the g_s and P_N started to decline in fifth leaf and continued for an hour after burning. No changes in c_i were observed indicating that reduction of CO_2 concentrations due to stomatal closure were balanced by reduction of P_N. No changes in O-J-I-P transients were recorded, indicating that local burning did not influence the linear electron transport in PSII directly. Within 8 min after burning, ABA concentration increased and then decreased between 16 and 60 min in fifth leaf. However, JA concentration increased after 15 min of local burning and concentration of salicylic acid (SA) concentration did not change. It is known that both ABA and JA induce stomata closure and inhibition of photosynthesis (Herde et al. 1997). Because electrical signals VP precede the systemic increase in the endogenous concentration of ABA and JA, Hlaváčková et al. (2006) believe that electrical signals (induced by propagating hydraulic signals) first triggered stomata closure directly by ionic fluxes across the membrane of the guard cells with some participation of chemical signals ABA and JA may play a role in the continuing process of stomatal closure in the second phase. However, the signaling pathway leading from electrical signals to the accumulation of ABA and JA is unknown. The external application of an electrical current was able to initiate systemic ABA and JA accumulation, indicating the involvement of ion fluxes in triggering hormonal response (Herde et al. 1996, 1999b). Then ABA may have a stimulatory effect on JA accumulation (Peña-Cortéz et al. 1995), but local burning can apparently activate JA biosynthesis independently (Herde et al. 1996, 1999b). These results provide confirming evidences that electrical and chemical signals interactively participate in photosynthetic responses to local burning (Hlaváčková et al. 2006).

Inhibition of P_N after heat treatment in tomato plants is accompanied also by changes in pigment composition. Increase in zeaxanthin and decrease of violaxanthin concentrations 5 h after heat treatment indicates involvement of xanthophyll cycle in deexcitation of excess absorbed light by NPQ and protection of photosynthetic apparatus against photoinhibition. No changes in chlorophyll concentrations were found confirming that such photoprotection is sufficient and no photo-bleaching occurred (Herde et al. 1999a).

VP and AP are often generated together, but AP usually precedes the VP due to the faster rate of propagation, not to mention that the changes in membrane potentials referred as AP by some authors may be interpreted as VP by other researchers (Stahlberg and Cosgrove 1997; Mancuso 1999). Grams et al. (2009) were not able to affiliate electrical signals evoked by heating in maize leaf. Heat-induced electrical signals are usually VP; however, Grams et al. (2009) reported short repolarization period similar to AP. Also Pyatygin et al. (2008) reported generation of APs in response to burning in 2-week-old pumpkin seedlings. Heat stimulation of the leaf tip of maize plants resulted in transient decline of P_N in central part of leaf lamina. While P_N declined, g_s first rapidly increased and then declined. The changes in fluorescence yield were observed by simultaneous measurements of quantum yields of PSI (excitation at $\lambda = 715$ nm) and PSII (excitation at $\lambda = 440$ nm). Because of chloroplast dimorphism in C_4 plants with bundle sheath chloroplast lacking grana and PSII, simultaneous measurements of PSI and PSII chlorophyll fluorescence may provide further evidence that the electrical signals propagate via the veins. And this was confirmed, heat stimulation affected quantum yield of PSI before PSII. Simultaneously pH was measured using a microprobe. Acidification of cytoplasm and alkalization of apoplast during propagation of electrical signals may trigger the photosynthetic response, as isolated chloroplast showed strong dependence of Φ_{PSII} on pH (Grams et al. 2009).

Current injection and mechanical wounding resulted in two major responses in tobacco and potato plants: a transient stomatal closure within 2–3 min and a more pronounced closure at 10 min with simultaneous decrease in P_N. The membrane potential kinetics resembles the gas exchange kinetics (Peña-Cortéz et al. 1995; Herde et al. 1998). The authors suggested that the fast response is triggered by AP and the second more pronounced response is triggered by VP and subsequently by plant hormones (ABA, JA).

2.5 Possible Mechanism Underlying Photosynthetic Limitation upon Impact of Electrical Signals

Without a doubt, the electrical signals have significant impact on photosynthesis. Whereas the VPs have usually uniform negative impact on P_N, the response of APs range from stimulation, no response to inhibition of photosynthesis. With the current stage of knowledge it is difficult to explain such different reaction in response to APs, especially when it is believed that AP in plants is a nonspecific bioelectric signal which does not carry any stimulus-specific information, but it gives a signal to resting tissue and organs about the onset of adverse conditions in some region (Pyatygin et al. 2008). On the other hand, some results strongly suppose the view that different stimulus-dependent APs cause specific response (Fromm et al. 1995). The different responses to AP may lay in the restriction of AP propagation in phloem without reaching the mesophyll cells (no response), or stimulation of opening closed stomata and thus increase of P_N (e.g. in drought-stressed plants), but this will need further examination. In contrast, this is not the case in VPs which have ability the spread even over dead plant region (Mancuso 1999), thus there is no restriction in propagation and only the negative impact on photosynthesis has been recorded. It can be suggested that AP and VP may have similar mechanism of action on photosynthesis, although both signals depends on different ion translocation. Whereas the APs depends on Ca^{2+}, Cl^- and K^+ flows, the VPs depends on H^+ flow (Mancuso 1999; Fromm and Lautner 2007). However, active transport of H^+ across the plasma membrane could also play an essential role in generation of APs (Opritov et al. 2002; Bulychev and Kamzolkina 2006a, b) and may be responsible for downregulation of photosynthetic assimilation. The chlorophyll fluorescence trace in response to AP and VP is similar (compare Figs. 2.4d and 2.8b). Thus the mechanism underlying photosynthetic limitation on impact of AP and VP is may be similar, dependent on changes in pH in cytoplasm (Grams et al. 2009) although the role of Ca^{2+} can not be excluded. Taken together, the negative impact on photosynthesis is predominant and further I try to outline the possible mechanism underlying photosynthetic limitation on impact of electrical signals.

The first barrier in the pathway of CO_2 molecules into the leaf is stomata aperture. In most cases VPs induce stomata closure (Koziolek et al. 2003;

Hlaváčková et al. 2006; Kaiser and Grams 2006; Grams et al. 2007), but may also induce rapid hydropassive stomata opening in fixed leaflets of *Mimosa*; however, fixed leaflets are not typical for plants in natural habitat (Koziolek et al. 2003; Kaiser and Grams 2006). No response was also recorded (Lautner et al. 2005). On the other hand, APs have no effect on g_s (Pavlovič et al. 2010, 2011) or even induce stomata opening (Fromm and Fei 1998; Grams et al. 2007). Kaiser and Grams (2006) explained such different response by either the decrease of epidermal cells turgor (similar to the pulvinar motor cells) and subsequent rapid hydropassive stomata opening or by decrease of guard cell turgor and subsequent stomata closure. Stomata closure is triggered by ionic fluxes across the plasma membrane of guard cells and the ionic channels participating in this process are activated by membrane depolarization and increase Ca^{2+} concentration (Hlaváčková et al. 2006). Since guard cells are sensitive to CO_2 concentration and APs induce increase in c_i (Pavlovič et al. 2010, 2011) it can not be excluded that stomata closure is driven indirectly by increase CO_2 concentration (Kaiser and Grams 2006). Nevertheless, it seems that stomata limitation of photosynthesis does not occur, because c_i is usually constant during the course of stomata closure (Hlaváčková et al. 2006). The plant hormones (ABA, JA) play later also indispensable role in stomata closure (see discussion in Sect. 2.4.2., Hlaváčková et al. 2006).

It has been hypothesized that the primary target of electrical signals on photosynthesis is in dark reactions (Bulychev and Kamzolkina 2006a, b; Grams et al. 2009; Pavlovič et al. 2011); however, some authors do not exclude direct interference of electron transport chains through direct impact by the electrical signals (Koziolek et al. 2003). It seems that both sites are targeted by electrical signals, but the major impact is in the dark reaction of photosynthesis (Pavlovič et al. 2011). It has been hypothesized that changes in Ca^{2+} or H^+ concentration are involved in inhibition of photosynthesis. For example, a key Calvin cycle enzyme Fru-1,6-bisphosphatase, is Ca^{2+} dependent enzyme inhibited by high Ca^{2+} concentration in chloroplastidic stroma (Hertig and Wolosiuk 1983; Kreimer et al. 1988). During a day, the light-stimulated Ca^{2+}/H^+ antiporter pumps Ca^{2+} from stroma into the thylakoid lumen (where is necessary for OEC assembly) and prevents Ca^{2+}-mediated inhibition of CO_2 fixation in the light (Ettinger et al. 1999; Sai and Johnson 2002). It is possible that Ca^{2+} efflux into the cytoplasm during AP may disrupt these fluxes which maintain low stroma Ca^{2+} concentration, and thus inhibit the enzymatic reactions of photosynthesis. However, Grams et al. (2009) found no changes in Φ_{PSII} with increasing Ca^{2+} concentration in external solution in isolated chloroplasts, but the Φ_{PSII} was sensitive to changes in pH. They supposed that, for example, carbonic anhydrase, a pH dependent enzyme, may be strongly involved in photosynthetic limitation. However, one may suppose that changes in pH are caused by the increase in Ca^{2+} level in cytoplasm. This rise may initiate the reaction sequences that perturb H^+ flows across the plasma membrane (Bulychev and Kamzolkina 2006b). Whatever ion (Ca^{2+} or H^+) is responsible for altering the photosynthetic reactions its major primary target seems is in enzymatic reaction of photosynthesis. This is supported by observations, that first traffic jam of electrons is generated and increase excitation pressure at PSII (1−qP) occurred, and later acidification of thylakoid lumen triggered NPQ and subsequently the excess excitation energy is quenched by harmless manner (Fig. 2.5, Pavlovič et al. 2011).

Besides dark reaction of photosynthesis, the electrical signals also interfered with photochemical reaction in thylakoid membrane. The changes in ionic environment and photoelectrochemical field may influence the charge separation and recombination reactions in PSII (Dau and Sauer 1991, 1992; Meiburg et al. 1983; Pavlovič et al. 2011). In comparison to dark enzymatic reactions, this effect seems to be rather minor than substantial; however, provide important evidence that variable chlorophyll fluorescence is under photoelectrochemical and photoelectrical control (Vredenberg and Bulychev 2002, 2010; Vredenberg 2004; Pavlovič et al. 2011).

2.6 Effect of Electrical Signalling on Respiration

Effect of electrical signalling on respiration has been much less studied than effect on photosynthesis. Since photosynthesis and respiration occur simultaneously, gas exchange measurements can not partition between these two processes. Carbon dioxide flux in gas exchange system in the light corresponds to net photosynthesis (P_N): the rate of CO_2 fixation during photosynthesis (gross photosynthesis, P_G) minus the rate of CO_2 simultaneously lost during respiration (R_D). However, by measuring the gas exchange in response to electrical signals in the dark, it is possible to determine the CO_2 lost by respiration. Figure 2.4a, b document that the R_D is the major contributor to the CO_2 lost in response to APs at least in Venus flytrap (R_D before stimulus in the dark is ∼9 nmol CO_2 g^{-1} DW s^{-1} and after stimulus at low light intensities dropped to ∼50 nmol CO_2 g^{-1} DW s^{-1}, and even to ∼60 nmol CO_2 g^{-1} DW s^{-1} in the dark, indicating rapid stimulation of R_D). This was also documented in the work published by Pavlovič et al. (2010, 2011). This documents the importance of using chlorophyll fluorescence technique in studying the effect of electrical signals on photosynthesis, because only changes in chlorophyll fluorescence indicate that besides R_D also P_N is affected. The separated tentacles of carnivorous *Drosera prolifera* showed many times greater R_D in comparison with photosynthetic lamina. If we take into account extensive electrical activity in *Drosera* tentacles (Williams and Pickard 1972a, b, 1974, 1980; Williams and Spanswick 1976), it is tempting to assume that the electrical irritability is responsible for such high R_D (Adamec 2010).

Besides Venus flytrap, rapid increase of R_D was documented in *Conocephalum conicum* in response to cut or electrical stimulus (Dziubinska et al. 1989). Filek and Kościelniak (1997) showed increased shoot R_D in response to wounding the roots by high temperature in *Vicia faba*. The response was hindered by local cooling of the stem or by treatment with sodium azide which blocked the change in the electric potential. Fromm et al. (1995) found increased R_D in response to stigma stimulation by pollen but decreased R_D in response to cooling and heating in *Hibiscus* plant. It is interesting that pollen as well as cooling trigger APs (heating triggered VPs), which have different effect on R_D. This indicates that stimulus-dependent electrical signals cause specific responses in ovarian metabolism. Application of killed pollen, which was heated at 150°C or pollen of other species trigger no electrical signal and did not affect R_D. Ten minutes after pollination the amount of ATP increased by 12% but cold shocked and wounded flowers have lower concentration of ATP by 19 and 22%, respectively. This is consistent with the data from respiration measurements. Moreover starch level increased 15-fold after pollination, indicating that female reproductive system prepares for the following fertilization by increasing its metabolism.

Propagation of electrical signals in plants is costly. In *M. pudica*, the ATP level is much higher in pulvini than in the tissue between pulvini and stimulation causes consumption of ATP with simultaneous increase in ADP content (Lyubimova et al. 1964; Fromm and Eschrich 1988b). During the 1–3 s required for trap closure in *D. muscipula* 29% of ATP is lost (Jaffe 1973; Williams and Bennett 1982). Fromm and Eschrich (1988b) concluded that ATP is consumed for extrusion of positive charges during reestablishing or original resting potential and this is associated with the increase of R_D found by Dziubińska et al. (1989). Increase cellular level of ADP after AP may stimulate R_D. ADP initially regulates the rate of electron transport, which in turn regulates citric acid cycle activity, which, finally, regulates the rate of the glycolytic reactions. Thus, plant R_D is controlled from the "bottom up" by the cellular level of ADP (for overview see Taiz and Zeiger 2002), concentration of which is increased after APs.

2.7 Conclusions

Although plants have never developed the same degree of neuronal network complexity as in animals, electrical signaling plays an important role in their life. Numerous physiological responses of plant excitation have been reported (for review see Fromm and Lautner 2007) which help the plants survive in changing environment. If you look at carnivorous Venus flytrap growing in poor soil catching its prey rich in nutrients, without doubt you must think that electrical signaling is beneficial for plants (up to 75% of leaf nitrogen is taken up from insect prey, Schulze et al. 2001). Regarding carbon metabolism, we can just speculate whether the effect of electrical signals on photosynthesis and respiration has some purpose or it is just a negative consequence of ion interaction. Soma data may support the view about the usefulness of electrical signals in communication regarding photosynthesis (e.g. stomata opening after irrigation of roots in water-stressed plants, Fromm and Fei 1998; Grams et al. 2007) other seems to be without any meaning (e.g. inhibition of photosynthesis in response to trigger hair irritation in carnivorous plant Venus flytrap, Pavlovič et al. 2010, 2011). All three processes, photosynthesis, respiration and plant cell excitability, are dependent on ion fluxes. Therefore, it is tempting to assume that ion fluxes during propagation of electrical signals may interfere with electron transport chain directly by repulsion of charges or indirectly by changes in ionic environment and thus affecting enzymatic activity. Electrical signals are also costly in term of consumption of ATP and the changes of the level of this important molecule may regulate photosynthesis and respiration.

Further studies would be directed towards better understanding the mechanisms underlying photosynthetic response, i.e., which ion triggers the response, what are the primary targets of electrical signals on metabolism, how can different electrical signals triggered different response (e.g. stimulation vs. inhibition of photosynthesis and respiration). Recent studies also reveal that interference between electrical signals and photosynthesis is suitable model for studying the photoelectrochemical and photoelectrical component of variable chlorophyll a fluorescence—a parameter which is widely used in plant physiology. Further studies are needed for complex understanding of electrical signal-mediated response on the light and dark reaction of photosynthesis.

Acknowledgments This work was supported by the grant VEGA 1/0520/12 from the Scientific Grant Agency of the Ministry of Education of the Slovak Republic.

References

Adamec L (2010) Dark respiration of leaves and traps of terrestrial carnivorous plants: are there greater energetic costs in traps. Cent Eur J Biol 5:121–124

Affolter JM, Olivo RF (1975) Action potentials in Venus's-flytraps: long term observations following the capture of prey. Am Midl Nat 93:443–445

Beilby MJ (2007) Action potential in charophytes. Int Rev Cytol 257:43–82

Blankenship RE (2002) Molecular mechanisms of photosynthesis. Blackwell Science, MPG Books Ltd, Cornwall

Bulychev AA, Kamzolkina NA (2006a) Differential effects of plasma membrane electric excitation on H^+ fluxes and photosynthesis in characean cells. Bioelectrochemistry 69:209–215

Bulychev AA, Kamzolkina NA (2006b) Effect of action potential on photosynthesis and spatially distributed H^+ fluxes in cells and chloroplasts of *Chara corallina*. Russ J Plant Physiol 53:1–9

Bulychev AA, Vredenberg WJ (1999) Light-triggered electrical events in the tylakoid membrane of plant chloroplast. Physiol Plant 105:577–584

Bulychev AA, Cherkashin AA, Rubin AB, Vredenberg VS, Zykov VS, Mueller SC (2001) Comparative study on photosynthetic activity of chloroplasts in acid and alkaline zones of *Chara corallina*. Bioelectrochemistry 53:225–232

Bulychev AA, Kamzolkina NA, Luengviriya J, Rubin AB, Müller SC (2004) Effect of a single excitation stimulus on photosynthetic activity and light-dependent pH banding in *Chara* cells. J Membr Biol 2:11–19

Burdon-Sanderson JS (1873) Note on the electrical phenomena which accompany stimulation of the leaf of *Dionaea muscipula*. Proc R Soc 21:495–496

Dau H, Sauer K (1991) Electrical field effect on chlorophyll fluorescence and its relation to photosystem II charge separation reactions studied by a salt-jump technique. BBA-Bioenerg 1098:49–60

Dau H, Sauer K (1992) Electrical field effect on the picosecond fluorescence of photosystem II and its relation to the energetics and kinetics of primary charge separation. BBA-Bioenerg 1102:91–106

Davies E, Schuster A (1981) Intercellular communication in plants: evidence for a rapidly generated, bidirectionally transmitted wound signal. Proc Nat Acad Sci USA 78:2422–2426

Dziubińska H, Trębacz K, Zawadski T (1989) The effect of excitation on the rate of respiration in the liverwort *Conocephalum conicum*. Physiol Plant 75:417–423

Escalante-Pérez M, Krol E, Stange A, Geiger D, Al-Rasheid KAS, Hause B, Neher E, Hedrich R (2011) A special pair of phytohormones controls excitability, slow closure, and external stomach formation in the Venus flytrap. Proc Nat Acad Sci USA 108:15492–15497

Ettinger WF, Clear AM, Fanning KJ, Peck ML (1999) Identification of a Ca^{2+}/H^+ antiport in the plant chloroplast thylakoid membrane. Plant Physiol 199:1379–1385

Filek M, Kościelniak J (1997) The effect of wounding the roots by high temperature on the respiration rate of the shoot and propagation of electric signal in horse bean seedlings (*Vicia faba* L. minor). Plant Sci 123:39–46

Finazzi G, Johnson GN, Dall'Osto L, Joliot P, Wollman FA, Bassi R (2004) A zeaxanthin-independent nonphotochemical quenching mechanism localized in the photosystem II core complex. Proc Nat Acad Sci USA 101:12375–12380

Fisahn J, Herde O, Willmitzer L, Peña-Cortés H (2004) Analysis of the transient increase in cytosolic Ca^{2+} during the action potential of higher plants with high temporal resolution: requirement of Ca^{2+} transients for induction of jasmonic acid biosynthesis and PINII gene expression. Plant Cell Physiol 45:456–459

Fleurat-Lessard P, Bouché-Pillon S, Leloup C, Bonnemain JL (1997) Distribution and activity of the plasma membrane H^+-ATPase in *Mimosa pudica* L. in relation to ionic fluxes and leaf movements. Plant Physiol 113:747–754

Fromm J (1991) Control of phloem unloading by action potentials in *Mimosa*. Physiol Plant 83:529–533

Fromm J (2006) Long-distance electrical signalling and its physiological functions in higher plants. In: Volkov AG (ed) Plant electrophysiol. Springer, Berlin

Fromm J, Bauer T (1994) Action potentials in maize sieve tubes change phloem translocation. J Exp Bot 45:463–469

Fromm J, Eschrich W (1983) Electric signals released from roots of willow (*Salix viminalis* L) change transpiration and photosynthesis. J Plant Physiol 141:673–680

Fromm J, Eschrich W (1988a) Transport processes in stimulated and non-stimulated leaves of *Mimosa pudica*: I. The movement of ^{14}C-labelled photoassimilates. Trees 2:7–17

Fromm J, Eschrich W (1988b) Transport processes in stimulated and non-stimulated leaves of *Mimosa pudica*: II. Energesis and transmission of seismic stimulation. Trees 2:18–24

Fromm J, Eschrich W (1988c) Transport processes in stimulated and non-stimulated leaves of *Mimosa pudica*: III. Displacement of ions during seismonastic leaf movements. Trees 2:65–72

Fromm J, Fei H (1998) Electrical signaling and gas exchange in maize plants of drying soil. Plant Sci 132:203–213

Fromm J, Lautner S (2007) Electrical signals and their physiological significance in plants. Plant, Cell Environ 30:249–257

Fromm J, Hajirezaei M, Wilke I (1995) The biochemical response of electrical signalling in the reproductive system of *Hibiscus* plant. Plant Physiol 109:375–384

Graham JS, Hall G, Pearce G, Ryan CA (1986) Regulation of synthesis of proteinase inhibitors I and II mRNAs in leaves of wounded tomato plants. Planta 169:399–405

Grams TEE, Koziolek C, Lautner S, Matyssek R, Fromm J (2007) Distinct roles of electric and hydraulic signals on the reaction of leaf gas exchange upon re-irrigation in *Zea mays* L. Plant, Cell Environ 30:79–84

Grams TEE, Lautner S, Felle HH, Matyssek R, Fromm J (2009) Heat-induced electrical signals affect cytoplasmic and apoplastic pH as well as photosynthesis during propagation through the maize leaf. Plant, Cell Environ 32:319–326

Herde O, Atzorn R, Fisahn J, Wasternack C, Willmitzer L, Peña-Cortéz H (1996) Localized wounding by heat initiates the accumulation of proteinase inhibitor II in abscisic acid-deficient plants by triggering jasmonic acid biosynthesis. Plant Physiol 112:853–860

Herde O, Peña-Cortéz H, Willmitzer L, Fisahn J (1997) Stomatal responses to jasmonic acid, linolenic acid and abscisic acid in wild-type and ABA-deficient tomata plants. Plant, Cell Environ 20:136–141

Herde O, Peña-Cortéz H, Willmitzer L, Fisahn J (1998) Remote stimulation by heat induces characteristic membrane-potential responses in the veins of wild-type and abscisic acid-deficient tomato plants. Planta 206:146–153

Herde O, Peña-Cortéz H, Fuss H, Willmitzer L, Fisahn J (1999a) Effects of mechanical wounding, current application and heat treatment on chlorophyll fluorescence and pigment composition in tomato plants. Physiol Plant 105:179–184

Herde O, Peña-Cortéz H, Wasternack C, Willmitzer L, Fisahn J (1999b) Electric signaling and *pin2* gene expression on different abiotic stimuli depend on a distinct threshold level of endogenous abscisic acid in several abscisic acid-deficient tomato mutants. Plant Physiol 119:213–218

Hertig CM, Wolosiuk RA (1983) Studies on the hysteretic properties of chloroplast fructuse-1,6-bisphospatase. J Biol Chem 258:984–989

Hlaváčková V, Krchňák P, Nauš J, Novák O, Špundová M, Strnad M (2006) Electrical and chemical signals involved in short-term systemic photosynthetic responses of tobacco plants to local burning. Planta 225:235–244

Hoddinott J (1977) Rates of translocation and photosynthesis in *Mimosa pudica*. New Phytol 79:269–272

Hodick D, Sievers A (1988) The action potential of *Dionaea muscipula* ellis. Planta 174:8–18

Hunt S (2003) Measurements of photosynthesis and respiration in plants. Physiol Plant 117:314–325

Jaffe MJ (1973) The role of ATP in mechanically stimulated rapid closure of the Venus's-flytrap. Plant Physiol 51:17–18

Johnson MP, Pérez-Bueno ML, Zia A, Horton P, Ruban AV (2009) The zeaxanthin-independent and zeaxanthin-dependent qE components of nonphotochemical quenching involve common conformational changes within the photosystem II antenna in *Arabidopsis*. Plant Physiol 149:1061–1075

Juniper BE, Robins RJ, Joel DM (1989) The carnivorous plants. Academic, London

Kaiser H, Grams TEE (2006) Rapid hydropassive opening and subsequent active stomatal closure follow heat-induced electrical signals in *Mimosa pudica*. J Exp Bot 57:2087–2092

Koziolek C, Grams TEE, Schreiber U, Matyssek R, Fromm J (2003) Transient knockout of photosynthesis mediated by electrical signals. New Phytol 161:715–722

Kreimer G, Melkonian M, Latzko E (1985) An electrogenic uniport mediates light-dependent Ca2+ influx into intact spinach chloroplasts. FEBS Lett 180:253–258

Kreimer G, Melkonian M, Holtum JAM, Latzko E (1988). Stromal free calcium concentration and light-mediated activation of chloroplast fructose-1,6-bisphosphatase. Plant Physiol 86:423–428

Krupenina NA, Bulychev AA (2007) Action potential in a plant cell lowers the light requirement for non-photochemical energy-dependent quenching of chlorophyll fluorescence. Biochim Biophys Acta 1767:781–788

Lautner S, Grams TEE, Matyssek R, Fromm J (2005) Characteristics of electrical signals in poplar and responses in photosynthesis. Plant Physiol 138:2200–2209

Lazár D (2006) The polyphasic chlorophyll a fluorescence rise measured under high intensity of exciting light. Funct Plant Biol 33:9–30

Lazár D (2009) Modelling of light-induced chlorophyll *a* fluorescence rise (O-J-I-P transient) and changes in 820 nm-transmittance signal of photosynthesis. Photosynthetica 47:483–498

Lichtner FT, Williams SE (1977) Prey capture and factors controlling trap narrowing in *Dionaea* (Droseraceae). Am J Bot 64:881–886

Lucas WJ (1975) The influence of light intensity on the activation and operation of the hydroxyl efflux system of *Chara corallina*. J Exp Bot 26:347–360

Lyubimova MN, Demyanovskaya NS, Fedorovich IB, Homlenskite IV (1964) Participation of ATP in the motor function of the *Mimosa pudica* leaf. Transl Biokhim 29:774–779

Malone M, Stankovič B (1991) Surface potentials and hydraulic signals in wheat leaves following localised wounding by heat. Plant, Cell Environ 14:431–436

Mancuso S (1999) Hydraulic and electrical transmission of wound-induced signals in *Vitis vinifera*. Aust J Plant Physiol 26:55–61

Maxwell K, Johnson GN (2000) Chlorophyll fluorescence—a practical guide. J Exp Bot 51:659–668

Meiburg RF, Van Gorkom HJ, Van Dorssen RJ (1983) Excitation trapping and charge separation in photosystem II in the presence of an electrical field. BBA-Bioenerg 724:352–358

Müller P, Li X-P, Niyogi KK (2001) Non-photochemical quenching. A response to excess light energy. Plant Physiol 125:1558–1566

Nedbal L, Whitmarsh J (2004) Chlorophyll fluorescence imaging of leaves and fruits. In: Papageorgiou GC, Govindjee (eds) Chlorophyll a fluorescence a signature of photosynthesis. Springer, Dordrecht

Opritov VA, Pyatygin SS, Vodeneev VA (2002) Direct coupling of action potential generation in cells of a higher plant (*Cucurbita pepo*) with the operation of an electrogenic pump. Russ J Plant Physiol 49:142–147

Papageorgiou GC, Govindjee (2004) Chlorophyll a fluorescence: a signature of photosynthesis. Springer, Dordrecht

Pavlovič A (2010) Spatio-temporal changes of photosynthesis in carnivorous plants in response to prey capture, retention and digestion. Plant Signal Behav 5:1–5

Pavlovič A, Mancuso S (2011) Electrical signalling and photosynthesis: can they co-exist together? Plant Signal Behav 6:840–842

Pavlovič A, Demko V, Hudák J (2010) Trap closure and prey retention in Venus flytrap (*Dionaea muscipula*) temporarily reduces photosynthesis and stimulates respiration. Ann Bot 105:37–44

Pavlovič A, Slováková L', Pandolfi C, Mancuso S (2011) On the mechanism underlying photosynthetic limitation upon trigger hairs irritation in carnivorous plant Venus flytrap (*Dionaea muscipula* ellis). J Exp Bot 62:1991–2000

Peña-Cortés H, Fisahn J, Willmitzer L (1995) Signals involved in wound-induced proteinase inhibitor II gene expression in tomato and potato plants. Proc Nat Acad Sci USA 92:4106–4113

Pikulenko MM, Bulychev AA (2005) Light-triggered action potentials and changes in quantum efficiency of photosystem II in *Athoceros* cells. Russ J Plant Physiol 52:584–590

Plieth C, Tabrizi H, Hansen U-P (1994) Relationship between banding and photosynthetic activity in *Chara corallina* as studied by the spatially different induction curves of chlorophyll fluorescence observed by an image analysis system. Physiol Plant 91:205–211

Pyatygin SS, Opritov VA, Vodeneev VA (2008) Signaling role of action potential in higher plants. Russ J Plant Physiol 55:312–319

Ricca U (1916) Soluzione d'un problema di fisiologia: la propagazione di stimulo nella *Mimosa*. Nuovo G Bot Ital 23:51–170

Roháček K (2002) Chlorophyll fluorescence parameters: the definitions, photosynthetic meaning, and mutual relationship. Photosynthetica 40:13–29

Sai J, Johnson CH (2002) Dark-stimulated calcium ion fluxes in the chloroplast stroma and cytosol. Plant Cell 14: 1279–1291

Samejima M, Sibaoka T (1983) Identification of the excitable cells in the petiole of *Mimosa pudica* by intracellular injection of procion yellow. Plan Cell Physiol 24:33–39

Schansker G, Tóth SZ, Strasser RJ (2005) Methylviologen and dibromothymoquinone treatments of pea leaves reveal the role of photosystem I in the chl a fluorescence rise. Biochim Biophys Acta 1706:250–261

Schreiber U (2004) Pulse-amplitude-modulation (PAM) fluorometry and saturation pulse method: an overview. In: Papageorgiou GC, Govindjee (eds) Chlorophyll a fluorescence: a signature of photosynthesis. Springer, The Netherlands

Schulze W, Schulze ED, Schulze I, Oren R (2001) Quantification of insect nitrogen utilization by the Venus flytrap *Dionaea muscipula* catching prey with highly variable isotope signatures. J Exp Bot 52:1041–1049

Shiina T, Tazawa M (1986) Action potential in *Luffa cylindrica* and its effects on elongation growth. Plant Cell Physiol 27:1081–1089

Sibaoka T (1962) Excitable cells in *Mimosa*. Science 137:226

Sibaoka T (1991) Rapid plant movements triggered by action potentials. Bot Mag Tokyo 104:73–95

Srivastava A, Guissé B, Greppin H, Strasser RJ (1997) Regulation of antenna structure and electron transport in photosystem II of *Pisum sativum* under elevated temperature probed by the fast polyphasic chlorophyll a fluorescence transient: OKJIP. Biochim Biophys Acta 1320:95–106

Stahlberg R, Cosgrove DJ (1996) Induction and ionic basis of slow wave potentials in seedlings of *Pisum sativum* L. Planta 199:416–425

Stahlberg R, Cosgrove DJ (1997) The propagation of slow wave potentials in pea epicotyls. Plant Physiol 113:209–217

Stankovič B, Davies E (1998) The wound response in tomato involves rapid growth and electrical responses, systemically up-regulated transcription of proteinase inhibitor and calmodulin and down-regulated translation. Plant Cell Physiol 39:268–274

Stolarz M, Król E, Dziubinska H, Kurenda A (2010) Glutamate induces series of action potentials and a decrease in circumnutation rate in *Helianthus annuus*. Physiol Plant 138:329–338

Strasser RJ, Govindjee (1992) On the O-J-I-P fluorescence transient in leaves and D1 mutants of *Chlamydomonas reinhardtii*. In: Murata M (ed) Research in photosynthesis, vol 2. Kluwer Academic Publishers, Dordrecht

Strasser RJ, Tsimilli-Michael M, Srivastava A (2004) Analysis of the chlorophyll a fluorescence transient. In: Papageorgiou GC, Govindjee (eds) Chlorophyll a fluorescence: a signature of photosynthesis. Springer, Dordrecht

Taiz L, Zeiger E (2002) Plant physiology, 3rd edn. Sinauer Associates Inc., Publishers, MA, p 690

Trebacz K, Sievers A (1998) Action potentials evoked by light in traps of *Dionaea muscipula* ellis. Plant Cell Physiol 39:369–372

Volkov AG, Adesina T, Jovanov E (2007) Closing of Venus flytrap by electrical stimulation of motor cells. Plant Signal Behav 2:139–145

Volkov AG, Adesina T, Jovanov E (2008a) Charge induce closing of *Dionaea muscipula* ellis trap. Bioelectrochemistry 74:16–21

Volkov AG, Adesina T, Markin VS, Jovanov E (2008b) Kinetics and mechanism of *Dionaea muscipula* trap closing. Plant Physiol 146:694–702

Volkov AG, Foster JC, Ashby TA, Walker RK, Johnson JA, Markin VS (2010) *Mimosa pudica*: electrical and mechanical stimulation of plant movements. Plant, Cell Environ 33:163–173

Vredenberg WJ (2000) A three-state model for energy trapping and chlorophyll fluorescence in photosystem II incorporating radical pair recombination. Biophys J 79:26–38

Vredenberg WJ (2004) System analysis and photo-electrochemical control of chlorophyll fluorescence in terms of trapping models of photosystem II: a challenging view. In: Papageorgiou GC, Govindjee (eds) Chlorophyll a fluorescence: a signature of photosynthesis. Springer, Dordrecht

Vredenberg WJ (2011) Kinetic analyses and mathematical modeling of primary photochemical and photoelectrochemical processes in plant photosystems. Biosystems 103:138–151

Vredenberg WJ, Bulychev AA (2002) Photo-electrochemical control of photosystem II chlorophyll fluorescence in vivo. Bioelectrochemistry 57:123–128

Vredenberg WJ, Bulychev AA (2003) Photoelectric effects on chlorophyll fluorescence of photosystem II in vivo. Kinetics in the absence and presence of valinomycin. Bioelectrochemistry 60:87–95

Vredenberg WJ, Bulychev AA (2010) Photoelectrochemical control of the balance between cyclic- and linear electron transport in photosystem I. algorithm for $P700^+$ induction kinetics. Biochim Biophys Acta 1797:1521–1532

Vredenberg WJ, Durchan M, Prášil O (2009) Photochemical and photoelectrochemical quenching of chlorophyll fluorescence in photosystem II. Biochim Biophys Acta 1787:1468–1478

Wildon DC, Thain JF, Minchin PEH, Gubb IR, Reilly AJ, Skipper YD, Doherty HM, O'Donnel J, Bowles DJ (1992) Electrical signaling and systemic proteinase inhibitor induction in the wounded plant. Nature 360:62–65

Williams SE, Bennett AB (1982) Leaf closure in the Venus flytrap: an acid growth response. Science 218:1120–1121

Williams SE, Pickard BG (1972a) Receptor potentials and action potentials in *Drosera* tentacles. Planta 103:193–221

Williams SE, Pickard BG (1972b) Properties of action potentials in *Drosera* tentacles. Planta 103:222–240

Williams SE, Pickard BG (1974) Connections and barriers between cells of *Drosera* tentacles in relation to their electrophysiology. Planta 116:1–16

Williams SE, Pickard BG (1980) The role of action potentials in the control of capture movements of *Drosera* and *Dionaea*. In: Skoog F (ed) Plant growth substances. Springer, Berlin

Williams SE, Spanswick RM (1976) Propagation of the neuroid action potential of the carnivorous plant *Drosera*. J Comp Physiol 108:211–223

Yan X, Wang Z, Huang L, Wang C, Hou R, Xu Z, Qiao X (2009) Research progress on electrical signals in higher plants. Prog Nat Sci 19:531–541

Chapter 3
Mathematical Modeling, Dynamics Analysis and Control of Carnivorous Plants

Ruoting Yang, Scott C. Lenaghan, Yongfeng Li, Stephen Oi and Mingjun Zhang

Abstract The focus of this chapter is to analyze the sensing and actuating mechanism of the Venus flytrap (*Dionaea muscipula*) by developing a control model to explain these phenomena. The mathematical model captures the dynamic responses of the flytrap in different environmental conditions. In addition, this model emphasizes the existence of the threshold accumulation of the trigger signal and the semi-closed state, which are largely ignored in other models. Furthermore, a biomimetic robot was constructed to demonstrate the feasibility of the mathematical model. While the robot serves as a prototype to demonstrate the control model, future applications using this control could aid microsensors and microgrippers to reduce false alarms. In summary, this chapter uses the Venus flytrap as an example to illustrate the integration between biology, theoretical modeling, and engineering. Such integration and inspiration from the natural world will significantly contribute to advances in these various disciplines.

R. Yang
Institute for Collaborative Biotechnologies, University of California,
Santa Barbara, CA 93106-5080, USA

S. C. Lenaghan · S. Oi · M. Zhang (✉)
Department of Mechanical, Aerospace and Biomedical Engineering,
University of Tennessee, Knoxville, TN 37996-2210, USA
e-mail: mjzhang@utk.edu

Y. Li
Division of Space Life Science, Universities Space Research Association,
Houston, TX 77058, USA

3.1 Introduction

Unlike other plants, carnivorous plants have access to increased nutrient pools through the capture and digestion of insects. This alternative strategy to nutrient acquisition gives them a competitive advantage over plants that must acquire nutrients from the soil. To coordinate the capture of insects, carnivorous plants must overcome several challenges. First, they must have developed a sophisticated mechanism for sensing their prey. This mechanism must be sensitive enough to detect prey, but must also not provide false signals to the plant. Second, the carnivorous plant must have a trapping system that can reliably capture insects, without too many failed attempts. This is especially important due to the high energetic cost of trapping systems. Over the course of evolution many carnivorous plants have developed variable strategies to accomplish these goals, both in sensing and trapping. Perhaps the most well-known example of a carnivorous plant is Venus flytrap (*Dionaea muscipula*), which has long been regarded as "one of the most wonderful plants in the world" (Darwin 1875), due to the rapid closure of its trap. Scientific studies have shown that the flytrap has a feedforward control system. The so-called "feedforward" means that the system responds the external environment in a predefined manner, in contrast to "feedback" that actively takes account of the subsequent modification of the control (Brogan 1985). Figure 3.1 illustrates a simple comparison of feedforward control and feedback control. In a feedback control system, when the thermostat senses that the temperature has reached to a certain upper limit, the cooling system is switched on, and consequently, cool air flows into the room (Fig. 3.1a). In contrary, in a feedforward control system, the cooling system is directly turns on or off in response to the status of the toggle switch (Fig. 3.1b). The feedforward control has been widely employed in modern industrial applications, such as precise positioning (Leang et al. 2009) and gene delivery (Yang et al. 2010b). Actually, the feedforward mechanism naturally exists in all levels of biological systems (Mangan and Alon 2003; Milo et al. 2002), while the evolutionary process gradually optimizes the control coefficients in terms of energy and efficiency.

To illustrate this idea, we will take a closer look at the three components of the control system for the flytrap: the sensor, motor, and decision-making block (Fig. 3.2). First, the sensor of the flytrap is comprised of 3–5 trigger hairs on each lobe. The trigger hairs can transfer a mechanical stimulus into an electrical signal [namely an action potential (AP)], which rapidly spreads throughout the trap (Burden-Sanderson 1882; Stuhlman and Darder 1950). Second, the "motor" of the flytrap is believed to be movement of water within the lobes. Recent experiments with ion and water channel blockers and uncouplers suggested that a large number of water channels exist between two layers of the lobes to facilitate transcellular water transport (Fagerberg and Howe 1996; Volkov et al. 2008d). In particular, the leaf curvature creates a strong hydraulic force for the rapid closure process. When an AP opens the water channels, an ion gradient is created that allows water to flush through from the inner to the outer layers and results in the fast closure (Markin et al. 2008).

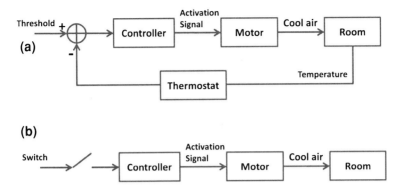

Fig. 3.1 Illustrative diagram of feedback (**a**) and feedforward (**b**) control mechanism for an air conditioner

Finally, the decision-making block of the flytrap consists of multiple steps and is of particular interest in terms of energy efficiency. The first decision is a toggle switch, when a total electrical charge is accumulated to a threshold. Experiments have suggested the threshold is approximately 14 μC (Volkov et al. 2008a). Interestingly, it requires two mechanical stimuli within ∼ 30 s at room temperature (Volkov et al. 2008a). Thus the double-trigger process can be seen as a natural error-proof design to prevent the trap initiating by a burst of wind or heavy rain. Our forced water/air stimulation experiments showed that the trap did snap if a strong stream water or air directly hit and bent one trigger hair twice within 30 s (Yang et al. 2010a). However, in the natural environment, the chance of doing so is slim.

Besides that, the flytrap has a second error-proof design: a semi-closed state to decide whether proceed with the digestion process, after being triggered to the semi-closed state (Fagerberg and Allain 1991; Fagerberg and Howe 1996). This decision-making process has been largely overlooked in literature (Yang et al. 2010a). However, it is critical for the flytrap to avoid false trapping. This false or inaccurate trapping results in high energetic losses from the secretion of digestive enzymes into an empty trap. This false trapping is also highly wasteful, since a single trap can only digest three meals in its lifetime (Yang et al. 2010a).

As shown Fig. 3.3, the motion of the Venus flytrap can be divided into three distinct states: (1) The fully open state (Fig. 3.3a), which occurs in the absence of prey, and is characterized by a convex curvature of the trap lobes; (2) The semi-closed state (Fig. 3.3b, d), which occurs immediately after the trap is triggered, and is characterized by interlocking cilia that restrict large prey but allow small prey to escape; and (3) The fully closed state (Fig. 3.3c, e), which occurs after prolonged stimulation, and is characterized by a tight appression and recurved bending of the trap margins (Yang et al. 2010a).

The open state is considered as a stable state, and does not spontaneously close, even with gusts of wind, or rain drops. Using a strong stream of water or a gust of air directed at a trigger hair, it is possible to initiate trap closure only when a

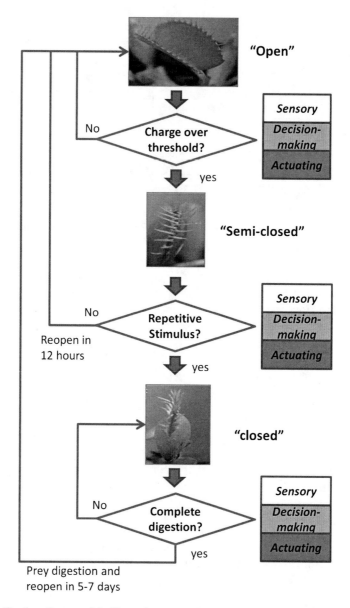

Fig. 3.2 The flow diagram of the Venus flytrap opening/closing mechanism. The trap opening is triggered by mechanical stimulation of the trigger hairs. The resulting action potentials are accumulated in the leaves. Once the charge accumulates above the trigger threshold, water rushes through open water channels and the trap is semi closed. If no further stimulus occurs, then the trap reopens in 12 h; otherwise the trap will finally be fully closed and proceeds to digestion. It requires 5–7 days to reopen after digestion of prey

3 Mathematical Modeling, Dynamics Analysis and Control of Carnivorous Plants 67

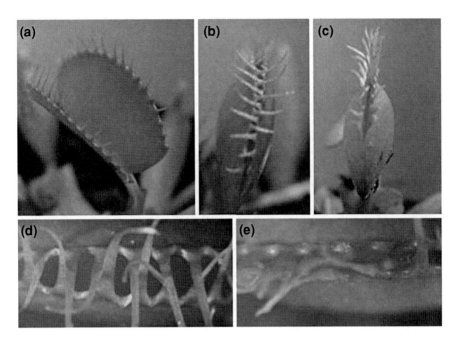

Fig. 3.3 Closing states of the Venus flytrap (*Dionaea muscipula*). **a** The open state of the trap characterized by the separation of the 2 lobes. **b** The semi-closed transition state, lobes brought together, but not tightly closed. **c** Tightly closed state, trap appearance changes and buckling occurs around outer edges of trap, lobes in direct contact. **d** and **e** Higher magnification of the trap lobes in the semi closed, and tightly closed states respectively. Images captured using a Sony HDR-HC9 handycam attached to a Meiji EMZ-13TR stereoscope. Images were processed using Pinnalce studio™ ultimate. Adapted with permission from (Yang et al. 2010a)

trigger hair is flexed twice in 30 s (Volkov et al. 2008a). When a trap is in the semi-closed state, two different outcomes are possible. The trap can proceed to the fully closed state on further stimulation, or the trap can return to the open state. This decision-making step operates as a failsafe, where if the prey escapes or other nonnutritive material is captured, then the semi-closed state will return to the fully open state and the energetic costs are minimized. If, however, there is a constant stimulus in the semi-closed state, such as the prey struggling and bending the trigger hairs, then the trap will proceed to the fully closed state. Considering the duration of digestion, 5–7 days (Stuhlman 1948; Yang et al. 2010a), false trapping is very expensive to the plant.

In light of these unique mechanisms, we proposed a mathematical model to explain the nonlinear dynamics and error-proof control mechanism of the flytrap (Yang et al. 2010a). In recent years, several other mathematical models have been proposed (Forterre et al. 2005; Markin et al. 2008; Volkov et al. 2008b). Forterre et al. suggested that leaf geometry creates a buckling instability generating the rapid snap force (Forterre et al. 2005). Bobji considered the Venus flytrap as a bistable vibrator, which is stable in both the open and closed states (Bobji 2005). Markin et al. put forward a hydroelastic curvature model stating that the lobes possess curvature elasticity and have inner and outer hydraulic layers with different hydraulic pressures (Volkov et al. 2008c). However, these models described only the poststimulation closure stage, and neglected the triggering process, reopening process, and the intermediate semi-closed transition state, while our model took all these under consideration (Yang et al. 2010a).

In this chapter, we will improve and analyze our mathematical model, and further explain the sensory, decision making, and actuating mechanisms of the Venus flytrap. Furthermore, a biomimetic robot was constructed to demonstrate the feasibility of the mathematical model. While the robot serves as a prototype to demonstrate the control model, future applications using this control could aid microsensors and microgrippers to reduce false alarms. In summary, this chapter uses the Venus flytrap as an example to illustrate the integration between biology, theoretical modeling, and engineering. Such integration and inspiration from the natural world will significantly contribute to advances in these various disciplines.

3.2 Mathematical Modeling

The Venus flytrap model consists of the three components, as also shown in Fig. 3.3.

1. **Sensory**: Action potential activated by mechanical stimuli
2. **Decision making**:

    a. The double-trigger process
    b. Semi-closed state

3. **Actuating**: Driven by water movement in the lobe, the trap transits between different states

Generally speaking, our model is based on the water movement driven assumption (Markin et al. 2008), which can be described in detail as follows. Once stimulation of the trigger hairs occurs, action potentials (APs) are generated that rapidly spread through the surface of the trap lobes. When the accumulated charge in the lobe exceeds to a threshold, a cascade of processes are initiated that are crucial for the transport of water between the inner and outer layers. ATP hydrolysis and fast proton transport leads to the creation of an ion gradient that initiates the opening of water pores (Volkov et al. 2008b). This rapid water transport between the inner and outer layers of the lobes causes the trap to transition to the semi-closed state within 0.3 s (Volkov et al. 2008b). If no further stimulation occurs, the lobes will slowly return to the open state within 12–24 h. Otherwise, the trap transitions to the fully closed state, which takes 5–7 days to

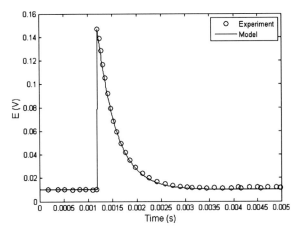

Fig. 3.4 Model of the action potential of the Venus Flytrap

digest and reopen (Yang et al. 2010a). In the following, we will lay out the flytrap model in two parts: double-trigger process and water kinetics.

3.2.1 Double-Trigger Process

The trigger process results from the action potential generated by the mechanical stimuli on the trigger hairs (Hodick and Sievers 1986, 1988, 1989; Stuhlman and Darder 1950). Volkov et al. (2007) discovered that a total charge of 14 μC is necessary to initiate closure. Moreover, the AP jumped to ∼0.15 V at 0.0012 s with duration time of 1.5 ms (Volkov et al. 2007). By observing this, this AP can be approximated by an exponential function as follows:

$$u_t = \begin{cases} 0.01 + 0.137 e^{-2,600 t}, & t \geq 0 \\ 0, & t < 0 \end{cases} \quad (3.1)$$

As illustrated in Fig. 3.4, in the above model ut evoked at 0.0012 s fits well with recent measurements (Volkov et al. 2007). Consequently, constant AP leads to a stepwise accumulation of a bioactive substance, which results in a charge accumulation (Ueda et al. 2007). The charge accumulation can be described by a linear dynamic system,

$$\dot{C} = -k_c (C - C_b) + k_a u_t \quad (3.2)$$

where C (unit: μC), k_c (s^{-1}) and k_a (μC · V^{-1} · s^{-1}) represent the accumulated charge in the lobes, charge dissipation rate, and accumulation rate, respectively. Without loss of generality, the basal charge C_b is assumed to be 4 μC.

Figure 3.5 shows how APs lead to a stepwise response of charge accumulation in the lobes. The first AP cannot accumulate enough charge; however, if the second AP occurs within 30 s (red dashed or green dot-dashed line), then the

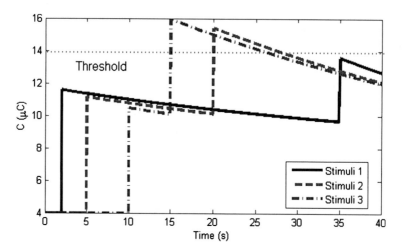

Fig. 3.5 Double stimuli occur within 30 s is necessary to trigger closure. The first AP cannot accumulate enough charge accumulation; however, if the second AP occurs within 30 s (*dashed* or *dot-dashed* line), then the charge rises above the activation threshold 14 μC and leads to a trigger impulse. If two APs occur over 30 s, the trigger threshold cannot be reached (*solid line*)

charge grows above the activation threshold 14 μC and leads to a trigger impulse. If two APs occur over 30 s, the trigger threshold cannot be reached (blue solid line). If the temperature in the environment rises above 35°C, the basal charge could significantly increase and thus only one stimulation would be required for the trap closure (Brown and Sharp 1910).

3.2.2 Water Kinetics

Many researchers believed that a large number of water channels exist between two layers of the lobes to facilitate rapid water transport (Fagerberg and Howe 1996; Volkov et al. 2008d). When the charge accumulates above the threshold, voltage-gated channels will be opened (Rea 1983; Volkov et al. 2007; Volkov et al. 2008b; Volkov et al. 2008d), and create an ion gradient drawing water through from the inner to the outer layer as quickly as 10^6 molecules per second (Schäffner 1998).

The mathematical model for the opening and closing mechanism of the Venus flytrap is based on the water movement between the outer and inner hydraulic layers of the lobes. As illustrated in Fig. 3.6, the open state of the Venus flytrap corresponds to a maximum/minimum water volume in the inner/outer layer (Fig. 3.6a), the opposite to the closed state (Fig. 3.6c). The semi-closed state has equal water volume in both layers (Fig. 3.6b).

Without loss of generality, the total water volume is assumed to be constant and normalized to 1. And the water volumes in the outer and inner layer of the lobes are

3 Mathematical Modeling, Dynamics Analysis and Control of Carnivorous Plants

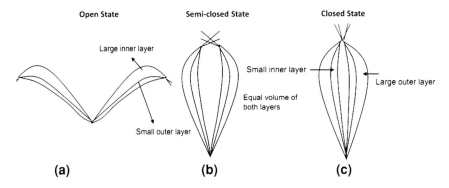

Fig. 3.6 Water volume in the outer and inner layers in the open, semi closed, and closed states. **a** The open state has large inner layer and small outer layer. **b** The semi-closed state has equal volume in both layers. **c** The closed state has large outer layer and small inner layer, in contrast to the open state. Revised with permission from (Yang et al. 2010a)

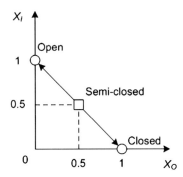

Fig. 3.7 Nonlinear dynamics of the Venus flytrap model. The total water volume has been normalized to one. The open state corresponds to $(X_I, X_O) = (1, 0)$ (*circle on the left*), the semi-closed state is $(X_I, X_O) = (0.5, 0.5)$ (*square*) and the closed state $(X_I, X_O) = (0, 1)$ (*circle on the right*). The open/closed states are two attractors and semi-closed state is a saddle point. The water dynamics moves along the line $X_O + X_I = 1$. It tends to the open state when $X_O > 0.5$, and converges to the closed state when $X_O < 0.5$. Adapted with permission from (Yang et al. 2010a)

denoted by X_O and X_I, respectively. The open and closed states are considered as stable states, while the semi-closed state is an intermediate state that can move towards either the open or closed states. Therefore, our basic idea is to construct a bistable system interlinking the three states. As shown in Fig. 3.7, there exist two stable equilibria, (0, 1) and (1, 0) (Circles) corresponding to the open and closed states, and an unstable equilibrium (0.5, 0.5) (Square) for the semi-closed state. Water transport can drive the water dynamics from stable states. Once the dynamics run across the semi-closed state, the dynamics will flip to another stable state; Otherwise, it will return to the original stable state when the water transport stops.

Following this idea, the mathematical model of the flytrap opening/closing mechanism can be described as follows:

$$\dot{X}_O = \frac{\alpha X_O^p}{X_O^p + X_I^p} + u_h + u_a - u_c - \mu X_O \qquad (3.3)$$

$$\dot{X}_I = \frac{\alpha X_I^p}{X_O^p + X_I^p} - u_h - u_a + u_c - \mu X_I \qquad (3.4)$$

$$X_I + X_O = \alpha/\mu \qquad (3.5)$$

where u_h, u_e and u_c are the water transport rates driven by a hydraulic gradient, electric gradient and chemical signal driven reverse gradient in different processes. Water supply rate and consumption rate denote as α and μ, respectively.

Without any stimuli, i.e., $u_h = u_e = u_c = 0$, the water dynamics becomes

$$\dot{X}_O = \frac{\alpha X_O^p}{X_O^p + X_I^p} - \mu X_O \qquad (3.6)$$

$$\dot{X}_I = \frac{\alpha X_I^p}{X_O^p + X_I^p} - \mu X_I \qquad (3.7)$$

$$X_I + X_O = c \qquad (3.8)$$

where $c = \alpha/\mu$. The system (X_O, X_I) exhibits bistability. It has three steady states: both the open state $S_1^0 = (0, c)$ and the closed state $S_3^0 = (c, 0)$ are stable, and the semi-closed state $S_2^0 = (c/2, c/2)$ is unstable. Moreover, the system allows an invariant manifold, $W_u = \{(X_O, X_I) \in \mathbb{R}^{2+}, X_O = X_I\}$, and an unstable submanifold

$$W_u^0 = \left\{(X_O, X_I) \in \mathbb{R}^{2+}, X_O = X_I < \frac{pc}{2}\right\} \subset W_u,$$

as well as two stability regions $W(S_1^0) = \{(X_O, X_I) \in \mathbb{R}^{2+}, X_O < X_I\}$, and $W(S_3^0) = \{(X_O, X_I) \in \mathbb{R}^{2+}, X_O > X_I\}$ (Li and Zhang 2011).

As shown in Fig. 3.8, the system shows bistability with two stable and one unstable steady states. The stability region $W(S_1^0)$ (solid line) is equivalent to $X_O < c/2$, because of $X_I + X_O = c$. And the stability region $W(S_3^0)$ (dash line) is equivalent to $X_O > c/2$. Furthermore, the higher the Hill coefficient p, the faster the rate of convergence is. In this chapter, we assume $p = 2$.

In fact, if the supply and consumption rate is balanced, i.e., $\alpha = \mu$, the above system can be reduced to be one dimension.

$$\dot{X}_O = \frac{\alpha X_O^2}{X_O^2 + (1 - X_O)^2} + u_h + u_a - u_c - \mu X_O \qquad (3.9)$$

where the hydraulic transport rate u_h is proportional to the hydraulic pressure induced by the volume difference. The hydraulic gradient is defined as follows:

$$u_h(t; t_0) = k_h(X_I - X_O)e^{-\lambda_h(t-t_0)}. \qquad (3.10)$$

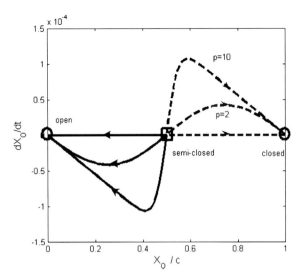

Fig. 3.8 Phase diagram of the flytrap system without stimulus. This system shows bistability with two stable and one unstable steady states. The stability region $W(S_1^0)$ (*solid line*) is $X_O < c/2$, while the stability region $W(S_3^0)$ (*dash line*) is $X_O > c/2$. Furthermore, the dynamics will be converged faster as the Hill coefficient p, increases

where k_h and λ_h are the hydraulic transport coefficient and time constant; respectively, while t_0 denotes the starting time. The rationale behind this equation is based on the assumption that the pressure inside the leaf layer is proportional to the amount of fluid confined in it, and thus the pressure transfer rate is proportional to the volume difference (Markin et al. 2008).

Continuous bending of the trigger hairs after semiclosure will still activate APs, opening the water channels, and pushing more water to the inner layer. The water transport caused by the electric gradient is much slower than the transport due to hydraulic pressure. The electric transport rate u_e is reversely proportional to the remaining volume in the inner layer due to the hydraulic pressure.

$$u_e(t;t_0) = k_e e^{\frac{\lambda_e(t-t_0)}{1-X_O}}, \qquad (3.11)$$

where k_e and λ_e are the electric transport coefficient and time constant, respectively.

The reopening from the sealing to the open state is due to the chemical signal as follows:

$$u_c(t;t_0) = k_c e^{-\lambda_c(t-t_0)}, \qquad (3.12)$$

where k_c and λ_c are the electric transport coefficient and time constant, respectively. The definition of the system parameters is summarized in Table 3.1.

3.2.2.1 Capture Process: (From the Open to the Semi-Closed State)

The driving force of this process is dominated by the hydraulic pressure, thus the electric gradient and the chemical signal are negligible, i.e., $u_a = 0$, and $u_c = 0$. Thus starting at the activating moment, the entire water dynamics can be represented by the following equations:

Table 3.1 Parameters of the Venus flytrap model

Symbol	Description	Value
k_c (s^{-1})	Dissipation rate of charge accumulation	0.02
k_a (μC/V · s)	Accumulation rate of charge	15.8
C_T (μC)	Trigger threshold	14
C_b (μC)	Basal charge	4
α (h^{-1})	Water supply rate	1
μ (h^{-1})	Water consumption rate	1
k_h (s^{-1})	Hydraulic transport coefficient	10
λ_h (s^{-1})	Hydraulic time constant	3
k_e (s^{-1})	Electric transport coefficient	0.8
λ_e (s^{-1})	Electric time constant	1
k_c (h^{-1})	Electric transport coefficient	0.36
λ_e (h^{-1})	Electric time constant	0.27

$$\dot{X}_O = \frac{\alpha X_O^2}{X_O^2 + (1 - X_O)^2} + k_h(1 - 2X_O)e^{-\lambda_h t} - \mu X_O \quad (3.13)$$

$$X_I = 1 - X_O$$

This system does not have any steady state until u_h vanishes. As shown in Fig. 3.9, $\dot{X}_O$ rapidly decreases and tends to zero when X_O tends to 0.5. The hydraulic transport coefficient k_h controls the speed of closure: the higher the faster. Additionally, the hydraulic time constant dominates the vanishing time of u_h, therefore it implies how close the lobes can be. Larger λ_h will cause the trap unable to reach the semi-closed state, and begin to move backward to the open state.

Figure 3.10 shows the water of the inner layer rapidly flushes to the outer layer and approaches an equal volume in 0.3 s. This implies that the lobes transition from the open state to the semi-closed state.

3.2.2.2 Release Process: (From the Semi-Closed to the Open State)

The release process is mainly dependent on the growth effect between two layers, thus, we can assume $u_h = 0$, $u_a = 0$, and $u_c = 0$. The water dynamics tend to the semi-closed state $(X_I, X_O) = (0.5, 0.5)$ during the capture process, but the inner layer is still slightly larger than the outer layer. However, without further water transport between the two layers, the water dynamics will return to the closest equilibrium, i.e., the open state $(X_I, X_O) = (1, 0)$. Figure 3.11 illustrates the reopening process; the inner volume is gradually refilled, while the outer volume slowly returns to the minimal level. It takes about 14 h to transition from the semi-closed state to the open state. The opening speed significantly decreases in the final 2 h.

3 Mathematical Modeling, Dynamics Analysis and Control of Carnivorous Plants

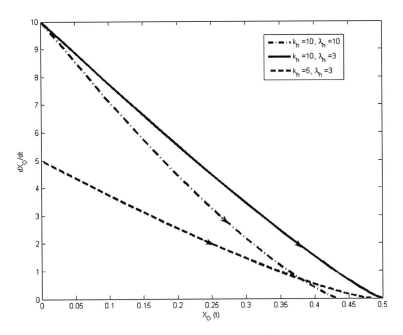

Fig. 3.9 Phase diagram of the model in the capture process. $\dot{X}_O$ rapidly decreases and tends to zero when X_O tends to 0.5. Larger hydraulic transport coefficient k_h (*solid*) shows faster closure than the smaller k_h (*dashed*). And larger λ_h (*dash-dotted*) will cause the trap unable to reach the semi-closed state, and begin to move backward to the open state

Fig. 3.10 Illustration of the capture process. The water rapidly flows into the outer layer, and the open trap $(X_I, X_O) = (1, 0)$ moves to the semi-closed state $(X_I, X_O) = (0.5, 0.5)$ in 0.3 s

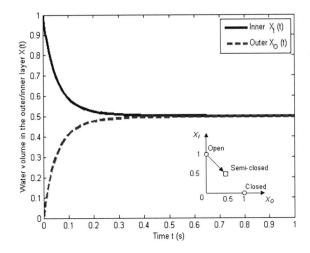

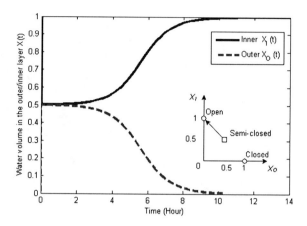

Fig. 3.11 Illustration of the release process. It takes about 14 h to transition from the semi-closed state $(X_I, X_O) = (0.5, 0.5)$ to the open state $(X_I, X_O) = (1, 0)$. The opening speed significantly decreases in the final 2 h

3.2.2.3 Sealing Process: (From the Semi-Closed to the Closed State)

The sealing process requires only electric gradients, i.e., $u_h = 0$, and $u_c = 0$. Suppose n stimuli occur at the time $t_0, \ldots, t_n$, then the water dynamics can be expressed as follows:

$$\dot{X}_O = \frac{\alpha X_O^2}{X_O^2 + (1 - X_O)^2} + k_e e^{-\frac{\lambda_e t}{1 - X_O}} \sum_{i=0}^{n} e^{\frac{\lambda_e t_i}{1 - X_O}} - \mu X_O \qquad (3.14)$$
$$X_I = 1 - X_O$$

Similar to the capture process, the water dynamics in the sealing process does not consist of steady states. Each stimulus u_{ei} leads to a small burst of water transport. The amount of water transport depends on the remaining volume of the inner layer and electric transport coefficient k_e as well as time constant λ_e. The transport coefficient k_e and time constant λ_e largely affect the transport rate at the beginning. However, small volume of the inner layer makes the water transport more and more difficult. As illustrated in Fig. 3.12, four trigger impulses occur at 0, 1, 2, and 3 s, respectively. Each small burst of water transport leads to a stepwise response accelerating the closing. Finally, the trap will reach the closed state due to the cell growth if no more stimuli drive the water transport.

3.2.2.4 Reopening Process: (From the Fully Closed to the Open State)

After digestion, the flytrap releases a chemical signal to initiate reopening. The first stage of the reopening process depends on slow water transport of the reverse chemical gradient, which pushes the trap across the semi-closed state. Following the action, the second stage of the reopening is the same as the release process, which depends on the cell growth. A detailed model can be presented in the following form:

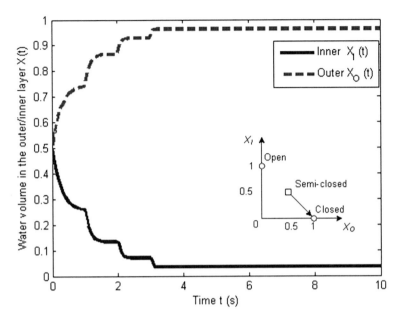

Fig. 3.12 Illustration of the sealing process. Stepwise stimuli accelerate the flytrap transition from the semi-closed state $(X_I, X_O) = (0.5, 0.5)$ to the closed state $(X_I, X_O) = (0, 1)$. Four trigger impulses occur at 0, 1, 2 and 3 s, respectively. Each small burst of water transport leads to a stepwise response to accelerate the closing. Finally, the trap will reach the closed state mainly due to the cell growth if the no more stimuli drive the water transport

$$\dot{X}_O = \frac{\alpha X_O^2}{X_O^2 + (1 - X_O)^2} - k_c e^{-\lambda_c t} - \mu X_O \qquad (3.15)$$

$$X_I = 1 - X_O$$

This process is controlled by the transport coefficient k_c and the time constant λ_c. As illustrated in Fig. 3.13, without mechanical stimuli, the dynamics move across the semi-closed state in 8 h driven by chemical energy and gradually move to the fully open state in 14 h. The maximum opening speed occurs near the position of the semi-closed state. The opening speed significantly decreases in the final 2 h. This result is consistent with the observation by Stuhlman (Stuhlman 1948).

3.2.3 Summary of Model

The above chapter generates a mathematical model to explain the complex transition processes between different states of the flytrap. The proposed model improves the original version in (Yang et al. 2010a), which comprehensively

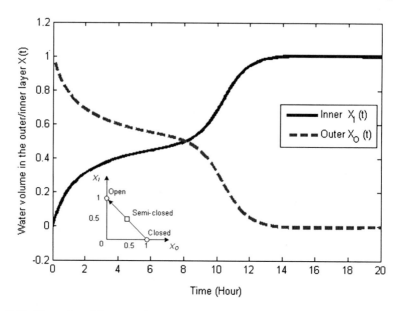

Fig. 3.13 Illustration of the reopening process. Electrochemical force drives the water dynamics across the semi-closed state $(X_I, X_O) = (0.5, 0.5)$ in 15 h. Then the Flytrap transits from semi-closed state $(X_I, X_O) = (0.5, 0.5)$ to open state $(X_I, X_O) = (1, 0)$ in 8 h. The maximum opening speed occurs in the neighborhood of the semi-closed state. The opening speed significantly decreases in the final 2 h

accounts for the effects of electrical charge accumulation, multiple mechanical stimuli, and the slow opening and fast closing states. Different with other models (Forterre et al. 2005; Markin et al. 2008), this model emphasizes two natural error-proof feedforward control designs—the double-trigger and semi-closed state—that are critical decision-making process to prevent closure and digestion in vain by false alarms. The closure and digestion process are energetically costly that a trap can snap for only a few times in its life time. This biological error-proof systems have lots of "biomimetic" engineering applications. In the next section, we will construct a flytrap robot based on our model in this section.

3.3 Flytrap Robot

Humans have always strived to learn special functions from the biological systems, since evolutionary tends to optimize these functions, simplify the underlying mechanism, and reduce false alarms. Biomimetics has fundamentally advanced the fields of materials science, mechanics, electronics, information technology, and many others (Bhushan 2009). Biomimetic robotics aims at physically reproducing the same functions of the biological creatures, thus often among the first to model and implement the biological mechanism (Bar-Cohen and Breazeal 2003; Trivedi

Fig. 3.14 The schematic of the Venus flytrap robot. The deflection sensor gives the stimuli signal to controller. The controller moves the necessary rotation angle of the center gear with the help of the internal angle sensor in the servomotor. Then the motor determines the open angle of the trap by the three-interlocking gears

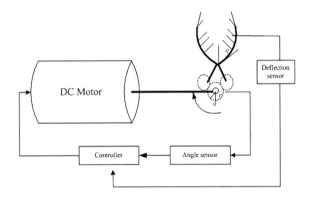

et al. 2008). The unique functions of the Venus flytrap certainly are of significant interests to biomimetic robotics. In robotics, the flytrap can serve as a deployable structure, such as gripper, that can be attached to a fixed structure to perform independent function and increase overall flexibility and adaptation (Vincent 2003).

The first stage of flytrap robots is to mimic the sensory and closure, similar to a real trap. This type of flytrap robots has been created as toys, such as Fly Catcher® and Artificial Venus Flytrap®, and smart materials (Shahinpoor and Thompson 1995). The second stage recently became to mimic the bistable curvature (Forterre et al. 2005) that creating buckling force to fast snap. Kim et al. (2011) built a flytrap robot using bistable unsymmetrically laminated carbon fiber reinforced prepeg (CFRP). The composite is bent in the same curvature configuration by shape memory alloy coil spring. The robot can snap in ~ 100 ms.

In this chapter, we takes a different aspect—error-proof decision-making mechanism—to develop the third generation of flytrap robots. In fact, this mechanism can be applied for a large number of engineering applications, such as microsensors and microgrippers. As illustrated in Fig. 3.14, a flytrap robot consists of four components: controller, sensors, motor, and lobes. The deflection sensor gives the stimuli signal to controller. The controller moves the necessary rotation angle of the center gear with the help of the internal angle sensor in the servo-motor. Then the motor determines the open angle of the trap by the three-interlocking gears.

The Venus flytrap robot is about one foot high, as shown in Fig. 3.15a and b. The Venus flytrap robot consists of two main parts, four leaves and the base. Each leaf is made of two 2×5 inch trapezoidal plastic lobes. The outer surface of the leaf is coated by green rubber films with finger-like projections, cilia, on the edge; the inner surface has six deflection sensors (Flexpoint sensor systems, Inc., Bend Sensor® K06), two in each lobe. The sensors are able to detect the deflection and translate the mechanical stimuli to a resistive value and feedback to the logic in the Arduino Duemilanove microcontroller (Arduino Inc.). The Arduino Duemilanove is a USB microcontroller board based on the ATmega328 with 14 digital input/output pins and 6 analog inputs. The control algorithm can be written into the board using a USB cable.

Fig. 3.15 An overview of the Venus flytrap robot. **a** The Venus flytrap robot consists of two main parts, four leaves and the base. Each leaf is made of two 2 × 5 inch trapezoidal plastic lobes. **b** The outer surface of the leaf is coated by green rubber films with finger-like projections, cilia, on the edge; the inner surface has six deflection sensors, which detects deflection and feeds back to microcontroller. **c** The Arduino Duemilanove microcontroller is used to implement the control algorithm for the Venus flytrap robot. A protoshield is mounted on the Arduino board for circuit design. **d** The logic, motor and gearbox are hidden inside the base. The motor drives a three-interlocking gear box, which further rotates the leaves through rotating shafts and rods

As shown in Fig. 3.15c and d, both logic and motors are hidden inside the base. The leaves are driven by a DC servomotor (Futaba S3001), which has 3.0 kg/cm torque and rotation speed up to 0.22 s/60° at 6 V working voltage. The motor is then attached to a gearbox with three-interlocking gears. The center gear is powered by the motor, while the side gears rotate in opposite direction but equal step size when the center gear is rotating. A rod is mounted in the center of each side gear, and connected to the rotating shaft. The rotating shaft is further mounted to a lobe. Thus, two rotating shafts rotate the lobes in opposite direction as the center gear rotates.

For the flytrap robot, the opening time was accelerated. The water supply rate α of the flytrap robot was thus chosen to be 100 fold of the real flytrap model.

3 Mathematical Modeling, Dynamics Analysis and Control of Carnivorous Plants

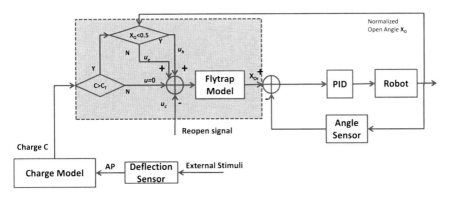

Fig. 3.16 The control diagram of the closing and opening mechanisms for the Venus flytrap. The APs and reopen chemical signal trigger different control events under various conditions. The activated control signal is sent to the flytrap system, resulting in the movement

In addition, we assumed that the maximum/minimum opening angles of the trap were 75 and 20°, thus the semi-closed state existed at 47.5°.

The movement of the flytrap robot can be generalized by Fig. 3.16. When the sensor detects the mechanical stimulation, an impulse is sent to the charge accumulation block. As the accumulated charge C exceeds the trigger threshold C_T, either the capture process or sealing process will start. If not, the dynamics follow the growth process. If the flytrap is at the fully closed state, a reopening signal will initiate. All the controls are delivered to the DC motor, which rotates the gear to manipulate the open angle of the flytrap. Specifically, we use the following model to define the reference trajectories that are tracked by the DC motor.

$$\dot{X}_O = \frac{\alpha X_O^2}{X_O^2 + (1 - X_O)^2} + u_h + u_a - u_c - \mu X_O \tag{3.16}$$

$$u_h(t; t_0) = k_h(X_I - X_O)e^{-\lambda_h(t-t_0)}, \tag{3.17}$$

$$u_e(t; t_0) = k_e e^{-\frac{\lambda_e(t-t_0)}{1-X_O}}, \tag{3.18}$$

$$u_c(t; t_0) = k_c e^{-\lambda_c(t-t_0)}, \tag{3.19}$$

$$X_0 = \frac{\eta - \eta_0}{\eta_1 - \eta_0}, \tag{3.20}$$

$$\eta = k\theta. \tag{3.21}$$

The movement of the flytrap robot is illustrated in the video (see http://web.utk.edu/~mjzhang/Projects.html). If the prey stimulates the sensor on the inner surface of the flytrap, the trap snaps to the semi-closed state. The trap will proceed to the sealed state if the prey keeps struggling and stimulates more sensor deflection signals; otherwise the trap will slowly reopen.

3.4 Conclusions

Inspiration from the natural world has long intrigued engineers, leading to advances in aerospace, biomedical, and mechanical engineering fields. To translate biological principles into engineering applications it is necessary to first apply engineering principles, such as mathematical modeling, dynamics analysis, and control to understand the biological system. Oftentimes, by understanding the biological system in terms of engineering, more advanced engineered systems can be developed. In this chapter we have demonstrated some of the interesting characteristics of the Venus flytrap that have drawn both biologists' and engineers' interest. To explain these biological phenomena in engineering terminology, a control systems model was developed to explain these processes. The proposed model improves on the original version in (Yang et al. 2010a), which accounts for the effects of electrical charge accumulation, multiple mechanical stimuli, and the slow opening and fast closing states. Differing from other models (Forterre et al. 2005; Markin et al. 2008), this model emphasizes two natural error-proof feed-forward control designs—the double trigger and semi-closed state—that are critical decision-making processes that prevent closure and digestion in cases of inappropriate trapping or false alarms. These decision-making steps are of particular interest in engineered systems, where similar energetic demands are encountered. The design of biomimetic robots, micro-sensors, and many other manufacturing processes could significantly benefit from a thorough understanding of these principles.

In addition to theoretical modeling, we have built a DC-motor driven flytrap robot that physically demonstrates the unique mechanisms incorporated in our model. This robot demonstrates how the theoretical contributions can be applied to a realistic system. This study provides a complete example that illustrates the transition from bioinspiration to an engineered system, and lays down the groundwork for future studies aimed at learning from nature.

References

Bar-Cohen Y, Breazeal CL (2003) Biologically inspired intelligent robots. SPIE Publications, WA

Bhushan B (2009) Biomimetics: lessons from nature: an overview. Phil Trans R Soc A 367(1893):1445–1486

Bobji M (2005) Springing the trap. J Biosci 30:143–146

Brogan WL (1985) Modern control theory, 3rd edn. Prentice Hall, NJ

Brown WH, Sharp LW (1910) The Closing response in dionaea. Bot Gaz 49(4):290–302

Burden-Sanderson J (1882) On the electromotive properties of the leaf of Dionaea in the excited and unexcited states. Philos Trans R Soc Lond B Biol Sci 173:1–55

Darwin C (1875) Insectivorous plants. Murray, London

Fagerberg WR, Allain D (1991) A quantitative study of tissue dynamics during closure in the traps of Venus's Flytrap Dionaea muscipula ellis. Am J Bot 78(5):647–657

Fagerberg WR, Howe DG (1996) A quantitative study of tissue dynamics in Venus's Flytrap Dionaea muscipula (droseraceae). II. Trap Reopening Am J Bot 83(7):836–842

Forterre Y, Skotheim JM, Dumais J, Mahadevan L (2005) How the Venus Flytrap snaps. Nature 433(7024):421–425

Hodick D, Sievers A (1986) The infulence of Ca^{2+} on the action potential in mesophyll Cells of Dionaea muscipula ellis. Protoplasma 133:83–84

Hodick D, Sievers A (1988) The action potential of Dionaea muscipula ellis. Planta 174(1):8–18

Hodick D, Sievers A (1989) On the mechanism of trap closure of Venus Flytrap (Dionaea muscipula ellis). Planta 179(1):32–42

Kim S-W, Koh J-S, Cho M, Cho K-J (2011) Design and amp; analysis a flytrap robot using bi-stable composite. In: IEEE international conference on robotics and automation (ICRA), Shanghai, 9–13 May 2011, pp 215–220

Leang K, Zou Q, Devasia S (2009) Feedforward control of piezoactuators in atomic force microscope systems. IEEE Control Syst Mag 29(1):70–82

Li Y, Zhang M (2011) Nonlinear dynamics in the trapping movement of the Venus Flytrap. In: American control conference, San Francisco, CA, pp 3514–3518

Mangan S, Alon U (2003) Structure and function of the feed-forward loop network motif. PNAS 100(21):11980–11985

Markin VS, Volkov AG, Jovanov E (2008) Active movements in plants: mechanism of fly catching by Venus Flytrap. Plant Sign Behav 3:778–783

Milo R, Shen-Orr S, Itzkovitz S, Kashtan N, Chklovskii D, Alon U (2002) Network motifs: simple building blocks of complex networks. Science 298(5594):824–827. doi:10.1126/science.298.5594.824

Rea PA (1983) The dynamics of H+ efflux from the trap lobes of Dionaea muscipula ellis (Venus's Flytrap). Plant Cell Environ 6(2):125–134

Schäffner AR (1998) Aquaporin function, structure, and expression: are there more surprises to surface in water relations? Planta 204(2):131–139

Shahinpoor M, Thompson M (1995) The Venus Flytrap as a model for a biomimetic material with built-in sensors and actuators. Mat Sci Eng C Biomim 2:229–233

Stuhlman O Jr (1948) A physical analysis of the opening and closing movements of the lobes of Venus' Flytrap. Bull Torrey Bot Club 75(1):22–44

Stuhlman O Jr, Darder E (1950) The action potentials obtained from Venus's Flytrap. Science 111:491–492

Trivedi D, Rahn CD, Kier WM, Walker ID (2008) Soft robotics: biological inspiration, state of the art, and future research. Appl Bionics Biomech 5(3):99–117

Ueda M, Nakamura Y, Okada M (2007) Endogenous factors involved in the regulation of movement and "memory" in plants. Pure Appl Chem 79(4):519–527

Vincent JF (2003) Deployable structures in biology. In: Hara F, Pfeifer R (eds) Morpho-functional machines—the new species: designing embodied intelligence. Springer, Tokyo, pp 23–40

Volkov AG, Adesina T, Jovanov E (2007) Closing of Venus Flytrap by electrical stimulation of motor cells. Plant Sign Behav 2:139–144

Volkov AG, Adesina T, Jovanov E (2008a) Charge induced closing of Dionaea muscipula ellis trap. Bioelectrochemistry 74(1):16–21

Volkov AG, Adesina T, Markin VS, Jovanov E (2008b) Kinetics and mechanism of Dionaea muscipula trap closing. Plant Physiol 146(2):694–702

Volkov AG, Carrell H, Adesina T, Markin VS, Jovanov E (2008c) Plant electrical memory. Plant Sign Behav 3:490–492

Volkov AG, Coopwood KJ, Markin VS (2008d) Inhibition of the Dionaea muscipula ellis trap closure by ion and water channels blockers and uncouplers. Plant Sci 175(5):642–649

Yang R, Lenaghan SC, Zhang M, Xia L (2010a) A mathematical model on the closing and opening mechanism for Venus Flytrap. Plant Sign Behav 5(8):968–978

Yang R, Tarn T-J, Zhang M (2010b) Data-driven feedforward control for electroporation mediated gene delivery in gene therapy. IEEE Trans Control Syst Technol 18(4):935–943

Chapter 4
The Telegraph Plant: *Codariocalyx motorius* (Formerly Also *Desmodium gyrans*)

Anders Johnsson, Vijay K. Sharma and Wolfgang Engelmann

Abstract The telegraph plant (*Codariocalyx motorius*) has drawn much interest among plant physiologists because of its peculiar movements of the leaflets. While the terminal leaflets move from a horizontal position during the day and downward during the night, the lateral leaflets display rhythmic up and down movements in the minute range. The period length of the lateral leaflets is temperature dependent, while that of the terminal leaflet is temperature compensated. The movements of both the leaflets are regulated in the *pulvini,* a flexible organ between the leaflets and the stalk. Electrophysiological recordings using microelectrodes have revealed the physiological mechanisms underlying the leaflet movements. Early experiments related to effect of mechanical load, light, electric and magnetic fields on the leaflet oscillations by the Indian physicist Bose, and followed up by others, are presented. Experimental approaches are discussed and indicate, that Ca^{2+}, various membrane channels, electric and osmotic mechanisms participate in the oscillating system. Modelling the pulvinus tissue would certainly aid in understanding the signal transduction during the movements. New approaches of modelling the mechanisms could further help in understanding the oscillations in the leaflet movements. Such oscillations might be of much broader relevance than known so far, although not as conspicuous as in the leaflet movements.

A. Johnsson (✉)
Department of Physics, Norwegian University of Science and Technology,
NO-7491 Trondheim, Norway
e-mail: anders.johnsson@ntnu.no

V. K. Sharma
Chronobiology Laboratory, Evolutionary and Organismal Biology Unit,
Jawaharlal Nehru Centre for Advanced Scientific Research,
Jakkur, P.O. Box 6436Bangalore, 560064 Karnataka, India

W. Engelmann
Botanisches Institut, Universität Tübingen, Auf der Morgenstelle 1,
D-72076 Tübingen, Germany

4.1 Introduction

The plant that will be discussed in the present chapter is popularly known as 'The Telegraph Plant' (or the 'semaphore plant') because of the typical movements of its leaflets. Its present name is *Codariocalyx motorius*.[1] Figure 4.1 illustrates different shapes of its two types of leaves, one larger (terminal) leaflet and two (one or sometimes none) smaller lateral leaflets. The terminal leaf makes up and down movement in a daily fashion, while the lateral leaflets resort to more rapid movements, noticeable to the naked eye. This movement is rhythmic with a period of 3–4 min at 25°C has been associated with the flags in a semaphore telegraph, thus giving rise to its present name.

The first mention of the leaflet movements in the telegraph plant was made in 1790 (Hufeland 1790; Aschoff 1991) by the physician Hufeland (who named it as *Hedysarum gyrans* motitans L.f., L.f. stands for Linné's son). The plant is indigenous to South-East Asia. Its seeds from Professor Groschke i Mitau were brought from England (originally from Bengal) to Hufeland.

The Indian biophysicist Bose (1858–1937) carried out extensive pioneering studies on this plant to understand the physiology of leaflet movements. He is presumably the first to discover electrical pulsations and its connection with the leaflet movements (Shepherd 1999, 2005).

In the present chapter we will primarily focus on the *lateral* leaflet movements of *Codariocalyx*, although time to time we will also compare it with the movement of the *terminal* leaf. We will discuss the physiology of the tiny organ between the leaflets and rhachis ('midrib') of the leaf, called the *pulvinus*,[2] the organ which controls the leaflet movements. Its structure is shown and described in Fig. 4.2. The tissue between the epidermis and central core of the pulvinus consists of *motor cells*. These cells elongate and shrink due to internal, osmotically driven water transport system, which causes changes in their volume. These volume changes occur at varying sites of the pulvinus and determine the direction of the movements of the leaflet. In fact, functioning of this organ is considered analogous to an 'Osmotic Motor'. The pulvinus is also thought to be the site of the pacemaker that controls the oscillatory movements of the leaflets.

We will first discuss the motor cells of the pulvinus, their transport of ions, mainly Ca^{2+}, K^+ and H^+, across the membranes (the last ones are important in the osmotic water transport), and then concentrate on oscillations in the motor cells and in the leaflet movements. It is obvious that for a coordinated movement of the

---

[1] The present scientific name of the plant is *Codariocalyx motorius* (Houtt.) (Ohashi 1973), but for a very long time older names such as *Hedysarum gyrans* (L.f.), *Desmodium gyrans* (L.f.) DC, and *Desmodium motorium* (Houtt.) Merril., were used. In this chapter we will consistently use the present official name viz. *Codariocalyx*, but readers looking for relevant literature should also use '*Desmodium*' as the key word.

[2] *Pulvinus:* correctly *pulvinule*, since it is the joint of a leaflet, not a leaf; for simplicity and since it is commonly used, we will use pulvinus (plural: pulvini).

4 The Telegraph Plant

Fig. 4.1 Leaf systems of *C. motorius*, each consisting of one larger terminal leaflet and two (occasionally one or none) small lateral leaflets. The terminal leaflets exhibit daily up and down movements, while the lateral leaflets display rhythmic movements in the minute range (illustrated here by the photographic multiexposures). The *pulvini* (cylindrical brownish structures at leaflet bases, indicated by white arrows) are responsible for the movements. Courtesy of Bernd Antkowiak, Tübingen

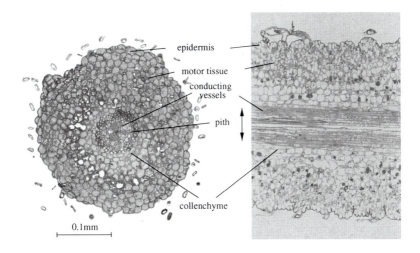

Fig. 4.2 Cross section (*left*) and longitudinal section (*right*) of the pulvinus of a lateral leaflet. Between the central core (*double headed arrow*, *right* panel) consisting of pith, conducting vessels and sclerenchyme, and the epidermis is a parenchymatous tissue consisting of collenchyme (two to three cell layers of inner part) and motor cells (outer part). The latter are about 15 µm long and have a diameter of about 10 µm. They can change their volume in a coordinated and reversible manner leading to the movements

motor cells of the leaflets a *coupling* between them is necessary. We will discuss and model this coupling between the pulvinus cells. We will also discuss systems of models to achieve such coordinated movement of motor cells in the pulvinus.

The pulvinus functions not only as an *executor* of leaflet movements, controlled by water volume changes in the motor cells, but also as a *sensor* for environmental light, touch, and heat signals, as well as a *transducer* of signals to its osmotic motor, which works as the executor to move the leaf.

Much knowledge has been gained from studies of pulvini and motor cells of related plants such as *Samanea* (Satter et al. 1990), *Mimosa* (Fromm and Eschrich 1990), *Albizzia* (Satter and Galston 1971) and *Phaseolus* (Mayer 1990). There exists a number of reviews on this topic by Coté (1995), Moran (2007), Engelmann and Antkowiak (1998), and a book by Satter et al. (1990). However, only the third one is devoted uniquely to *Codariocalyx* studies. Furthermore, intensive efforts to understand the reactions and volume changes of the *guard cells* of plants and their control of plant transpiration have contributed a great deal to the field (e.g. Gorton 1990; Pandey et al. 2007).

The content of the present chapter will roughly follow this sequence: Structure and function of the *Codariocalyx* pulvinus and its motor cells → ion transport leading to osmotic water transport → swelling and shrinking leading to positional changes of pulvinus → pulvinus bending leading to positional change of leaflets → rapid leaflet movements are possible to observe.

4.2 Anatomy and Physiology of the *Codariocalyx* Pulvinus

The movements of the lateral leaflets, visualised by multiple exposures in Fig. 4.1, are due to volume changes of the motor cells in the *pulvinus*. The pulvinus of the lateral leaflets has a cylindrical structure—about 3 mm long. Its transversal and longitudinal sections are illustrated in Fig. 4.2. The cross section shows the central core (vascular bundles and support tissue) and the tissue of 'motor cells' lying like a cylinder around the central core of the pulvinus. An epidermal layer, finally, defines the outer boundary of the pulvinus. The longitudinal section, likewise, demonstrates the central core and the motor cells, which are more or less cylindrical in structure (see also Whitecross 1982). This cylinder like structure of the motor cell tissue at the outer part of the pulvinus allows circular or elliptical movements of leaflets, as often found in addition to up- and down movements.

In contrast to the pulvini of the terminal leaflets and to pulvini of other species of plants with circadian movements there is no anatomic difference between the 'upper' (or *adaxial*, towards the stalk) and the 'lower' (or *abaxial*, away from the stalk) tissue of the motor cells (Weber 1990). When treating the mechanisms for the lateral leaflet movements we, therefore, would avoid the concepts of *flexor* and *extensor* cells.

Figure 4.3 illustrates schematically how the bending of the pulvinus comes about. The scheme of the cells in the left panel, forming a ring (simplified cross section through the pulvinus, showing just one layer of motor cells and no central core) show shrunken cells on the top and elongated cells at the bottom. Many of those rings next to each other would form a tube (=pulvinus) pointing slightly upward in the situation shown since the cells underneath are lengthened relative to the upper ones. The leaf would point downward when the cells in the 'upper' part are extended and the cells in the 'lower' part shrunken and upwards when the upper cells are shrunken and lower cells are extended.

The longitudinal cell length variations would lead to the curvature of the whole (cylinder) tissue if the cells are synchronised in an appropriate manner. Assuming the volume changes to occur in a longitudinal plane cutting through the pulvinus, a to-and-fro bending of the pulvinus cylinder is likely, causing a corresponding to-and-fro movement of the leaflets. If the pulvinus is more or less rotationally symmetric the leaflet movements would be elliptical or circular. The direction would be up and down if the cylinder axis is horizontal, but sidewise if it is directed vertically. If the pulvinus regions are not well synchronised, the movement is likely to be irregular. Occasionally recorded movements of the leaflets of *Codariocalyx* demonstrate all types of patterns (for examples see Chen 1996).

Two oppositely placed motor cells in the pulvinus are sketched in the right panel of Fig. 4.3. The upper one shows the shrunken state (contains less water), while the lower one the swollen state (higher water content). With increasing water uptake, only the *length* of the motor cells changes, while the radial diameter remains more or less unaffected. This is due to the arrangement of the cellulose microfibrills along the cell walls (oriented only tangentially to the cell wall, not longitudinally, Weber 1990).

There are further differences between pulvini of the lateral and terminal leaflets and their behaviour besides the ultradian and circadian length of the period. The terminal leaf contains a starch sheet between the central core and the motor cell tissue, which is thicker and partly double layered, containing much more starch at the lower (abaxial) part compared to the upper (adaxial) part. This starch sheet is missing in the lateral leaflet pulvini; there are only a few cells of the innermost parenchyma tissue containing single starch grains. The radial xylem elements of the terminal leaflet pulvinus are dorsoventrally asymmetric, containing more tracheids on the abaxial part. This is not so in the lateral leaflet pulvinus. The largest motor cells of the terminal leaflet pulvinus have a diameter of 32 µm, whereas the size of the lateral leaflet pulvini varies between about 11 and 18 µm. A further difference is that the circadian period of the terminal leaflets is modulated by alcohols, while the ultradian period of the lateral leaflets remains largely unaffected (Kastenmeier et al. 1977).

4.2.1 Pulvinus Shape and Bending

The increase in the *Codariocalyx* motor cell length during leaflet movement is estimated to be about 20% (Engelmann and Antkowiak 1998; citing Mitsuno 1987). A simple estimate of the curvature of the pulvinus can be made by

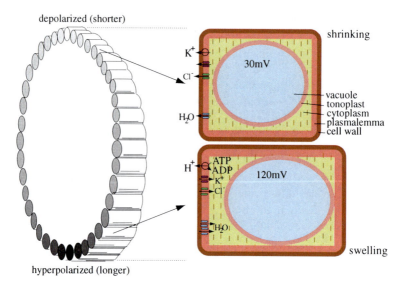

Fig. 4.3 Function of *Codariocalyx* pulvinus—schematic. *Left* panel shows schematics of one of the many motor cell rings in a lateral pulvinus of *C. motorius*. The cells at the top are in a shrunken state, while those at the bottom in a swollen state. This leads to the bending of the pulvinus. Half a period later the upper cells become larger, and the lower cells smaller, which leads to reversal in the direction of bending. The upper and lower cells are electrically depolarised and hyperpolarised, respectively. At the right are schematically shown a shrinking motor cell with cell wall (*brown*), outer membrane (plasmalemma, *pink*) and tonoplast (*pink*) around the vacuole (*light blue*). Microfibrils are drawn as dotted vertical lines onto the cytoplasm (*yellow*). Proton- and ion- pumps as well as water channels (aquaporins) which regulate the volume changes of the motor cell are indicated. In the shrunken state of the motor cells, the plasmalemma membrane is depolarised and outwardly directed K^+ and Cl^- channels open (and inward K^+ channels close). As a result, osmotically active particles leave the cell, the water potential changes, and the cell shrinks. After a while—due to the inward transport of H^+ by ATPases—the cell gets hyperpolarised, and K^+ re-enters the cell through voltage activated K^+ channels. The motor cells thus start swelling once again

considering its geometrical dimensions. The pulvinus length, L, is roughly about 3 mm, the total pulvinus radius, r, about 0.15 mm and the radius of the motor cell tissue, r', about 0.075 mm. We assume that of the motor cells, one half of the pulvinus is swollen, and the other half is shrunken.

Since $L = \alpha \times R$ (where α is the angle of curvature and R the radius of curvature of the pulvinus), simple calculations show that the difference in the lengths of the cells of outer and inner curvatures of the pulvinus would be $\Delta L = 2\alpha \times r'$ (where r' is the radius of the motor cell tissue).

For a leaflet movement cycle, α can be about 120° (=$2\pi/3$ radians; based on empirical observation). With r' of about 0.075 mm and L of about 3 mm, this bending can be achieved with a relative length increase $\Delta L/L$ of only about 0.1 or 10%. The calculation thus illustrates that the pulvinus structure allows very large curvatures and bending, as often found both in spontaneous ultradian movements and in externally stimulated leaflets.

4.2.2 Pulvinus Curvature and Water Transport

In the case of *Codariocalyx*, the pulvinus movements are reversible, since the motor cells take up and expel water. Thus, no irreversible lengthening of the cells would be present here, in contrast to other types of movements in plants (discussed e.g. for circumnutations around the plumb line, Johnsson 1997). The pulvinus movements are, therefore, not based on growth. Whether expansin is involved in the bending of the pulvinus, as is the case in growth, is unknown. During the growth of, for example, the oat coleoptile, auxin activates H^+-pumps in the plasma membrane, by which the cell wall solution is acidified. This activates expansin, a cell wall loosening protein, allowing the cell wall to yield to the wall tension created by turgor pressure. As a result the cells enlarge (Cosgrove 2005; Cosgrove et al. 2002).

Using the example above with a relative increase (or decrease) in the length of the motor cell during a leaflet oscillation by about 10%, one can estimate the water flux in and out of motor cells.

Water volume transported into and out of a cylindrical motor cell at the rim of the tissue will be $\Delta V = \pi r^2 \times \Delta l$, where r is the radius of one motor cell and l is the length of the cylindrical motor cell.

The surface area A (curved area of the cylinder + two circular end areas) of the motor cell will be

$$A = 2\pi r \times l + 2\pi r^2 \tag{4.1}$$

This might be regarded as the maximum estimate of the area—if the water is expelled from the circular end areas into the neighbour cells the second term on the right hand side of the above Eq. 4.1 should be ignored. In case it streams into the extracellular space it should be included.

The volume flux of water (i.e. volume water, ΔV, passing the plasmalemma per unit area per second) J_w will then be $J_w = \Delta V/(A \times$ (time for transport of ΔV)).

The volume change occurs during roughly half the period of time for one *Codariocalyx* leaflet oscillation, i.e., for about 150 s. This would correspond to a volume flux of about 3×10^{-7} cm^3/cm$^2 \times$ s. Downward movement of leaflet is, however, often more rapid and therefore can be estimated to be shorter than about 50 s (from data presented in Antkowiak 1992, p. 65, see also Fig. 4.4, short versus long interval between the vertical lines) which increases the value to about 10^{-6} cm^3/cm$^2 \times$ s. The volume flow will further increase if the circular cell end areas are not included in the above estimation.

4.2.3 Pulvinus Water Transport

The water flux through the motor cell membrane is mainly through aquaporin channels (Maurel et al. 2008; Kaldenhoff and Fischer 2006; Moshelion et al. 2002a). It is assumed to be driven by the difference in the *water potential*, as summarised

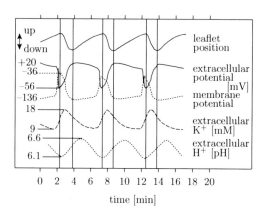

Fig. 4.4 Time course of physiological events in the motor cells of the lateral leaflet of *Codariocalyx*. Up and down movement of a lateral leaflet for three cycles. The vertical lines mark the onset of down and up movements, and serve as a temporal reference for other events. Extracellular potential (*solid curve*, varying between +20 and −56 mV), and membrane potential (*dotted curve*, varying between −36 and −136 mV), run opposite to each other. Extracellular K⁺ (*dashed curve*, varying between 9 and 18 mM), and extracellular pH (*dotted curve* at the *bottom*, varying between 6.1 and 6.6). After Antkowiak (1992)

in the following formula (e.g., Noble 1974): $J_w = -L_p \times \Delta\Psi$. L_p is the permeability of the membrane, $\Delta\Psi$ the water potential difference over the membrane.

The formula describes that water is transported due to gradients in the water potential, Ψ. The basic components of Ψ are the physical pressure in the cells P (including the matrix potentials), and the osmotic pressure, Π, in the cells. The water potential difference is therefore written as:

$$\Delta\Psi = (\Delta P - \sum \sigma_j \times \Delta\Pi_j) \qquad (4.2)$$

A difference in physical pressure ΔP between the two compartments can be created due to the physical stress built up in the two parts via different 'stiff' cell walls. In the case of the pulvinus tissue, an external physical load can be applied to alter the P values.

The relevant $\Delta\Pi$ is the difference in osmotic water potential which depends on the number of molecules in the compartments. The indices j denote the different osmotically active constituents (uncharged or charged) in the system, while σ_j denotes the corresponding reflection coefficients (see e.g. Noble 1974).

The water conductivity constant L_p describes how the water flux between the compartments will (to a first approximation) depend on the difference in water potential. It depends on the membrane permeability and the tissue permeability between the compartments. In this way the flux resistance in the cell wall as well as the membrane conductivities will affect the water transport between the cells.

The simplified reasoning above shows that the water transport in motor cells of the pulvinus can be affected via changes in P and Π—caused by, for example, a physical load on the leaflets, touch, osmotic changes caused by adding uncharged solutes, by external changes in ion concentration, internally generated osmotic changes or by ion transport via illumination.

While we are not going to discuss here, at least not in details, the water transport and water binding to cellular molecules (see Kramer and Boyer 1995), providing a rough estimate of the quantitative flux processes of the motor cells might be useful. With J_w about 10^{-6} cm^3/cm^2 × s we need an estimate of L_p to obtain an estimate of $\Delta\Psi$. The motor cell permeability of *Codariocalyx* is not available but tabulated values (see e.g. Kramer and Boyer 1995) indicate that a value of $L_p = 10^{-5}$ cm/bar × s is reasonable (the same value was provided by Noble (1974) for *Nitella* and *Chara*). We then arrive at a $\Delta\Psi$ of about 10^{-1} bar, which would be required to shrink the motor cell in the time specified. Assuming this to be due to changes in Π alone, it would correspond to an estimated concentration difference of about 4 mM. Thus, the water efflux from the motor cells is likely to be driven by a concentration difference of about 5 mM. An increase in apoplastic K$^+$ of this order during the depolarising phase of the adaxial motor cells has already been measured (Antkowiak 1992).

Measurements of Π and P of the pulvinus tissue have already been published for the circadian movements of *Samanea* leaves (Gorton 1987). The structure of the more voluminous pulvinus tissue seems to be different from that of the lateral pulvinus of *Codariocalyx,* since they have (as already mentioned) distinct flexor and extensor structure. The water transport parameters might therefore be qualitatively different.

4.3 *Codariocalyx*: Experiments on Leaflet Movements

In the previous section we have mentioned that the pulvinus of the telegraph plant works like a sensor, as a signal transducer and as an effector of volume changes. The *Codariocalyx* leaflet movements are generated by synchronised, rhythmic osmotic changes in the motor cells of the pulvinus. As a consequence water flows in and out of cells in a synchronous manner.

This synchronisation is depicted in Fig. 4.3 left panel. The upper part of the ring of the motor tissue has cells with smaller volume, while the lower cells have larger volume. This facilitates the bending of the pulvinus. Half a period later the upper cells become larger, and the lower cells become smaller and thus the direction of bending is reversed.

Figure 4.3, right panel, shows in a schematics ion pumps and membranes which regulate the volume changes of the motor cells. Although we will discuss this in some details later, very briefly it can be mentioned here that in the shrunken phase of the motor cells, the plasmalemma membrane is depolarised and outward directed K$^+$ and Cl$^-$ channels are opened (and inward K$^+$ channels closed). This implies that osmotically active particles would leave the cell, leading to water potential changes (Eq. 4.2) and eventually the cell will shrink. After a while the plasmalemma ATPases will start inward transport of H$^+$, the cell will be hyperpolarised and K$^+$ will re-enter the cell through voltage activated K$^+$ channels. The motor cell thus increases its volume once again. Aquaporin channels are likely to be involved too, as shown by Moshelion et al. (2002a) for *Samanea*.

4.3.1 Background: J.C. Bose

The eminent Indian biophysicist Bose (1858–1937) pioneered in the scientific studies of the leaflet movements of *Desmodium* (as the plant was then referred to and we use this name in honour of Bose through some sections of the present subchapter). Some approaches and results by Bose will be mentioned and commented on in the following sections.

Much before biologists learnt to appreciate electric phenomena and oscillations in plants Bose finished his studies on the Indian telegraph plant and came to a number of insightful conclusions. Bose worked as Professor of Physics at the Presidency College, Calcutta (now Kolkata) after having studied physics under Rayleigh in Cambridge. He was a polymath and a pioneer in radio science, microwave optics (and preceded Marconi in many fields). After his retirement in 1917 he founded the Bose Institute (now a top research Institute in India). But he turned his main interest to biophysics and we give a brief account of his findings on physiology and behaviour of plants, with a focus on *Desmodium*.

As early as in 1901 Bose turned his attention to experimenting with bioelectric potentials and movements in plants. For little over 30 years he performed hundreds of experiments on electrical signalling in plants. Bose chose to study plants which displayed intrinsic oscillations. He discovered that the leaflets of the sensitive touch-me-not plant *Mimosa pudica* folds its leaflets when touched. Likewise, the Indian telegraph plant *Desmodium* showed spontaneous oscillations or gyration of the lateral leaflets, as well as up-and-down movements. He examined a host of other plant species including those capable of leaflet movements such as beans (*Phaseolus sp*) and *Biophytum, Impatiens,* and *Chrysanthemum, Ficus,* and so on. Most of his experiments required sophisticated apparatus, some of which he adapted from his earlier research in physics and for others he decided to device his own. With such equipments Bose was able to measure very small leaf movements with high temporal resolution (<1–2 s). He was also able to measure mechanical responses of *Mimosa* and *Desmodium*, and electric potentials associated with responses in plant tissues using a device analogous to the later chart-recorder. He recorded tropic movements in plants, twining of tendrils, thermonastic phenomena, photonastic responses, geotropism and day/night movements of petals and leaves of plants. For his later studies he stimulated plants electrically to mimic effects of mechanical stimulation such as scratching, touching, heating and chilling.

In this review of the experiments on the *Codariocalyx* leaflet movements we will first discuss experiments involving changes in some environmental parameters. They were also in focus during the early stages of the *Codariocalyx* studies, foremost by Bose.

4.3.2 Leaflet Movements and Temperature

In contrast to the period of the circadian movements of the terminal *Codariocalyx* leaf, the period of the movements of the lateral leaflets depends strongly on temperature. It decreases with increasing temperature with a Q_{10} value of 2.0 (Cihlar 1965) or 2.5 (Lewis and Silyn-Roberts 1987). Typical period values can be about 8.7 min at 15°C and about 2 min at 32°C. Giving high temperature pulses for 30 s by applying a 45°C heating element at about 5 mm from the pulvinus produced advance phase shifts or no phase shifts at all (Lewis and Silyn-Roberts 1987). Under the same condition it was possible to phase lock the movements to regular temperature pulses (entrainment) when the period of the pulses was shorter than that of the leaflet rhythm (data on limits of this entrainment not provided). Cihlar (1965) studied the effects of 'cold' and 'warm' temperature pulses and obtained both phase delays and advances, depending on the phase of perturbation. Experiments with small amplitude signals (every 5 min for 1 min at 20.3°C, otherwise 19.8°C) showed partial synchronisation (Engelmann, 'personal communication'). The experiments also indicate that it is likely that pulvinus functions as a temperature sensor.

4.3.3 Leaflet Movements and Mechanical Load

The water potential Ψ (Eq. 4.2) contains a term that depends on the physical pressure P. An external load on the leaflet is likely to change the pressure distribution in the motor cells of the pulvinus and could affect Ψ if sufficiently large. The presence of stretch-activated ion channels in a wide variety of cells (e.g. *Samanea* protoplasts, reviewed in Moran 2007; in guard cells: Cosgrove and Hedrich 1991; Zhang et al. 2007) makes mechanical handling of the pulvinus pertinent. Bose's studies of possible effects on the oscillation when applying a load to a *Codariocalyx* leaflet were thus quite logical and much ahead of his time.

The *Codariocalyx* lateral leaflet movement is faster in its downward position as compared to the upward position. Antkowiak (1992, p 65) specifies the time for the down movement to be about 50 s out of a mean period of about 250 s, i.e., the leaflets stay for about 20% of their cycle in downward position. Bose (1913) and Dutt and Guhathakurta (1996) and others have observed a greater speed in the downward movement. Further changes under loading conditions were studied by Bose (1913), applying weights of 5, 10 and 20 mg to the leaflets (the weight of a leaflet itself is typically 1.5 mg). The oscillatory period then increased and became about 16% longer at the highest load (judged from Fig. 14.7 in the book just cited), amplitude was reduced to roughly half, while a 25 mg load arrested the oscillations altogether.

While comparing circadian and ultradian leaf movements Pedersen et al. (1990) confirmed a similar decrease in amplitude of *Codariocalyx* leaflet movements with increasing load. The amplitude decreased drastically at 6.9 mg loading (amplitude decreased by about 65%) and continued in the downward position. The increase in period length due to this load was about 10% (typically from about 210–230 s). Contrary to the effect of loading on ultradian oscillations in *Codariocalyx*, a loading of the circadian leaf movements in *Oxalis* leaves did not change the period (although the amplitude was affected). To the best of our knowledge, the effect of loading of the *Codariocalyx* terminal leaf has never been reported.

If in *Codariocalyx*, loading of the terminal leaflet is without any effect on the period, as observed in *Oxalis,* it would support the idea that ultradian leaflet movements have a separate underlying mechanism than the circadian leaf movements. Loading experiments indicate that the 'rhythm mechanism' controlling the ultradian movement is sensitive to external pressure, while mechanisms underlying circadian leaf movements may be pressure insensitive or pressure compensated.

A clear demonstration of ultradian, endogenous leaf movements in *weightlessness* conditions, was reported by Solheim et al. (2009). Under space conditions *Arabidopsis* plants showed rhythmic rosette leaf movements, i.e., also when load from the leaf itself was absent. Different ultradian periods were also reported for the movements, for example, 80–90 min, about 45 min and even shorter periods; image sampling was done every 5 min, therefore periods of 10 min and shorter could not be analysed (data collected under white light conditions). When a centrifugal acceleration of about 1 g was turned on, gravitropic leaf movements occurred, but ultradian oscillations were present (e.g. with 100 min period, under white light conditions). Generally, rhythms under darkness had a longer period.

4.3.4 Leaf Movements and Light

The terminal leaf of *Codariocalyx* shows circadian movements (Engelmann and Antkowiak 1998; Antkowiak 1992) which are not connected with the movements of the lateral leaflets (Sect. 4.2). However, circadian modulation of lateral leaflet movement occurs. For example, a cut lateral leaflet continues to oscillate longer (about 20 h or longer) if cut off from the plant in the first half of the night (under 12:12 h light/dark conditions) but shorter (below 10 h) if cut off in early morning (Engelmann, 'personal communication').

In complete darkness (with infrared recording photographs and a strobe lamp with infrared filter), the lateral leaflet oscillations continue for more than 24 h (Mitsuno and Sibaoka 1989), but the energy status of the tissue might have influenced the results as Sharma et al. (2003) reported them to gradually die out after 6 h in darkness (but they could be restarted when light was switched on; see also Bose 1913). The movements are often remarkably regular under darkness, describing either an elliptical path or regular up- and down movements (but totally irregular movements may also occur, Antkowiak 1992).

In experiments by Sharma et al. (2003) white LEDs were used (partly to avoid any temperature influence) in the flux region between 0.5 and 75 µmol/m^2/s. For fluxes up to 20 µmol/m^2/s the period did not change much (being about 175 s, but also higher values recorded). Higher intensities caused a decrease in the period to about 130 s. Strong abaxial white light (50 W/m^2, which corresponds to about 250 µmol/m^2/s) arrested the leaflet movement, and the motor cells became *depolarised* (Antkowiak 1992). However, before reaching the frozen state, the amplitude and period of the oscillations reduced significantly.

Leaflet movements studied under white light shining from above often display movements of the up and down type. Light might not only be perceived by the leaflets but could also (or mainly) be absorbed by the pulvinus. When the adaxial part of the pulvinus was irradiated by white light, the leaflet moved upwards, while irradiation of the abaxial part caused a downward movement (Antkowiak 1992). The pulvinus may thus be the site of a light direction sensor as well as an executor of phototropic movements.

An absorption spectrum for the photoreceptor responsible for the leaflet oscillation is still lacking for *Codariocalyx*. The effects of blue, red and far-red light has been studied in species with larger pulvini, especially in motor cell protoplasts obtained from *Samanea saman* pulvini (Kim et al. 1992 etc.). Moran (2007) provided an extensive overview of these key studies. The K$^+$-channels of pulvinus cell protoplasts are sensitive to blue, red and/or far-red lights and they are known to react to circadian signals (see also Kim et al. 1993). Phytochromes constitute the relevant photopigments in the *Samanea* system (as well as in *Phaseolus*) and blue-light photoreceptors are indicated to be the same as the ones involved in phototropism reactions. The *Samanea* pulvinar motor cells are the first system described to combine light and circadian regulation of K$^+$ channels at the level of transcript and membrane transport (Moshelion et al. 2002b). In *Phaseolus*, blue-light osmoregulation of motor cell protoplasts has been studied by Wang et al. (2001) where K$^+$ efflux channels in flexor cells of *Samanea* are also activated by blue light (Suh et al. 2000).

Even if the ultradian oscillations might be modulated by circadian clocks, studies of phase locking of the movements to repetitive light signals are yet to be performed on the *Codariocalyx* leaflet oscillations. It would be worthwhile to scan in such experiments the whole period range from minutes to several hours. To the best of our knowledge, issues related to the localisation of photoreceptors (pulvinus and/or leaflet with signal transport to pulvinus) are yet to be resolved (see Engelmann and Antkowiak 1998). In this situation one can infer about the quality of light signal cascade in *Codariocalyx* only from studies of leaf movements on other species.

In a series of experiments done on *Codariocalyx* between 1914 and 1919 (Bose 1919) Bose showed that under constant light and constant darkness, a diurnal (circadian) rhythm of leaflet oscillations continues. This, to the best of our knowledge, is the first ever demonstration of persistence of circadian rhythm under constant conditions, which today is described as circadian (or free-running) under constant light or constant dark conditions.

4.4 *Codariocalyx* Experiments: Contributions from Electro-Physiology and Biochemistry

In the present section, a cellular approach to the *Codariocalyx* motor cells will be taken. Two main developments will be followed. Results from microelectrode and patch-clamp studies of pulvinus and cells will be discussed and Ca^{2+} regulation of the pulvinus motor cells—and the leaflet movements—will be outlined and experimental outcomes presented. The discussion will continue in Sect. 4.5 with emphasis on electrical perturbations of the pulvinus. Modelling of the pulvinus tissue, i.e., of coupled pulvinus cells will be the theme of Sect. 4.6.

4.4.1 Microelectrode Electrophysiology

Periodic potential changes can be recorded from the pulvinus surface and from its apoplast (Misuno and Sibaoka 1989; Guhathakurta and Dutt 1961; Antkowiak 1992; Antkowiak and Engelmann 1998; Antkowiak et al. 1989); the temporal order of some variables is depicted in Fig. 4.4. The membrane potential oscillates between -36 and -136 mV (mean values), and the peak-to-peak value thus amounts to 100 mV. Leaflet position and the extracellular potential in the abaxial part of the pulvinus vary synchronously. When the extracellular potential has reached its minimum (-56 mV), downward movement commences and the upward movement starts.

Microelectrodes were successfully inserted into the motor cells of the pulvinus during leaf movements, and the measurements revealed several key features of the cellular activity (Antkowiak 1992; Antkowiak and Engelmann 1989; Engelmann and Antkowiak 1998).

The extracellular potentials are likely to be generated from the membrane potentials of the abaxial motor cells (as recorded in simultaneous extra- and intracellular recordings in the pulvinus; Antkowiak and Engelmann 1995). They are about 180° out of phase with the membrane potential.

During the downward leaflet movement, the membrane potential of the motor cells in the abaxial part of the pulvinus was depolarised, cf. Fig. 4.3, right panel. When the membrane potential of the abaxial motor cells were depolarised, K^+ and H^+ were expelled from the motor cells, which increased the apoplastic K^+ and H^+ content. Water was then transported out of the motor cells. The K^+ and H^+ activities and the extracellular voltage were also monitored simultaneously using ion-selective microelectrodes (Antkowiak et al. 1991). Once again, oscillations with a period in the minute range and a stable phase relationship with each other and with other measured parameters were detected. The change in the K^+ content amounted to about 5–10 mM extracellularly.

If the abaxial motor cells swell and take up water, the leaflet moves upward. During the upward movement the membrane potential of abaxial cells is negative and K$^+$ (via selective ion channels) and probably Cl$^-$ ions are accumulated (Moran 2007). As the motor cells spontaneously depolarise, ion and water fluxes are reversed. This process is in accordance with the current models for leaf movements of other species, such as *S. saman* (Moran 1990; Satter and Galston 1981; Satter et al. 1988), *Phaseolus* (Mayer 1990) and *M. pudica* (Fromm and Eschrich 1990).

The concept outlined has been tested in several ways on the rhythm of the *Codariocalyx* lateral leaflets:

- Strong white light (50 W/m^2) applied to the abaxial pulvinus cells caused leaflets to move downward and they remained in a lower position until the light was switched off (Antkowiak et al. 1992). At the same time, the membrane potential of the abaxial motor cells was depolarised, as expected.
- *Vanadate*, which blocks H$^+$-ATPases in the plasma membrane (and, with different affinities, all kinases and phosphatises; Serrano 1990; Engelmann and Antkowiak 1998), slowed leaflet movement rhythm and finally arrested the leaflets in the lowermost position. Again, the measurements showed that abaxial motor cells were depolarised.
- The volatile anaesthetic *enflurane* arrested the leaflet movements in the uppermost position (Engelmann and Antkowiak 1998). Enflurane was shown to prevent spontaneous depolarisation of the motor cells, and caused depletion of H$^+$ in the extracellular space.

The concept of the processes in the leaflet rhythm discussed so far is consistent, but the basic mechanisms underlying leaflet oscillation are not specified. The literature on endogenous, rapid oscillations in plant cells and tissue is primarily concentrated on the cellular regulation of Ca^{2+} and its control. It is therefore appropriate to discuss its possible role in the *Codariocalyx* leaflet movements.

4.4.2 Ca^{2+} Regulation in Plant Cells

Calcium ions play a fundamental role in cellular metabolism and signalling (e.g. Chrispeels et al. 1999). It is also of interest in different types of cellular and tissue oscillations (e.g. Schuster et al. 2002), for example, in plants systems like guard cells (Pandey et al. 2007) and coleoptiles (Felle 1988) and in mammalian cells (Kraus et al. 1996). Therefore, it should also be considered while modelling the motor cell oscillations in the *Codariocalyx* pulvinus and the concomitant leaflet movements.

Schuster et al. (2002) identified, in a general model approach to cellular Ca^{2+} oscillations, six types of concentration variables (inositol triphosphate IP$_3$, Ca^{2+} in cytoplasm, in endoplasmic reticulum, in mitochondria, binding sites of calcium buffers and, finally, an active IP$_3$ receptor of Ca^{2+} through release channels). Simple models of Ca^{2+} oscillations contain two of these variables while the 'extended models' have three or more variables. The approaches used and the results obtained by Goldbeter and coworkers (see Goldbeter et al.1996), Dupont et al. (1991) and Goldbeter et al. (1990) on Ca^{2+} regulation in cells have been of general interest and importance. They have also been applied to the *Codariocalyx* leaflet movements, as will be discussed later, where we will direct our focus on simulations and experimental tests of the models for Ca^{2+} regulation of the rhythmic *Codariocalyx* leaflet movements.

4.4.3 Ca^{2+} and the Phosphatidyl Inositol Signalling Chain

Considerable experimental work has focused on understanding the possible role of the phosphatidylinositol signalling chain in the *Codariocalyx* pulvinus. The volume regulation of plant cells involves this signalling chain (Chen et al. 1997; Moran 2007), and therefore shrinking of motor cell might be due to Ca^{2+} triggered plasmalemma depolarisation, which in turn opens up the outward-directed K^+ channels (Kim et al. 1996).

Figure 4.5 illustrates such Ca^{2+}-loops in a cell (here it is assumed to be a motor cell in the pulvinus; schematic after Dupont et al. 1991). Inositol-1,4,5-triphosphate (IP_3) triggers Ca^{2+} release from the internal stores (Z in figure) of cells by binding to a receptor. This opens Ca^{2+} channels in the membranes of the organelles. IP_3 is affected by external signals (S) via plasmalemma-based receptors (R). The cytosolic Ca^{2+} concentration (X) is important in the model and can be pumped into an IP_3—insensitive, intracellular store (Y). When the cytosolic Ca^{2+} concentration reaches a certain level it is believed that it (or an efflux of Cl^- ions) triggers depolarisation of the cell which opens up outward-directed K^+ channels (Moran 2007). Concomitantly, the motor cell shrinks due to ion driven osmotic water transport out of the cell. The Ca^{2+} concentration supposedly increases again in the cytoplasm either due to an influx from the apoplast or an influx from intracellular stores as shown in Fig. 4.5. This increase continues until cells are depolarised once again.

Agonists and antagonists of IP_3 have been administered via the transpiration stream to pulvini of *Codariocalyx* to record possible changes in the period length of leaflet movements. For example, Li^+ ions, known for its inhibiting effect on the formation of inositol lengthened the period of the rhythm (Weber et al. 1992; Chen 1996; Chen et al. 1997)—this effect was partially reversed by simultaneously adding inositol.

Ca^{2+} channels in the plasmalemma and antagonists of calmodulin do not seem to be a part of the underlying oscillator since the Ca^{2+} ionophore A23187 (inhibitor of Ca^{2+} channels in the plasmalemma), and certain ions, as well as verapamil, nifedipin and antagonists of calmodulin (W7) were not found to affect the lateral leaflet rhythm (Menge 1991).

The momentary increase in the intracellular Ca^{2+} concentration that triggers depolarisation may stem from intracellular storages since inhibition of intracellular Ca^{2+} channels with 3,4,5-trimethyoxybenzoic acid 8-(diethylamino) octyl ester (TMB-8) affects the period (Chen et al. 1997). Neomycin (like lithium, an inhibitor

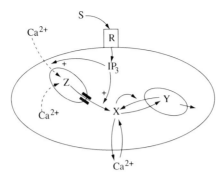

Fig. 4.5 Schematics of a minimal model of Ca^{2+} signalling. A stimulus S is sensed by the receptor R located at the cell membrane, which via inositol-1,4,5-triphosphate (IP_3) influences the Ca^{2+} uptake and release from IP_3-sensitive Ca^{2+} storage Z. The intracellular Ca^{2+}, X, is determined by the IP_3-insensitive Ca^{2+} storage, Y. The rate of change of Ca^{2+} is described by two differential equations. They can be extended to couple one cell to its neighbouring cells by the diffusion of Ca^{2+}. (According to Dupont et al. 1991; cited by Chen 1997)

of IP_3 formation), increased the period length, but only at lower concentrations (0.2 mM). At higher concentrations (0.8 and 2.0 mM) it shortened the period (Chen et al. 1997). However, this shortening of period might be due to the activation of a G protein (known to increase IP_3 formation; Aridor and Sagi–Eisenberg 1990;[3]) Mastoparan activates G proteins (Ross and Higashijima 1994; cf. Takahashi et al. 1998), and therefore it could increase IP_3 formation. One would therefore expect a shortening of period length, which has indeed been reported in the literature (Chen et al. 1997; Table 2). Since neomycin also affects Ca^{2+} channels, kanamycin was used in these studies as a control. In rat brain slices, it inhibits Ca^{2+} channels, but had no effect on the PI signalling chain and as expected, if the PI chain is involved in the leaflet rhythm, it did not affect the period length of the lateral leaflet movement rhythm (Chen et al. 1997).

A possible involvement of the intracellular mediator cAMP (cyclic adenosine monophosphate) in the ultradian leaflet movement was tested by adding the phosphodiesterase (PDE) inhibitor theophylline as well as the PDE activator imidazole via the transpiration stream. Both substances lengthened the period, whereas the PDE inhibitor caffeine had no effect. These results suggest that cAMP does not affect the ultradian mechanism of *Codariocalyx* via PDE because its inhibition should lengthen the period and its activation should shorten it.

In summary, the results on the biochemistry of the Ca^{2+} regulation and control, and the PI signalling chain in the *Codariocalyx* pulvinus provide evidence supporting that the lateral leaflet movement is closely linked to the rhythmic changes occurring in the Ca^{2+} and PI signalling chain.

---

[3] *G proteins* (guanine nucleotide-binding proteins) are a family of proteins involved in transmitting chemical signals from outside the cell causing intracellular changes.

Modelling of the leaflet movements, both at the cellular and the tissue levels have already been performed and some approaches and results will be discussed later.

4.5 *Codariocalyx* Experiments: Contributions from Electromagnetic Perturbations of Rhythmic Leaflet Movements

Section 4.4 concentrated on the experiments and results concerning the processes in single motor cells involved in the leaflet movement. However, leaflet movement can only occur due to a combined and coordinated action of all the motor cells in a pulvinus. They must be synchronised in a way that a combined increase or decrease in the volume of the motor cells can occur, leading to the leaflet movement (see Fig. 4.3a). Modelling the movements must account for the interaction between the motor cells and their water regulation, which determines the movements of the leaflets.

Such coupling between the individual motor cells can in principle be of different kinds. The type of coupling that has been used most frequently is coupling by *diffusion* with signal molecules diffusing between the cells. In this chapter, and in the literature, models for the *Codariocalyx* movements prefer the Ca^{2+} ions to play a central role in such communication. Other types of coupling such as electric coupling are certainly a possibility (see e.g. overview by Gradmann 2001; Gradmann and Buschmann 1996) and can be intimately coupled to Ca^{2+} transport and signalling. A good example of which is the pacemaker cells of the heart, which constitute an oscillating centre, where electric signals are coupled in a synchronised way to the heart muscle tissue causing controlled regulation of the heart tissue (Winfree 1987b).

A short description of the oscillatory systems will be followed by an extension of the electrical approach to perturbations of the *Codariocalyx* lateral leaflet rhythm.

4.5.1 Interlude: Oscillations and Singularities

In a simplified picture, Fig. 4.6, several general features of biological oscillations are illustrated. The leaflet movements and thus the state of pulvinus can be represented in a *phase plane*: The state of the oscillator is described by the path of successive values forming a *limit cycle* (here we will not restrict our discussion to the two-variable system as shown in the figure). During one cycle of the oscillator it runs through a complete loop. A sudden perturbation of the oscillation changes the state of the oscillator. A new position on the limit cycle defines a phase change

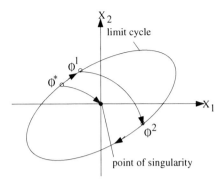

Fig. 4.6 A two-dimensional illustration of an oscillation, plotted as as a *limit cycle*. The arrowheads on the cycle show the direction of cycling; one run from the phase Φ^1 to Φ^1, corresponding to one oscillatory cycle with its characteristic period length. If the oscillator is purturbed by a light pulse at the phase Φ^1, the oscillation will be shifted to the phase Φ^2, provided a suitable, strong enough pulse is applied. There is a special phase Φ^*, at which a pulse of a particular strength (neither too strong and nor too weak) throws the oscillator into the point of singularity, and the oscillation stops (examples shown in Fig. 4.7)

in the oscillation. It can be a delay or an advance, as exemplified by the arrow from phase ϕ^1 to ϕ^2 in the Fig. 4.6.

The complexity of biological oscillations, involving quite a number of variables and numerous cooperating cells in a tissue (as in the pulvinus) can not be represented satisfactorily by a simple 2-dimensional model. However, it can still be useful and illustrative. Limit cycle representations and biological examples can be found in the literature dealing with both circadian and ultradian rhythms (e.g. Winfree 1987a, b; Engelmann 1996; Winfree 2001). The circadian *Kalanchoë* petal movement rhythm (Johnsson and Karlsson 1972; Karlsson and Johnsson 1972), and the ultradian transpiration rhythm of *Avena* leaves (Johnsson et al. 1979) are among the plant rhythms discussed from these aspect. Finally, the phase plane concept is also used in connection with studies of the electro-mechanical mechanisms of the heart beat rhythm (see e.g. Winfree 1987b; Guevara and Glass 1982; Glass and Mackey 1988).

Perturbations may also cause amplitude changes and the oscillator might be thrown out of the limit cycle, but often returns back to the limit cycle. The simplifications one uses in this representation of biological oscillations are quite extensive, while treatments of limit cycle behaviour are restricted to simple mathematical models (e.g. Strogatz 1994). Mathematical oscillators display under certain special circumstances a *singular point* within the limit cycle (Fig. 4.6). This is characterised by a halt of the oscillation. The singularity point can be reached by applying a perturbation of a precise magnitude at a precise phase (ϕ^* in Fig. 4.6).

The singularity can be *stable* or *unstable*. Winfree (1970, 1971) found a stable singularity for the circadian adult emergence rhythm of *Drosophila*. A suitable light pulse perturbation sent the oscillatory system into a singularity point and circadian clocks 'stopped' permanently. Likewise, in case of an ultradian rhythm, namely the water transpiration rhythm of young *Avena* leaves, the system could be frozen to a stable state by light pulses or by pulses of changing water potential (Johnsson et al. 1979). Subsequent pulses could restore the rhythm to its normal state.

In *Codariocalyx*, a singular state of the lateral leaflet rhythm can be reached by the application of a singular electric current pulse, sent through the pulvinus (Johnsson et al. 1993).

4.5.2 Applications of Electric Currents to Pulvinus

In Sect. 4.1 we outlined the results concerning the close association between leaflet movements, membrane potential of the motor cells, pulvinus surface potential and transport of several types of ions across the plasmalemma. The electric voltages, ion currents, and the voltage gated ion channels of the pulvinus can be perturbed by externally applied currents, since the K^+ transport system are triggered by electric pulses (Antkowiak and Engelmann 1995). The reaction of the leaflet movements to electric currents were observed already by Hufeland (1790), and in great details by Bose (1913) and others (Guhartakhurta and Dutt 1961).

The oscillatory leaflet movements provide a sensitive and a suitable system to detect small effects on the system, which controls the rhythmic movements. Johnsson et al. (1993) studied the phase and amplitude effects of the leaflet oscillations when pulses of direct current (DC) were applied to the leaflet tip via a (copper) spring system; the current passed the pulvinus and the stalk, which was placed in electrically grounded water. Changes in phase and amplitude were fairly easy to detect which could be observed once the movements became stable. Small DC current pulses (<10 µA) produced negligible effect on the oscillations, while large pulses diminished the amplitude and delayed the phase of the oscillation. The product of average current and exposure time was found to be relevant and the amplitude dampening was about 50% for 75 µA. Fostad et al. (1997) used a current clamp technique to stabilise and control the currents to 2, 5 or 10 µA currents for 1 s. Only delays were obtained in this case (the same results if the direction of the current was reversed; Fostad 1994; Fostad et al. 1997). The lowest charge which caused a physiological response was of the order of 2–10 µC (=2–10 µA × s). This charge compares well with the electric charge required for the closure of *Venus* flytrap leaf, which is about 14 µC (Volkov et al. 2007; Chaps. 1 and 7). Closure of the flytrap leaf occurred in about 300 ms (same for mechanical stimuli).

As mentioned earlier, phase changes in the rhythm consisted only of delays—possibly due to the fact that current through the pulvinus gets evenly distributed, which then stimulated the pulvinus uniformly, independent of the phase of leaflet movement. Current density in the pulvinus was estimated to be about 5×10^4 mA/m^2.

In Sect. 4.5.1 it was mentioned that a singular DC current pulse could stop the oscillations in leaflet movements. The perturbations causing such an arrhythmic state were in most cases given in the lower position of the leaflets and the current dose (current × time) amounted to 150–200 µA × s. Strong pulses stopped the rhythm permanently, leaving the leaflets in a lowered position. After such a perturbation the leaflet attained an intermediate position between the previous maximum and minimum. Figure 4.7 shows two recordings, the DC pulses were given in the position denoted '*'.

In this case the singular state seems to be unstable, since oscillations usually start spontaneously after some time (about 35 min, i.e. after 8–10 times the period). The system was, evidently, not damaged by the perturbation. The centre part of Fig. 4.7 shows the phase plane representations (velocity vs. position of movement) of the same experiments. For convenience they are separated in the presentations as before and after their stay in the singular point. The figure illustrates how the leaflets move rapidly into singularity and move out to the limit cycle after a while.

It should be noted that the *overall* leaflet movement is halted after the perturbations: This could imply that the rhythm of the individual motor cells is also stopped after the perturbation, or, alternatively, the cells might still oscillate, but in random phases. In the latter case, arrhythmic pulvinus could resemble the situation of a fibrillated heart (where individual cells or tissue areas oscillate but out of sync and with an increased frequency—see Winfree 1987b). To the best of our knowledge, the *Codariocalyx* pulvinus system is the only rhythmic plant system which has been shown to reach a singularity point after a brief electric perturbation.

Sharma et al. (2001) extended the phase studies by applying DC current pulses *asymmetrically* onto the pulvinus—the upper (adaxial) side of the pulvinus being the electrode contact area. In this way the current pulses could be applied at different phases of the adaxial motor cells (i.e. during their swollen and shrunken phase), and a phase response curve could be obtained. Significant phase delays were obtained when the pulses were applied in the downward position of the leaflets (swollen adaxial motor cells) or during their upward movements (adaxial motor cells shrinking). In the upward position (shrunken adaxial motor cells) small phase advances could be observed but during their downward movements no phase shifts were found. The action of the current pulses can be qualitatively explained by the changes in the extracellular potentials inside the pulvinus by the application of external current signals and its effects on the voltage gated channels.

Radiofrequency electromagnetic fields have also been applied on oscillating leaflets (Ellingsrud and Johnsson 1993). Sine-wave fields with a frequency of 27.12 MHz (industry frequency) were applied and stalks and leaflets were placed in a Crawford exposure box. This exposure box allows application of homogenous, perpendicular, controlled electric and magnetic fields. After perturbations (e.g. 0.53 W/cm^2 for 30 s) plants showed only phase advances in their leaf movement rhythms. The magnitude of phase shifts increased as a function of the irradiation dose. It is unlikely that temperature effects were responsible for the phase shifts, since the intensity was kept very low (cf. Cihlar 1965, Sect. 3.2). The smallest radiofrequency perturbations that caused phase shifts could have increased the temperature by not more than about 0.2°C. High frequency electromagnetic waves have been used in rather few plant studies but discussion of the results are concentrating on Ca^{2+} effects (e.g. Roux et al. 2008; and Pazur and Rassadina 2009).

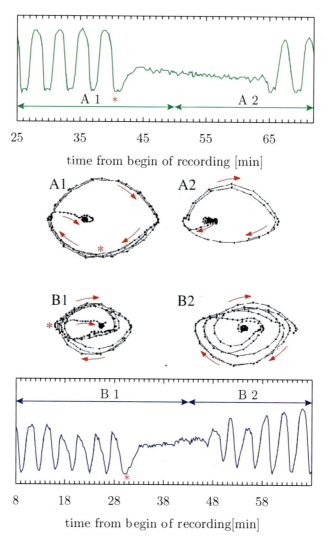

Fig. 4.7 The lateral leaflet oscillations in A and B were stopped by an electric current pulse of about 30 and 40 μA, applied for 5 and 4 s; respectively, when the leaflets were close to the lowermost position (marked by *). The leaflets regained their oscillatory movements once again after 25 and 20 min, respectively. The *middle* panel show phase plane plots of the perturbation experiments. Speed of movement plotted vs position (arbitrary units), both for movements before (A1, B1) and after reaching the singularities (A2 and B2). From Johnsson et al. (1993)

4.5.3 Static Magnetic Fields

A magnetic field, static or time varying, homogenous or inhomogenous, might affect the leaflet rhythms. It is interesting to note that a *static* magnetic field of about 125 mT (milli-Tesla) was reported to cause the inhibition of Ca^{2+} channel activation in the GH3 cells (Rosen 1996, see also Rosen 2003; GH3 is a cell line from rat pituitary tumor cells). This signalling pathway was shown to influence the *Codariocalyx* leaflet movements: Period as well as phase changes under the influence of fairly moderate static fields were observed (Sharma et al. 2000). Small permanent magnets placed close to the pulvinus (horizontal magnetic flux density around 50 mT (30–150 mT) caused prolonged period reversibly if a field was applied for a long time (typical values 6.1 ± 0.1 min without field and 7.1 ± 0.4 min with field at 20°C). Delay and advance phase shifts were induced (Fig. 2 in the paper cited) if magnets were placed close to the leaflet pulvinus for a brief period of time.

Static magnetic fields have been reported to modulate rotational movements of cucumber tendrils. These movements were accelerated under homogenous magnetic fields (flux density between 1 and 16 mT Ginzo and Décima 1995). Okada et al. (2005) report that circumnutation of *Arabidopsis thaliana* around the plumb line (period about 2.5 h, i.e. about 50 times slower than the leaflet movements) showed a decreased speed under a 50 Hz magnetic field with a magnitude of 1 mT.

Magnetic field homogeneity is difficult to control, since the leaflet movements imply position changes, and if the magnetic field is spatially inhomogenous, the *gradient field* can cause movements, so-called magnetotropic bending reactions (Kuznetsov and Hasenstein 1996). This is due to the field of attraction of diamagnetic anisotropic particles or molecules—for example, starch grains in gravisensing cells or ordered phospholipid layers in biological membranes.

Further electromagnetic experiments on *Codariocalyx* leaflet oscillations have never been reported. The same is true for experiments on entrainment by regularly repeated current pulses, although the presence of phase delays (stable phase change after about 4–5 periods) indicates that at least in a certain period range, entrainment should be possible to achieve. There is a wide interest regarding the effects of low intensity electromagnetic fields on biological systems in general and on their intracellular Ca^{2+} regulation in particular. Publications mostly concentrate on 50 Hz fields but have not primarily been focused on plant cells (Lindström 1995; McCreary et al. 2006; Grassi et al. 2004). That very low magnetic fields can affect Ca^{2+} transport in a biological system of highly purified plasma membrane vesicles was demonstrated by Bauréus Koch et al. (2003).

4.6 The "Heart of the Matter": Modelling the Pulvinus Tissue

The pulvinus is an electro/chemical/mechanical system, and modelling of the *Codariocalyx* leaflet movements would require contributions from several scientific fields—ranging from electrophysiology and biochemistry to systems analysis. The multitude of motor cells in the pulvinus must coordinate their physiology and osmotic pumps if rhythmic leaflet movement should result. In other words, the motor cells within a certain area should be 'coupled', or synchronised in time and space for the coordinated leaf movements to occur. This will be discussed in the section below.

4.6.1 Diffusion Coupling

One obvious way of coupling between the cells in the pulvinus is by structural/mechanical mode, which would occur via the cell walls. The modelling of the *Codariocalyx* leaflet movements has thus far been based on the assumption that the cells are *diffusively* coupled and we will focus on this approach. Such a coupling is possible via plasmodesmata and plasmalemma. One can also imagine an electric coupling between the cells, much as in the animal nerve systems where cells are 'coupled' via neurotransmitters (Gradmann and Buschmann 1996; Baluska et al. 2005, 2008; Brenner 2006).

In the case of diffusive coupling, the models often take their starting point in Turing's basic paper (Turing 1952). He showed that on a ring of diffusively coupled cells, it is possible to get 'concentration waves'. He assumed that in each cell there is a two-variable enzyme system and that one of the substances could diffuse into the neighbouring cells. The dynamics for the diffusion followed the 'Fick's law' (transport determined by a proportionality constant and the negative gradient of the substance). The diffusible substance is usually assumed to be Ca^{2+} ions, at least in the case of the leaf movements.

The dynamics of diffusible cytosolic Ca^{2+} ions with a concentration of X_i in the cell #i with neighbour cells #($i-1$) and #($i+1$) arranged in a ring (Fig. 4.8) can be expressed as

$$dX_i/dt = f(X_i, Y_i) + D_x[(X_{i-1} - X_i) + (X_{i+1} - X_i)] \qquad (4.3)$$

The Eq. 4.3 describes how the cytosolic concentration X_i changes according to the cellular dynamics given by the function f (first term in the right hand side) as well as by Ca^{2+} diffusion from the neighbouring cells, determined by the product of the coefficient D_x, and the concentration differences relative to the neighbour cells (second term in the right hand side). The constant D_x is >0 and can be thought of as a diffusion (or transport) constant.

Turing's cellular model was rather abstract, albeit with a two-dimensional (2-D) autocatalytic feedback involved (described by the function f). The concentration X in the cells could attain maximum values that 'travelled around' the ring. The approach was successful and flexible and was later extended to several biological systems in studies of morphology, and patterns in development.

In a series of papers Goldbeter and collaborators (Goldbeter et al. 1990; Goldbeter 1996) specified the content of the cellular dynamics to be that of a Ca^{2+} based oscillator as described in Sect. 4.4. Thus the ring structure had specific cellular mechanisms that could describe oscillatory Ca^{2+} regulation in the individual cell as well as the Ca^{2+} 'waves' on the ring of cells.

4 The Telegraph Plant

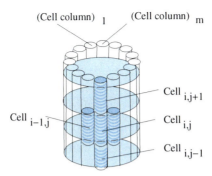

Fig. 4.8 Model of coupled cells arranged in a cylinder. In this cylindrical (but in reality 2D) matrix—consisting of n cells in a ring, with m such rings in a stack—each cell is coupled to its neighbours. Apart from the cells in the uppermost and lowermost rings all cells were coupled to four neighbours (two neighbour cells in the same, horizontal, ring and one neighbouring cell in each ring immediately over and under the cell; all darker blueish in the figure). Thus, cell #(i, j) is coupled to cells #(i − 1, j), #(i + 1, j), #(i, j + 1), and #(i, j − 1), respectively

4.6.2 Modelling Ca²⁺ Oscillations Applied to Leaflet Oscillations

Chen (1996) and coworkers (Chen et al. 1995, 1997) extended the Ca^{2+} models to a 2-D network of cells. The individual cells were assumed to have a dynamics of the 'Goldbeter type' and were coupled in a ring. Several rings were stacked up one on another. The individual cells were coupled not only to the 'horizontal' neighbours on a ring but also to the neighbours above and below the ring (Fig. 4.8). If the rings were to be 'cut open', they could be treated as one layer and studied as a first approximation of the cylindrical motor cell structure in the *Codariocalyx* pulvinus (and thus form a substrate for understanding the leaflet movements).

In this cylindrical (but in reality 2-D) matrix—consisting of n cells in a ring multiplied with m rings in a stack—each cell (apart from those in the uppermost and lowermost rings) was coupled to four neighbours (two neighbour cells in the same, horizontal, ring and one neighbour cell in each of the rings immediately over and under the cell (see legend of Fig. 4.8).

Chen (1996), Chen et al. (1995), Antonsen (1998) and others have simulated the behaviour of this cylindrical structure in models with 50×50 cells with various coupling conditions. The transport coefficients in 'up' and 'down' direction can of course differ from those in the sidewise directions of the cylinder. Simulations have also been performed with asymmetric transport coefficients in a clockwise and anticlockwise direction (G. Aanes, Department of Physics, Trondheim, 'personal communication').

Simulations were able to produce waves of Ca^{2+} in the matrix/around the cylinder (e.g. Antonsen 1998). The time development of waves on a ring is exemplified in the Fig. 4.9 where the Ca^{2+} concentration was varied for diffusive coupling according to the differential Eq. 4.3.

Some features shown by the systems analysis are:

- waves of Ca^{2+} can be found in both clock- and anti-clockwise directions (see Fig. 4.9)
- 'islands' of high Ca^{2+} concentration can be found, sometimes combining to waves which move around the cylinder (simulated in Fig. 4.10)
- up-and-down movements can be simulated

The simulations show an interesting range of patterns and time sequences that can be looked for in the leaflet movements (or in plant circumnutation movements).

4.6.3 From Concentration Variations to Movements

The Ca^{2+} patterns simulated in the cellular cylinder must be translated into pulvinus Ca^{2+} patterns and finally into leaflet movements. This is a modelling task that depends on steps partly discussed in the section above and comprises a tentative sequence:

Ca^{2+} concentration oscillations

- other cellular concentration/potential oscillations
- ion transport oscillations
- water transport oscillations of individual cells
- oscillations of volume and length in the pulvinus motor cells
- angular bending of the pulvinus tissue, and finally,
- leaflet movements.

The 'translation' of the cellular processes, for example, the increase in length after water uptake, to overall bending of the pulvinus and leaflet, requires a summation of differential length changes along the 'columns' or 'cell columns' of the cylinder as sketched in Fig. 4.8.

The dynamics of the leaflet movements and oscillations varies to a large extent (and are influenced by growth conditions, signals into the pulvinus etc.). The regular circular and regular up-and-down movements are predominant. However, movements can also be highly irregular or even absent. The cylinder simulations reported in the literature can describe many of these patterns but must be more elaborated, as will be discussed later.

4.7 Discussion

While discussing leaflet movements in *Desmodium* it is only proper to turn again to the contributions made by Bose. His main approaches and results on the 'Indian Telegraph Plant' although from a century ago, have gained appreciation and been

4 The Telegraph Plant

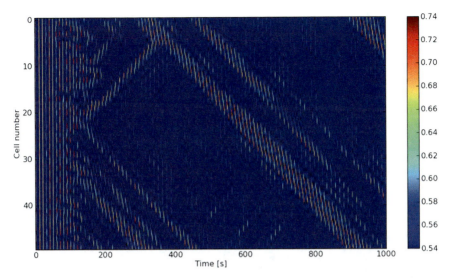

Fig. 4.9 Simulations of Ca^{2+} concentration on a ring with 50 cells, coupled according to the model described in the text. Cell numbered from $n = 1$ to 50, vertical axis; time development along horizontal axis (unit dependent on parameter values chosen). Ca^{2+} concentration indicated by colour scale to the right. Coupling is the same among all neighbouring cells, see text. At time $t = 0$, Ca^{2+} concentrations are randomly distributed in a defined concentration interval. After an initial period, cells couple to each other and Ca^{2+} patterns indicate waves in both clockwise and counter clockwise direction on the ring (clockwise when high concentration moves in direction of increasing cell number, counter clockwise when high concentration moves in direction of decreasing cell number). Simulation G. Aanes, Trondheim

followed up and largely extended—but are still of great interest. Bose's studies started with *qualitative* observations. The methods have been refined much in the meantime due to newer and more advanced techniques such as microelectrode recordings, patch-clamp analyses, progress in biochemistry, physiology, data recording and handling and video techniques. But many of the basic issues which fascinated Bose are still challenging and are in the centre of physiological research in plants. Among the many experimental plants that he experimented with we can imagine that *Desmodium* was his favorite—and in our days the observable oscillatory behaviour of this plant is still a fascinating object with many unsolved riddles.

The main conclusion that Bose came up with was far-reaching for that time. He concluded that much alike animals, plants too have components that are equivalent to receptors where stimuli are received, a conductor ('nerve') which electrically propagates the stimulus and an effector, or the terminal motor organ. In both *Codariocalyx* and *Mimosa*, a stimulus is transmitted electrically to the motor organ, the pulvinus, which has been the focus of this chapter. It has now been established beyond any doubt that the osmotic motor of the leaflet movements is governed by the cyclic depolarisation and hyperpolarisation events in the motor

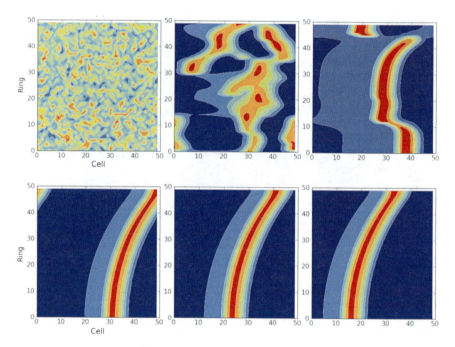

Fig. 4.10 Simulations of Ca^{2+} concentration in a 50 × 50 matrix of cells according to the model described in text. Ca^{2+} plots (colour scale as in Fig. 4.9) at different time points after start (A:0, B:32, C:81, D:119, E:120, and F:121 s after start of simulation). Coupling between cells was stronger in the clockwise than in the anti-clockwise direction, and weaker in the vertical direction. Patterns demonstrate: *panel A*—the Ca^{2+} concentrations are randomly distributed at time zero; *panels B* and *C*—after a while patchy distribution of Ca^{2+} in larger, high concentration areas; *panels D, E* and *F*—now only one wave of high Ca^{2+} concentration, moving from *right* to *left* in figure (note short time intervals between the three patterns). Simulation G. Aanes, Trondheim

cells of the pulvinus, a mechanism which Bose had predicted long ago. Ironically Bose had always argued that electrical signalling are key events in plant physiology, a view to which plant physiologist have come only recently (see e.g. Baluska 2010; Brenner et al. 2006 and references given in the different subsections).

The experiments by Bose were, in their simple, direct approach, important for the studies of *Codariocalyx* and other plants but also for general bioelectrical studies of plants. This has been appreciated only in the last few decades (e.g. Picard 1973; Davies and other contributions in Volkov 2006; Chaps. 1 and 7). Bose demonstrated that in several plants mechanical stimulation could be mimicked by electrical stimulation (Bose 1913, 1926, 1928) which thus emphasises the importance of the electrical systems in plants. Despite the fact that some prominent plant physiologists have been arguing that multi-functional electric signals are primarily responsible for co-ordinating plant responses to the environment, until not too long ago, most plant biologists were reluctant to view action potentials as of primary significance for plants. For example, as first proposed by Pickard

(1973) and Davies (1987a, b) and subsequently confirmed by Wildon et al. (1992), the 'protease inhibitor inducing factor', a wound signal, is electrical and not chemical. This has brought about a paradigm shift in the attitude of plant physiologists from that espousing the predominance of chemical signalling in plants to one emphasising electrical signalling—the very principle Bose sought to establish long ago. In fact, action potentials were recorded in plants in the 30s of the last century (microelectrodes in *Nitella* cells; Umrath 1930) and thus earlier than similar recordings in animal cells, as pointed out by Fromm and Lautner (2007).

We now shift our focus briefly to some experimental and modelling aspects of the *Codariocalyx* leaf movements in the light of the possible role of Ca^{2+} in the pulvinus controlled leaflet movements.

4.7.1 Experimental; Observations and Physiology

Empirical studies on the *Codariocalyx* leaflet movements have mainly concentrated on the underlying cellular biochemical/physiological mechanisms and the concomitant electrical reactions. The starting point has been the turgor volume changes of the pulvini cells, the 'motor pump'. As shown in Sect. 4.4, the findings support the participation of cellular Ca^{2+} ions and their oscillations in the leaflet movement. For example, several compounds targeting different sites in the discussed sequence of 'Ca^{2+} oscillations → ion pumps → turgor changes' have been tested and all seem to support the cellular model.

Newer optical fluorescence techniques to follow possible Ca^{2+} waves over the pulvinus surface would be an important step forward in future studies. This has been done with reasonable resolution in gravitropism studies by studying Ca^{2+} redistributions in whole plants like, for example, *Arabidopsis* (Toyota et al. 2008). Applying this method to leaflet movements would provide valuable tests of Ca^{2+} based models.

A few aspects that apparently have not yet been studied in the *Codariocalyx* system might be worth mentioning here: Aquaporins and their role in plant water relations are now attracting much interest (Agre 2004; Nobel lecture, reviews, e.g. by Johansson et al. 2000; Maurel et al. 2008; Maurel 2007; Kaldenhoff and Fischer 2006). Their possible effects on the period and amplitude as well as joint action of other substances seem worthwhile to study in the context of leaflet oscillations.

To study a possible role of *auxin* (indoleacetic acid) in the pulvinus mechanisms and in the Ca^{2+} oscillations would also be important. In our approach we have mainly focussed on Ca^{2+} based oscillations in the motor cells. It has been claimed that auxin oscillations are the basic mechanism for Ca^{2+} oscillations (e.g. Felle 1988). Furthermore, auxin has been assumed to be a key substance in the electric oscillations in plants roots (Scott 1962; Shabala et al. 1997). The period of these oscillations is about 5 min, viz. in the same range as the *Codariocalyx* leaflet movements. The effect of auxin is also of considerable interest from several points of view. Its action is connected with H^+-ATPase activity (e.g. Hager 2003; Robert-Kleber et al. 2003 and others) which is a part of the model discussed for the leaflet movements (cf. Fig. 4.3). Auxin concentration changes and auxin signalling with subsequent pH changes can be rapid as shown in several systems (e.g. Monshausen 2011) and they might, therefore, participate in the mechanisms of the *Codariocalyx* leaflet oscillations. In *Phaseolus*, auxin effects on osmoregulation in pulvini protoplasts have been studied (K^+ and Cl^- transport affected; Iino et al. 2001).

The pulvinus oscillations are reversible, so the *growth promoting* action of auxin is probably not relevant in the present context, since the pulvinus does not grow. However, a specific 'loose' cell wall structure of the motor cells is necessary for their oscillatory volume changes. This indicates a possible connection between auxin concentrations and the 'acid-growth' action of auxin as well as to the presence of expansins (for discussion on auxin, expansins and cell wall loosening see e.g. Durachko and Cosgrove 2009). Without going into details, it seems that both Ca^{2+} and auxin play a key role in the oscillations even if the growth mediated effects of the plant hormone is unlikely in the pulvinus reactions. Therefore, it would certainly be of interest to study the effects of auxin on the leaflet movements.

Other substances have also been tested on the leaflet oscillations but will only be briefly mentioned here. Bose (1913, p. 341, summary), reported that acids arrest the leaflets in the upper position, and alkalis do that in the lower position. Copper sulfate was reported to stop the leaflet movements in an upper position (Bose 1913, p. 338). It is likely that the sulfate ion is responsible for this effect, since potassium sulfate also modulated the period length in *Codariocalyx*. Chen et al. (1997) tested neomycin sulfate (neomycin inhibits IP_3) and found that sulfate lengthens the period of the rhythm. Drugs such as ether, chloroform (Bose 1913, p. 336), and modern anaesthetics (Antkowiak 1992) stop oscillations in its upper position (see Das 1932). Several of these investigations could preferably be extended.

4.7.2 Experimental: Electrophysiology

The pulvinar motor cells in *Codariocalyx* are small compared to those of many other plants possessing leaf pulvini (*Samanea, Mimosa, Phaseolus* and others). This makes protoplast studies and single cell physiological investigations cumbersome. However, it would be valuable to push forward investigations in this direction.

The electrophysiology of *Cordiocalyx* movements in vivo has been mapped (Antkowiak 1992; Engelmann and Antkowiak 1998). Electrical perturbation experiments have revealed that the phase of the lateral leaflet rhythm can be perturbed much in the same way as many other biological oscillatory systems. Several perturbation effects such as phase changes turn up as features, common to many oscillating systems. Finer studies are needed to dissect out details of the system, such as specification of photopigments mediating light-induced phase changes. We have already pointed out that electric DC pulses can cause phase shifts in the leaflet movement oscillation, and that short pulse perturbations can also elicit arrhythmicity in the pulvinus and the leaflet movements. This is analogue to the results of a fibrillation of the mammalian heart and its standstill (Winfree 1987b, 2000 and 2002;

Janse 2003); a recent paper on cardiac arrest induced by physical blow is (Solberg et al. 2011) despite the differences to the pulvinus and the motor cells.

The electrically induced phase changes point towards the possibility of entraining the leaflet oscillations to cyclic electric DC pulses (in a certain frequency range). This remains to be studied in the light of Bose's observation that repeated electric stimuli 'cause fatigue and an eventual loss of excitatory response'.

How does electric voltage affect Ca^{2+}-based oscillations in single cells? Electric pulses could affect specific voltage gated ion pumps (e.g. Moran 2007). The magnitude of pulses used in the *Codariocalyx* experiments is likely to be of sufficient magnitude. If the pulvinus has a length of about 3 mm and every motor cell has a length of about 30 µm, and about 100 cells (=3,000/30 cells) are arranged in a row in the pulvinus. If the voltage applied between the leaflet tip and the lower part of leaf stalk, about 100 V, is divided into three parts so that about 50 V fall over the row, this means that about 50/100 V will apply over each cells, i.e., about 500 mV over each cell. This can be regarded as a voltage capable of exciting voltage gated channels in the cells and lead to Ca^{2+} induced modulation of the oscillations.

A mapping of the time sequences in the perturbation experiments seems to be of some value to be considered in future experiments. DC pulse perturbations of the movements were administered in the millisecond to second range to stop the oscillations. Experiments on the time dependence of the subsequent effects on the ion transports, the osmotic pump and the leaflet movements would provide valuable information on the dynamics of the mechanisms.

Many physiological mechanisms in the *Codariocalyx* pulvinus cells remain to be investigated and specified. The leaf movements also show a great variety of movements (up/down, elliptical, circular, turning, twisting and irregular etc.), which are likely to be studied at the systems level as well. This will ultimately require models for the pulvinus that are flexible and allow different swelling patterns to arise and change dynamically. The pulvinus and the guard cell system stand out as focus systems for investigations of rapid plant reactions. Results from guard cell studies might be relevant also for the pulvinus leaflet cells, cf. results on stretch activated (e.g. Ca^{2+}) channels (Moran 2007; Zang et al. 2007; also see Cosgrove and Hedrich 1991; and the review by MacRobbie 1998).

It is anticipated that newer techniques will play a crucial role in meeting needs of future studies. Optical techniques should be adopted to design for mapping of Ca^{2+} oscillations, electric field variations should be possible to be detected by voltage-sensitive dyes (e.g. Konrad 2008) and by modern integrated electrode technique (cf. Ul Haque et al. 2007; Baikie et al. 1999; Porterfield 2007) etc.

4.7.3 Systems Approach and Modelling

The model approach discussed is based on diffusion coupled cellular oscillators. The details of the oscillators are not critical for the oscillations to occur (cf. Turings general approach to results on a ring of cells Turing 1952; or Fostad 1994). It would be important to find out which features of the leaflet movements are due to cellular mechanisms and which ones are due to transport between the individual cells. Features such as rotation of leaflets and the oscillatory period, entrainment and kind of patterns generated may result from the multi-oscillatory cooperation of the motor cells.

We have used some kind of Ockham's razor approach and chosen to test the published Goldbeter Ca^{2+} oscillator, combined in cellular matrices with simple coupling dynamics. Further work should probably study possibilities for other types of coupling, for example, via electric signals alone or in addition to Ca^{2+} coupling. The plasmodesmata and the plasmalemma are natural pathways for these types of coupling. The possible auxin participation in the oscillations could be of interest, but does not seem to be necessary for the time being.

In the oscillations, coupling via ionic transport will exist as well as via turgor regulation (see e.g. Gradmann and Buschmann 1996; Findlay 2001; Shabala et al. 2006) and will be part of a complete and complex picture of the leaf oscillatory system. In addition to the types of coupling mentioned, a mechanical cell to cell coupling via the cell walls might also be of interest. Cell wall loosening and acid effects will be natural refinements to introduce (see Sect. 4.7.2) and even a feedback from the cell walls onto the cytoplasm Ca^{2+} might be relevant (as reported for pollen tubes; Hepler and Winship 2010). Again, here it is noteworthy to mention that the Ca^{2+} ions may play an important role.

Extensions of the model and its features which would probably not change the main results of the present simulations are the following:

- The model cylinder should consist of several, not just one concentric layer of cells
- The conditions at the top and the bottom ring of the cylinder should be studied separately
- The coupling conditions in the cylinder should be varied longitudinally and radially
- The sequence leading from Ca^{2+} concentration waves and oscillations to bending of different parts of the cylinder should be studied
- A comparison with the circumnutations of organs such as the *Helianthus* and *Arabidopsis* hypocotyls
- Existence of patches of increased lengthening on the cylinder surface should be looked for and studied
- Occurrence of multiple frequencies, addition of noise, chaotic swelling/shrinking patterns on the pulvinus should be studied and simulated

In the circumnutating sunflower hypocotyls, Berg and Peacock (1992) studied the 'growth landscape' of the hypocotyl cylinder and revealed different patterns in oscillating and non-oscillating plants. Such patterns on the surface of the pulvinus can be predicted (see parts of simulation in Figs. 4.9 and 4.10) and might be detected with different techniques. Further comparisons with other types of circumnutations might be worthwhile (see e.g. reports on hypocotyls Berg and Peacock 1992; Engelmann et al. 1992; Neugebauer 2002; Johnsson 1997; Johnsson et al. 2009; Solheim et al. 2009 and on tendrils Engelberth 2003).

We would like to point out here that a chapter of this nature involves covering material published in several hundreds of papers. In the interest of keeping the list of references to a reasonable length, we have not made citations to the literature exhaustive. Instead, we have tried to be representative, and have therefore favoured, as far as possible, citing reviews, books or relatively recent articles. Omission of papers from the list of references, thus, does not imply that we consider those studies to be less important, it rather reflects our ignorance.

The modelling of the leaflet movement will by necessity provoke a large number of relevant questions as to the basic cellular mechanisms, the signalling and the cellular coupling and so on in *Codariocalyx*. We hope that our chapter (despite its obvious caveats, e.g. concerning coupling alternatives) has induced enough interest in some readers to a degree that they would be stimulated to start looking into the behaviour of the *Codariocalyx* leaflet movements, which have fascinated many of us who saw this plant in action–including Johann Wolfgang von Goethe, key figure in German literature, who described his meeting with *Codariocalyx* (*alias Desmodium, alias Hedysarum*):

'Still higher rises that feeling, to which I will give no name, at the sight of the *Hedysarum gyrans*, which lifts up and down its little leaves, without any visible outward occasion, and seems to play with itself as with our thoughts. Imagine a *Pisang* (=banana) to which this gift were imparted, so that it could let down and lift up again by turns its huge leafy canopy all by itself; whoever should see it for the first time, would step back for terror'. (from translation by Dwight 1839).

****

A short movie will be available under

http://tobias-lib.uni-tuebingen.de/index.php?la=en

References

Agre P (2004) Aquaporin water channels (Nobel lecture). Angew Chem Int Ed 43:4278–4290

Antkowiak B (1992) Elektrophysiologische Untersuchungen zur Seitenfiederblattbewegung von *Desmodium motorium*. PhD thesis, University of Tübingen, Germany

Antkowiak B, Engelmann W (1989) Ultradian rhythms in the pulvini of *Desmodium gyrans*: an electrophysiological approach. J Interdiscip Cycle Res 20:164–165

Antkowiak B, Engelmann W (1995) Oscillations of apoplasmic K^+ and H^+ activities in *Desmodium motorium* (Houtt) merril. Pulvini in relation to the membrane potential of motor cells and leaflet movements. Planta 196:350–356

Antkowiak B, Engelmann W, Herbjørnsen R, Johnsson A (1992) Effects of vanadate, N_2 and light on the membrane potential of motor cells and the lateral leaflet movements of *Desmodium motorium*. Physiol Plant 86:551–558

Antkowiak B, Mayer W-E, Engelmann W (1991) Oscillations of the membrane potential of pulvinar motor cells in situ in relation to leaflet movements of *Desmodium motorium*. J Exp Bot 42:901–910

Antonsen F (1998) Biophysical studies of plant growth movements in microgravity and under 1 g conditions. Doctoral thesis, Norwegian University of Science and Technology, Trondheim, Norway

Aridor M, Sagi-Eisenberg R (1990) Neomycin is a potent secretagogue of mast cells that directly activates a GTP-binding protein involved in exocytosis. J Cell Biol 111:2885–2891

Aschoff J (1991) Hufeland's interest in plant movements. Chronobiol 18:75–78

Baikie ID, Smith PJS, Porterfield DM, Estrup PJ (1999) Mulitple scanning bio-Kelvin probe. Rev Sci Instrum 70:1842–1850

Baluska F (2010) Recent surprising similarities between plant cells and neurons. Plant Signal Behav 5:87–89

Baluska F, Schlicht M, Volkmann D, Mancuso S (2008) Vesicular secretion of auxin: evidences and implications. Plant Signal Behav 3:254–256

Baluska F, Volkmann D, Menzel D (2005) Plant synapses: actin-based domains for cell-to-cell communication. Trends Plant Sci 10:106–111

Bauréus Koch CL, Sommarin M, Persson BR, Salford LG, Eberhardt JL (2003) Interaction between weak low frequency magnetic fields and cell membranes. Bioelectromagnetics 24:395–402

Berg AR, Peacock K (1992) Growth patterns in nutating and nonnutating sunflower (*Helianthus annuus*) hypocotyls. Am J Bot 79:77–85

Bose JC (1913) Researches on irritability of plants. Longmans, Green and Co. London, NY, Bombay, Calcutta, London

Bose JC (1919) Life movements in plants. Trans Bose Inst, pp 255–597

Bose JC (1926) The nervous mechanisms of plants. Longmans, Green and Co., London

Bose JC (1928) The motor mechanism of plants. Longmans, Green and Co., London

Brenner ED, Stahlberg R, Mancuso S, Vivanco J, Baluska F, Volkenburgh EV (2006) Plant neurobiology: an integrated view of plant signaling. Trends Plant Sci 1:413–419

Chen J-P (1996) Untersuchungen zur ultradianen Seitenfiederbewegung von *Desmodium motorium* und zu diffusiv gekoppelten Ca^{2+}-Oszillatoren. PhD thesis, University of Tübingen, Germany

Chen J-P, Eichelmann C, Engelmann W (1997) Substances interfering with phosphatidyl inositol signalling pathway affect ultradian rhythm of *Desmodium motorium*. J Biosc 22:465–476

Chen J-P, Engelmann W, Baier G (1995) Nonlinear dynamics in the ultradian rhythm of *Desmodium motorium*. Z Naturf 50:1113–1116

Chrispeels MJ, Holuigue L, Latorre R, Luan S, Orellana A, Peña-Cortes H, Raikhel NV, Ronald PC, Trewavas A (1999) Signal transduction networks and the biology of plant cells. Biol Res 32:35–60

Cihlar J (1965) Der Einfluss vorübergehender Temperaturänderungen auf Erregungsvorgänge bei Staubgefässen und bei *Desmodium gyrans*. PhD thesis, University of Tübingen, Germany

Cosgrove DJ (2005) Growth of the plant cell wall. Nat Rev Mol Cell Biol 6:850–861

Cosgrove DJ, Li LC, Cho HT, Hoffman-Benning S, Moore RC, Blecker D (2002) The growing world of expansins. Plant Cell Physiol 43:1436–1444

Cosgrove DJ, Hedrich R (1991) Stretch-activated chloride, potassium, and calcium channels coexisting in plasma membranes of guard cells of *Vicia faba* L. Planta 186:143–153

Coté GG (1995) Signal transduction in leaf movement. Plant Physiol 109:729–734

Das GP (1932) Comparative studies of the effect of drugs on the rhythmic tissues of animal and plant. Trans Bose Res Inst 8:146

Davies E (1987a) Action potentials as multifunctional signals in plants: a unifying hypothesis to explain apparently disparate wound responses. Plant Cell Environ 10:623–631

Davies E (1987b) The biochemistry of plants. Academic 12:243–264

Dupont G, Berridge MJ, Goldbeter A (1991) Signal-induced Ca^{2+} oscillations: properties of a model based on Ca^{2+}-induced Ca^{2+} release. Cell Calcium 12:73–85

Durachko DM, Cosgrove DJ (2009) Measuring plant wall extension (creep) induced by acidic pH and by alpha-expansin. J Vis Exp 25:1263

Dutt BK, Guhathakurta A (1996) Effect of application of load on the pulsatory movement of the leaflet of *Desmodium gyrans*. Trans Bose Res Inst 29:105–117

Dwight JS (1839) Select minor poems from the German of Goethe and Schiller with notes (specimens of foreign standard literature). Hilliard, Gray and Company, Boston, p 403

Ellingsrud S, Johnsson A (1993) Perturbations of plant leaflet rhythms caused by electromagnetic radiofrequency radiation. Bioelectromagnetics 14:257–271
Engelberth J (2003) Mechanosensing and signal transduction in tendrils. Adv Space Res 32:1611–1619
Engelmann W (1996) Leaf movement rhythms as hands of biological docks. In: Greppin H, Degli Agosti R, Bonzon M (eds) Vistas on biorhythmicity. University of Geneva, Geneva, pp 51–76
Engelmann W, Antkowiak B (1998) Ultradian rhythms in *Desmodium* (minireview). Chronobiol Internat 15:293–307
Engelmann W, Simon K, Phen CJ (1992) Leaf movement rhythm in *Arabidopsis thaliana*. Z Naturf 47C:925–928
Felle H (1988) Auxin causes oscillations of cytosolic free calcium and pH in *Zea mays* coleoptiles. Planta 174:495–499
Findlay GP (2001) Membranes and the electrophysiology of turgor regulation. Aust J Plant Physiol 28:617–634
Fostad OK (1994) Konstruksjon av strømpulsgenerator. Strømperturberingseksperiment og matematisk modellering/simulering i studier av oscillative bladbevegelser. Master thesis, University of Trondheim, Norway
Fostad OK, Johnsson A, Engelmann W (1997) Effects of electrical currents on *Desmodium gyrans* leaflet movements. Experiments using a current clamp technique. Biol Rhythm Res 28:244–259
Fromm J, Eschrich W (1990) Seismonastic movements in *Mimosa*. In: Satter RL, Gorton HL, Vogelmann TC (eds) The pulvinus: motor organ for leaf movement. American Society of Plant Physiologists, Rockville, pp 25–43
Fromm J, Lautner S (2007) Electrical signals and their physiological significance in plants. Plant Cell Environ 30:249–257
Ginzo HD, Decima EE (1995) Weak static magnetic fields increase the speed of circumnutation in cucumber (*Cucumis sativus* L.) tendrils. Experientia 51:1090–1093
Glass L, Mackey MC (1988) From clocks to chaos: the rhythms of life. Princeton University Press, Princeton
Goldbeter A (1996) Biochemical oscillations and cellular rhythms. The molecular bases of periodic and chaotic behaviour. Cambridge University Press, Cambridge
Goldbeter A, Dupont G, Berridge MJ (1990) Minimal model for signal-induced Ca^{2+}- oscillations and for their frequency encoding through protein phosphorylation. Proc Nat Acad Sci USA 87:1461–1465
Gorton HL (1987) Water relations in pulvini from *Samanea saman*. I. Intact pulvini. Plant Physiol 83:945–950
Gorton GL (1990) Stomates and pulvini: a comparison of two rhythmic turgor-mediated movement systems. In: Satter RL, Gorton HL, Vogelmann TC (eds) The pulvinus: motor organ for leaf movement. American Society of Plant Physiologists, Rockville, pp 223–237
Gradmann D (2001) Model for oscillations in plants. Aust J Plant Physiol 28:577–590
Gradmann D, Buschmann P (1996) Electrocoupling causes oscillations of ion transporters in plants. In: Greppin H, Degli Agosti R, Bonzon M (eds) Vistas on biorhythmicity. University of Geneva, Geneva, pp 239–269
Grassi C, D'Ascenzo M, Torsello A et al (2004) Effects of 50 Hz electromagnetic fields on voltage-gated Ca^{2+}channels and their role in modulation of neuroendocrine cell proliferation and death. Cell Calcium 35:307–315
Guevara MR, Glass L (1982) Phase locking, period doubling bifurcations and chaos in a mathematical model of a periodically driven oscillator: a theory for the entrainment of biological oscillators and the generation of cardiac dysrhythmias. J Math Biol 14:1–23
Guhathakurta A, Dutt BK (1961) Electrical correlate of the rhythmic pulsatory movement of *Desmodium gyrans*. Trans Bose Res Inst 24:73–82
Hager A (2003) Role of the plasma membrane H^+-ATPase in auxin-induced elongation growth: historical and new aspects. J Plant Res 116:483–505
Hepler PK, Winship LJ (2010) Calcium at the cell wall-cytoplast interface. J Integr Plant Biol 52:147–160

Hufeland W (1790) Über die Bewegung des *Hedysarum gyrans* und die Wirkung der Elektrizität auf dasselbe. Magazin für das Neueste aus der Physik und Naturgeschichte 6:5–27 (was published as anonymous, but traced to Hufeland)

Iino M, Long C, Wang XJ (2001) Auxin- and abscisic acid-dependent osmoregulation in protoplasts of *Phaseolus vulgaris* pulvini. Plant Cell Physiol 42:1219–1227

Janse MJ (2003) A brief history of sudden cardiac death and its therapy. Pharmacol Ther 100:89–99

Johansson I, Karlsson M, Johanson U, Larsson C, Kjellbom P (2000) The role of aquaporins in cellular and whole plant water balance. Biochim Biophys Acta 1465:324–342

Johnsson A (1997) Circumnutations: Results from recent experiments on earth and in space. Planta 203(Suppl.):S147–S158

Johnsson A, Bostrøm AC, Pedersen M (1993) Perturbation of the *Desmodium* leaflet oscillation by electric current pulses. J Interdisc Cycle Res 24:17–32

Johnsson A, Brogårdh T, Holje Ø (1979) Oscillatory transpiration of *Avena* plants: perturbation experiments provide evidence for a stable point of singularity. Physiol Plant 45:393–398

Johnsson A, Karlsson HG (1972) A feedback model for biological rhythms. I. Mathematical description and basic properties of the model. J Theor Biol 36:153–174

Johnsson A, Solheim GB, Iversen T-H (2009) Gravity amplifies and microgravity decreases circumnutations in *Arabidopsis thaliana* stems: results from a space experiment. New Phytol 182:621–629

Kaldenhoff R, Fischer M (2006) Aquaporins in plants. Acta Physiol (Oxf) 187:169–176

Karlsson HG, Johnsson A (1972) A feedback model for biological rhythms. II. Comparisons with experimental results, especially on the petal rhythm of *Kalanchoë*. J Theor Biol 36:175–194

Kastenmeier B, Reich W, Engelmann W (1977) Effect of alcohols on the circadian petal movement of *Kalanchoë* and the rhythmic movement of *Desmodium*. Chronobiol 4:122

Kim HY, Coté GG, Crain RC (1992) Effects of light on the membrane potential of protoplasts from *Samanea saman* pulvini. Involvement of K^+ channels and the H^+ ATPase. Plant Physiol 99:1532–1539

Kim HY, Coté GG, Crain RC (1993) Potassium channels in *Samanea saman* protoplasts controlled by phytochrome and the biological clock. Science 260:960–962

Kim HY, Coté GG, Crain RC (1996) Inositol 1,4,5-triphosphate may mediate closure of K^+ channels by light and darkness in *Samanea saman* motor cells. Planta 198:279–287

Konrad KR, Hedrich R (2008) The use of voltage-sensitive dyes to monitor signal-induced changes in membrane potential-ABA triggered membrane depolarization in guard cells. Plant J 55:161–173

Kramer PJ, Boyer JS (1995) Water relations of plants and soils. Academic, San Diego. ISBN 0-12-425060-2

Kraus M, Wolf B, Wolf B (1996) Cytoplasmic calcium oscillations. In: Greppin H, Degli Agosti R, Bonzon M (eds) Vistas on Biorhythmicity. University of Geneva, Geneva, pp 213–237

Kuznetsov OA, Hasenstein KH (1996) Intracellular magnetophoresis of amyloplasts and induction of root curvature. Planta 198:87–94

Lewis RD, Silyn-Roberts H (1987) Entrainment of the ultradian leaf movement rhythm of *Desmodium gyrans* by temperature cycles. J Interdiscipl Cycle Res 18:193–203

Lindström E, Lindström P, Berglund A, Lundgren E, Hansson Mild K (1995) Intracellular calcium oscillations in a T-cell line after exposure to extremely-low-frequency magnetic fields with variable frequencies and flux densities. Bioelectromagnetics 16:41–47

MacRobbie EAC (1998) Signal transduction and ion channels in guard cells. Philos Trans R Soc London B Biol Sci 353:1475–1488

Maurel C (2007) Plant aquaporins: novel functions and regulation properties. FEBS Lett 581:2227–2236

Maurel C, Verdoucq L, Luu DT, Santoni V (2008) Plant aquaporins: membrane channels with multiple integrated functions. Annu Rev Plant Biol 59:595–624

Mayer WE (1990) Walls as potassium storage reservoirs in *Phaseolus* pulvini. In: Satter RL, Gorton HL, Vogelmann TC (eds) The pulvinus: motor organ for leaf movement. American Society of Plant Physiologists, Rockville, pp 160–174

McCreary CR, Dixon SJ, Fraher LJ et al (2006) Real-time measurement of cytosolic free calcium concentration in Jurkat cells during ELF magnetic field exposure and evaluation of the role of cell cycle. Bioelectromagnetics 27:354–364

Menge C (1991) Die Wirkung von Ca^{2+}, Ca^{2+}-Chelatbildern, Ca^{2+}-Kanalblockern, Calmodulinantagonisten und des Ca^{2+}-Ionophors A23187 auf die ultradiane Rhythmik der Seitenfiederbewegung von *Desmodium motorium*. Diploma Thesis, Universität Tübingen, Germany

Mitsuno T (1987) Volume change in the motor cells of pulvinule of lateral leaflets of *Codariocalyx motorius*. Bull Kyoritsu Woman's Univ 33:115–124

Mitsuno T, Sibaoka T (1989) Rhythmic electric potential change of motor pulvinus in lateral leaflet of *Codariocalyx motorius*. Plant Cell Physiol 30:1123–1127

Monshausen GB, Miller ND, Murphy AS, Gilroy S (2011) Dynamics of auxin-dependent Ca^{2+} and pH signaling in root growth revealed by integrating high-resolution imaging with automated computer vision-based analysis. Plant J 65:309–318

Moran N (1990) The role of ion channels in osmotic volume changes in *Samanea* motor cells analyzed by patch-clamp methods. In: Satter RL, Gorton HL, Vogelmann TC (eds) The pulvinus: motor organ for leaf movement. American Society of Plant Physiologists, Rockville, pp 142–159

Moran N (2007) Osmoregulation of leaf motor cells. FEBS Lett 581:2337–2347

Moshelion M, Becker D, Biela A et al (2002a) Plasma membrane aquaporins in the motor cells of *Samanea saman*: diurnal and circadian regulation. Plant Cell 14:727–739

Moshelion M, Becker D, Czempinski K et al (2002b) Diurnal and circadian regulation of putative potassium channels in a leaf moving organ. Plant Physiol 128:634–642

Neugebauer A (2002) Dreidimensionale Registrierung circadianer und ultradianer Wachstumsvorgänge des Hypokotyls von *Arabidopsis thaliana* und *Cardaminopsis arenosa*. PhD thesis University of Tübingen, Germany

Nobel PS (1974) Biophysical plant physiology. Freeman and Company, San Fransisco. ISBN 0-7187-0592-3

Ohashi H (1973) The Asiatic species of *Desmodium* and its allied genera. Ginkgoana 1:1–318

Okada T, Miyazaki T, Ishii N, Fukushima T, Honda N (2005) Effect of the magnetic field of 50 Hz on the circumnutatiomn of the stem of *Arabidopsis thaliana*. Bull Maebashi Inst Technol 8:137–142

Pandey S, Zhang W, Assman SM (2007) Roles of ion channels and transporters in guard cell signal transduction. FEBS Lett 581:2325–2336

Pazur A, Rassadina V (2009) Transient effects of weak electromagnetic fields on calcium ion concentration in *Arabidopsis thaliana*. BMC Plant Biol 9:47

Pedersen M, Johnsson A, Herbjørnsen R (1990) Rhytmic leaf movements under physical loading of the leaves. Z Naturf 45c:859–862

Pickard BG (1973) Action potentials in plants. Bot Rev 39:172–201

Porterfield DM (2007) Measuring metabolism and biophysical flux in the tissue, cellular and subcellular domains: recent developments in self-referencing amperometry for physiological sensing. Biosens Bioelectron 22:1186–1196

Rober-Kleber N, Albrechtovà JTB, Fleig S et al (2003) Plasma membrane H^+-ATPase is involved in auxin-mediated cell elongation during wheat embryo development. Plant Physiol 131:1302–1312

Rosen AD (1996) Inhibition of calcium channel activation in GH3 cells by static magnetic fields. Biochim Biophys Acta 1282:149–155

Rosen AD (2003) Effect of 125 mT static magnetic field on the kinetics of voltage activated Na^+ channels in GH3 cells. Bioelectromagnetics 24:517–523

Ross EM, Higashijima T (1994) Regulation of G-protein activation by mastoparans and other cationic peptides. Methods Enzymol 237:26–37

Roux D, Faure C, Bonnet P et al (2008) A possible role for extra-cellular ATP in plant responses to high frequency, low amplitude electromagnetic field. Plant Signal Behav 3:383–385

Satter RL, Galston AW (1971) Potassium flux: a common feature of *Albizzia* leaflet movement controlled by phytochrome or endogenous rhythm. Science 174:518–520

Satter RL, Galston AW (1981) Mechanisms of control of leaf movements. Annu Rev Plant Physiol 32:83–110

Satter RL, Gorton HL, Vogelmann TC (1990) The pulvinus: motor organ for leaf movement. American Society of Plant Physiologists, Rockville

Satter RL, Morse MI, Lee Y, Crain RC, Cote G, Moran N (1988) Light-and clock-controlled leaflet movements in *Samanea saman:* a physiological, biophysical and biochemical analysis. Bot Acta 101:205–213

Schuster S, Marhl M, Höfer T (2002) Modelling of simple and complex calcium oscillations. From single-cell responses to intercellular signalling. Eur J Biochem 269:1333–1355

Scott BIH (1962) Feedback induced oscillations of five-minute period in the electric field of the bean root. Ann N Y Acad Sc 98:890–900

Serrano R (1990) Plasma membrane ATPases. In: Larsson C, Moller JM (eds) The plant plasma membrane. Springer, Berlin, pp 127–152

Shabala SN, Newman IA, Morris J (1997) Oscillations in H^+ and Ca^{2+} ion fluxes around the elongation region of corn roots and effects of external pH. Plant Physiol 113:111–118

Shabala S, Shabala L, Gradmann D et al (2006) Oscillations in plant membrane transport: model predictions, experimental validation, and physiological implications. J Exp Bot 57:171–184

Sharma VK, Bardal TK, Johnsson A (2003) Light-dependent changes in the leaflet movement rhythm of the plant *Desmodium gyrans*. Z Naturf 58c:81–86

Sharma VK, Engelmann W, Johnsson A (2000) Effects of static magnetic field on the ultradian lateral leaflet movement rhythm in *Desmodium gyrans*. Z Naturf 55c:638–642

Sharma VK, Jensen C, Johnsson A (2001) Phase response curve for ultradian rhythm of the lateral leaflets in the plant *Desmodium gyrans*, using DC current pulses. Z Naturf 56c:77–81

Shepherd, VA (1999) Bioelectricity and the rhythms of sensitive plants—the biophysical research of Jagadis Chandra Bose. Curr Sci 77:189–195

Shepherd VA (2005) From semi-conductors to the rhythms of sensitive plants: the research of J.C Bose. Cell Mol Biol 51:607–619

Solberg EE, Embra BI, Börjesson MB et al (2011) Commotio cordis—under-recognized in Europe? A case report and review. Eur J Cardiov Prev R 18:378–383

Solheim BGB, Johnsson A, Iversen TH (2009) Ultradian rhythms in *Arabidopsis thaliana* leaves in microgravity. New Phytol 183:1043–1052

Strogatz SH (1994) Non-linear dynamics and chaos. Addison-Wesley Publishing Company, Reading MA. ISBN 0201543443

Suh S, Moran N, Lee Y (2000) Blue light activates potassium-efflux channels in flexor cells from *Samanea saman* motor organs via two mechanisms. Plant Physiol 123:833–843

Takahashi K, Isobe M, Muto S (1998) Mastoparan induces an increase in cytosolic calcium ion concentration and subsequent activation of protein kinases in tobacco suspension culture cells. Biochim Biophys Acta 1401:339–346

Toyota M, Furuichi T, Tatsumi H, Sokabe M (2008) Cytoplasmic calcium increases in response to changes in the gravity vector in hypocotyls and petioles of *Arabidopsis* seedlings. Plant Physiol 146:505–514

Turing AM (1952) The chemical basis of morphogenesis. Philos Trans R Soc Lond B 237:37–72

Ul Haque A, Rokkam M, Carlo DAR et al (2007) A MEMS fabricated cell electrophysiology biochip for in silico calcium measurements. Sens Actuator B 123:391–399

Umrath K (1930) Untersuchungen über Plasma und Plasmaströmungen an Characeen. IV. Potentialmessungen an *Nitella mucronata* mit besonderer Berücksichtigung der Erregungserscheinungen. Protoplasma 9:576–597

Volkov AG, Adesina T, Jovanov E (2007) Closing of Venus flytrap by electrical stimulation of motor cells. Plant Signal Behav 2:139–145

Volkov AG (2006) Plant electrophysiology. Theory and methods. Springer, London. ISBN 978-3-540-32717-2

Wang XJ, Haga K, Nishizaki Y et al (2001) Blue-light-dependent osmoregulation in protoplasts of *Phaseolus vulgaris* pulvini. Plant Cell Physiol 42:1363–1372

Weber U (1990) Die Rolle von Ionenkanälen und Protonenpumpen bei der rhythmischen Seitenfiederbewegung von *Desmodium motoricum*. Diploma Thesis, Universität Tübingen, Germany

Weber U, Engelmann W, Mayer WE (1992) Effects of tetraethylammonium chloride (TEA), vanadate, and alkali ions on the lateral leaflet movement rhythm of *Desmodium motorium* (Houtt.) Merr. Chronobiol Int 9:269–277

Whitecross MI, Plovanic N (1982) Structure of the motor region of pulvinules of *Desmodium gyrans* leaflets. Micron 13:337–338

Wildon DC, Thain JF, Minchin PEH et al (1992) Electrical signalling and systemic proteinase inhibitor induction in the wounded plant. Nature 360:62–65

Winfree AT (1970) An integrated view of the resetting of a circadian clock. J Theor Biol 28: 327–374

Winfree A (1971) Corkscrews and singularities in fruitflies: resetting behaviour of the circadian eclosion rhythm. In: Menaker M (ed) Biochronometry. Natl Acad Sci, Washington

Winfree AT (1987a) The timing of biological clocks. Scientific American Books Inc, NY

Winfree AT (1987b) When time breaks down. The three-dimensional dynamics of electrochemical waves and cardiac arrythmias. Princeton University Press, Princeton NJ. ISBN 0-691-02402-2

Winfree AT (2000) Various ways to make phase singularities by electric shock. J Cardiovasc Electrophysiol 11:286–289

Winfree AT (2001) The geometry of biological time, 2nd edn. Springer, NY. ISBN 10: 0387989927

Winfree AT (2002) Chemical waves and fibrillating hearts: discovery by computation. J Biosci 27:465–473

Zhang W, Fan LM, Wu WH (2007) Osmo-sensitive and strech-activated calcium-permeable channels in *Vicia faba* guard cells are regulated by actin dynamics. Plant Physiol 143: 1140–1151

Chapter 5
Regulatory Mechanism of Plant Nyctinastic Movement: An Ion Channel-Related Plant Behavior

Yasuhiro Ishimaru, Shin Hamamoto, Nobuyuki Uozumi and Minoru Ueda

Abstract Leguminous plants open their leaves during the daytime and close them at night as if sleeping, a type of movement that follows circadian rhythms, and is known as nyctinasty. This movement is regulated by the drastic volume changes in two kinds of motor cell of the pulvinus, which is located at the bottom of the leaf stalk. The detailed mechanism of the ion channel-regulated volume change of the motor cells largely remains to be elucidated. In this chapter, we reviewed the mechanism of nyctinasty from two view points, electrophysiology of potassium channel and endogenous chemical substance triggering nyctinastic leaf closure. We focused on the nyctinasty of *Samanea saman* plant because almost all of physiological studies on nyctinasty have been carried out using this plant.

5.1 Ion Channel-Related Regulatory Mechanism on Plant Nyctinastic Movement

Samanea saman, a nyctinastic plant of legume family, opens and closes its leaves and leaflets during the diurnal cycle of light and dark, as well as moves its leaves under the control of circadian rhythm. This movement is regulated by the drastic volume changes in two kinds of motor cell of the pulvinus, which is located at the

Y. Ishimaru · M. Ueda (✉)
Faculty of Science, Tohoku University, 6-3 Aramaki-aza-Aoba, Aoba-Ku,
Sendai 980-8578, Japan
e-mail: ueda@mail.tains.tohoku.ac.jp

S. Hamamoto · N. Uozumi
Faculty of Engineering, Tohoku University, 6-6 Aramaki-aza-Aoba, Aoba-Ku,
Sendai 980-8578, Japan

bottom of the leaf stalk. The opening movement is brought about by the increased volume of the extensor cell, lower half of the pulvinus, and decreased volume of flexor cell, upper half of the pulvinus. The closing movement is caused by the inversely combined volume changes of extensor and flexor cells.

Measurement of the cationic composition of pulvinal extracts using atomic absorption revealed that K^+ is the major cation sufficiently concentrated in the pulvinus (Satter et al. 1974). This indicates that K^+ is the major osmolytes which may determine the turgor pressure of the motor cells. To determine whether the leaflet movements are correlated with K^+ flux, extensor cell area, flexor cell area, and entire secondary pulvini K^+ content were analyzed. K^+ content of entire secondary pulvini remained constant during the movement, whereas that of the extensor and flexor cells largely altered in K^+ content. During the open state, extensor K^+ content increased and flexor K^+ content decreased. At the closed state, K^+ content of the both cell was opposite. The effect of light on K^+ movement was also analyzed. After pinnae were incubated in dark, they were exposed to red light. K^+ concentrations in the extensor cell reduced, but that in the flexor cells increased. This suggested the phytochrome is correlated with the control of the K^+ flux (Satter et al. 1974).

To determine whether K^+ channel in the plasma membrane of the motor cell participate in the leaflet movement, Moran's group carried out patch clamp recording on the extensor and flexor cell protoplasts (Moran et al. 1988). Depolarization of the plasma membrane of flexor protoplast elicited whole-cell K^+ currents, which represented the efflux (outward rectification) of K^+ from the cytoplasm. The single channel conductance of K^+ channel was about 20 pS in the excised patch clamp recording. Likewise, extensor protoplast showed the outward current by membrane depolarization. The identification on K^+ transport activity across plasma membrane in motor cells was supported by the blockage of K^+ channel activity using tetraethylammonium (TEA), a commonly used K^+ channel blocker. TEA also reduced light-mediated leaflet opening and induced leaflet closure in dark. It may be inferred from these data that K^+ efflux through outwardly rectifying K^+ channel is necessary for the cell shrinkage resulting in the leaf movement.

The detailed mechanism of the volume regulation of the motor cells largely remains to be elucidated. On the other hand, the study on the movement of the stomatal guard cells has been relatively progressed. Stomatal guard cells provide excellent model for studying motor cell because these cells are similar to pulvinar motor cells in many aspects. The molecular mechanism on the volume changes of the guard cells has been extensively dissected. The size of protoplast from stomatal guard cells changes in response to light, drought stress phytohormone abscisic acid (ABA), and CO_2 concentrations, etc. (Albrecht et al. 2003; Israelsson et al. 2006; Acharya and Assmann 2009). The recent progress of the ABA signal transduction clarified the interaction of ABA with the ABA receptor, which led to the decrease of guard cell volume through several components (Umezawa et al. 2010). Several recent studies suggest that activities of channels in guard cells are regulated by protein phosphorylation and dephosphorylation (Sato, Gambale et al. 2010;

Geiger et al. 2009, 2010; Sato et al. 2009). *Arabidopsis thaliana* inward-rectifying K$^+$ channel KAT1 has been suggested to play a role for stomata opening by K$^+$ uptake to guard cells. Activity of KAT1 was reduced by the phosphorylation of C-terminal region mediated by the guard cell-specific ABA-activated protein kinase SnRK2.6/OST1/SRK2E (Merlot et al. 2002; Mustilli et al. 2002; Yoshida et al. 2002). Moreover, the calcium-dependent pathway involving possible calcium-dependent protein kinases has contributed to the guard cell volume regulation. Several evidences indicate that control of the shrinkage of the guard cells by K$^+$ channels and anion channels in guard cells. The recently identified guard cell anion channel SLAC1 was activated by calcium-dependent protein kinase 23, CPK23, as well as SnRK2.6/OST1/SRK2E (Geiger, Scherzer et al. 2009, 2010), whose modification also led to a downregulation of the channel activity. As described above, some kinases had been identified as the key component of the ABA-mediated signaling pathway, and control of the activity of KAT1 and SLAC1 for stomatal closure. Similar to guard cell ion channels, *S. saman* motor cell depolarization-activated K$^+$ channels are tightly associated by the kinase activity (Moran 1996). This study was described by the patch clamp method using protoplast prepared from the extensor region of the pulvini. In the absence of either Mg^{2+} or ATP in the pipette solution, corresponding to the cytosolic condition at the whole-cell configuration of the patch clamp method, the K$^+$ channel activity was diminished in few minutes. The diminished activity was recovered in excised inside-out patch clamp configuration by restoring MgATP to the cytosolic side of the plasma membrane. Moreover, depolarization-activated K$^+$ channel was tested using 1-(5-Isoquinolinesulfonyl)-2-methylpiperazine dihydrochloride (H7) which is a wide range kinase inhibitor, which was previously shown to block phosphorylation in plant cell (Hidaka et al. 1984; Raz and Fluhr 1993). H7 inhibited K$^+$ channel in both whole cell and excised patch configuration. The removal of the H7 from the solution recovered the K$^+$ channel activity. These results indicated that some kinases resulted in the activation of the outward-rectifying K$^+$ channel by the protein kinase in the extensor cell. The further study has shown that the protein phosphorylation also directly regulated the inward-rectifying K$^+$ channel in the motor cell. To test the dependence on the inward-rectifying K$^+$ channel activity on phosphorylation, a strong protein phosphatase inhibitor, okadaic acid, was applied into the cytosolic buffer and the patch clamp experiment was performed. The effect of okadaic acid on inward-rectifying K$^+$ channel in extensor cells was not significant, whereas in flexor cells, okadaic acid significantly enhanced the inward K$^+$ current. The other channels, SPICK1 and SPICK2, a homolog of the *Arabidopsis Shaker*-type K$^+$ channel AKT2 were found to be expressed in *Samaea saman* motor cell and controlled under the circadian rhythm (Yu et al. 2001). In addition, SPICK2 was phosphorylated by the cAMP-dependent protein kinase (PKA) (Yu et al. 2006). Considering the circumstances mentioned above, SPICK2 was likely to be the K$^+$ influx channel regulated by phosphorylation (Fig. 5.1).

Controlling the membrane potential and pH of extracellular space of the cell is crucial issue for regulating ion channel-mediated ion flux rapid changes of cell volume. Stomatal opening is guided by light, including blue and red light.

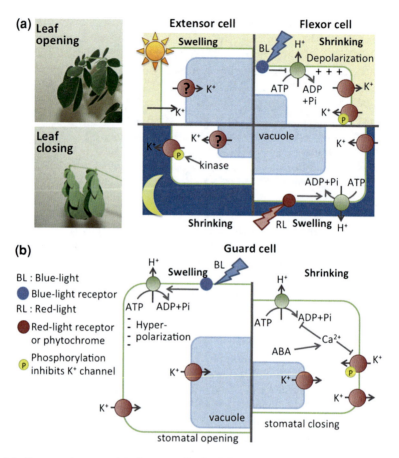

Fig. 5.1 Osmoregulatory model of motor cell of pulvinus and stomatal guard cell. **a** Extensor cell swell and flexor cell shrink in the leaf opening phase. Leaf closing phase is caused by opposite volume changes of extensor cell and flexor cell. **b** Swelling and shrinking of stomatal guard cell

Blue light activates plasma membrane H^+-ATPase and hyperpolarizing plasma membrane potential by releasing H^+ from guard cell (Kinoshita and Shimazaki 1999; Briggs and Christie 2002; Roelfsema and Hedrich 2005; Vavasseur and Raghavendra 2005). Hyperpolarization of the plasma membrane of guard cell induces gate opening of inward-rectifying K^+ channel. In *S. saman* flexor cell, illumination with blue light for a few minutes induces membrane depolarization-activated K^+ channel to release K^+ from the cytosol, which led to end up to a rapid cell shrinkage (Suh et al. 2000). This indicates that depolarization resulted from a blue light resulted in arrest of proton pump, and increase of depolarization-activated K^+ channel activity. In guard cell, gene product of phototropin *PHOT1 PHOT2* has been cloned and their protein products demonstrated to be the blue light receptor protein (Kinoshita, Doi et al. 2001). In *Phaseolus vulgaris*, a major food legume, phototropin was identified as the initial elements of shrinking pulvinar motor cell (Inoue et al. 2005). Moreover, cryptochrome and flavin-binding aquaporin are also involved in pulvinar blue light response (Lorenz et al. 2003). Blue light receptor of *S. saman* pulvinus is not yet identified, and these photoreceptors can be an excellent candidate for the first element in the phototransduction cascade of shrinking motor cell.

In the closed state of stomata, most of the vacuoles show particle shape which are formed by fragmentation of the large central vacuole (Diekmann et al. 1993). Interestingly, such small vacuoles fuse with one another to reform a large central vacuole during stomata opening (Gao et al. 2005). Prior to this process, huge amount of ions moves from cytoplasm to the vacuole via transport systems in the tonoplast. K^+ is accumulated by H^+/K^+ antiporter, and anion can be transported into vacuole via both anion channel and H^+/anion antiporter (Pei et al. 1996; Hafke et al. 2003; De Angeli et al. 2006; Kovermann et al. 2007). Among the solute release from the guard cell, more than 90% originate from the vacuole (MacRobbie 1998). During the stomatal closure, ABA triggers cytosolic Ca^{2+} increases and Ca^{2+} sensitivity, which activates both cation channel and anion channel to promote the ion efflux reducing the cellular volume. TPC1 channel was shown to be Ca^{2+}-activated, Ca^{2+}-permeable channel and found in vacuolar membrane of many different plant cell types (Hedrich and Neher 1987). TPC1 channel are also found in guard cell vacuolar membrane, leading to the hypothesis that TPC1 channel participates in Ca^{2+} signaling by releasing Ca^{2+} from the vacuole (Ward and Schroeder 1994; Ward et al. 1995; Peiter et al. 2005). Cytosolic Ca^{2+} which is released from vacuole activates vacuole membrane-localized K^+ channel TPK1 (Bihler et al. 2005; Gobert et al. 2007; Latz et al. 2007). Five TPK channels are conserved in *A. thaliana* genome and TPK channel family is widely conserved in other plant species, including *S. saman* (Moshelion et al. 2002; Hamamoto et al. 2008a, b). TPK1 is expressed in guard cell vacuolar membrane and shows all of the features of the previously demonstrated vacuolar K^+ channel in guard cell. When plant was treated with ABA, *tpk1* mutant plant showed slower stomatal closure, whereas stomatal closure in *TPK1* overexpressor plant was slightly faster than wild-type plant (Gobert et al. 2007). From these observations, TPK1 contribute to K^+ release from the vacuole of the guard cell (Ward and Schroeder 1994; Gobert et al. 2007). However, dynamic movement of intracellular organism has not been demonstrated in motor cell of *S. saman*. Considering the similarities between guard cell and motor cell, vacuole fragmentation and reconstitution might be involved in the alteration of the cellular volumes.

5.2 Chemical Studies on Nyctinastic Leaf Movement

In general, plants are rooted and unable to move from place to place by themselves. However, some plants are able to move in certain ways. Three plants shown in Fig. 5.2 open their leaves in the daytime and "sleep" at night with their

Fig. 5.2 Three nyctinastic plants in daytime (*left*) and at night (*right*) (from the *left*, *Senna obtsusifolia* L., *Phyllanthus urinaria* L., and *Mimosa pudica* L.)

leaves folded. This leaf movement is circadian rhythmic and regulated by a biological clock with a cycle of about 24 h. This phenomenon, known as nyctinasty, is widely observed in leguminous plants, and has been of great interest to scientists for centuries with the oldest records dating from the time of Alexander the Great.

It was Charles Darwin, well known for his theory of evolution, who established the science of plant movement in his later years. In 1880, Darwin published an invaluable book entitled "The Power of Movement in Plants" (Darwin 1880), based on experiments using more than 300 different kinds of plants, including nyctinastic species. However, despite the great advance in science that has been made since Darwin's time, it is still difficult to establish the molecular basis of these processes. Our study focuses on the chemical mechanisms of Darwin's observations using endogenous chemical factors.

Physiological mechanism of leaf movement has been extensively studied (for a review, Satter and Galston 1981; Satter et al. 1990). Nyctinastic leaf movement is induced by the swelling and shrinking of motor cells in the pulvinus, joint-like thickening located at the base of the petiole. Motor cells play a key role in plant leaf movement. A flux of potassium ions (K^+) across the plasma membrane of the motor cells is followed by massive water flux, which results in swelling or shrinking of these cells. An issue of great interest is the regulation of the opening and closing of the K^+ channels involved in nyctinastic leaf movement. The channels involved in the light-induced control of leaf movement are studied from various aspects (Moshelion and Moran 2000; Moshelion et al. 2002; Suh et al. 2000). However, no information was achieved on the circadian rhythmic regulation of the channel concerning nyctinsaty. Chemical studies on the nyctinasty have also been carried out and many attempts have been made to isolate the endogenous factors that is expected to be involved in nyctinasty (for a review, see Schildcknecht 1983). But, all of them containing known phytohormones were ineffective under physiological conditions.

Leaf Opening and Closing Substances in Nyctinastic Plants

We isolated two types of endogenous bioactive substances from several nyctinastic plants: leaf opening and closing factors, which possibly mediate nyctinastic leaf movement (for a review, see Ueda and Yamamura 2000c). When the leaves of a leguminous plant were disconnected from the stem, their leaflets continued to open and close according to the circadian rhythm: open in the daytime and closed in the night time. Artificial application of leaf closing factor made the leaflet folded even in the daytime, whereas leaf opening factor made the leaflet open even at night. To date, we have isolated five sets of leaf movement factors (**1–10**) from five nyctinastic plant species (Fig. 5.3, Miyoshi et al. 1987; Shigemori et al. 1989; Ueda et al. 1995a, b, 1997a, b, 1998a–c, Ueda and Yamamura 1999a, b, Ueda et al. 1999c, d, 2000a). All of these factors were effective at the concentrations of 10^{-5} to 10^{-6} M when applied exogenously. And the content of these factors in the plant body was estimated to be 10^{-5} to 10^{-6} M by HPLC analyses (Ueda et al. 1998c, 1999c), which showed that they were effective at physiological concentration. It was also shown that each nyctinastic plant has a specific set of leaf movement factors (Ueda et al. 2000a, b). None of the factors were effective in the plants belonging to other genuses, even at a 10,000-fold to 100,000-fold concentration (Miyoshi et al. 1987; Shigemori et al. 1989; Ueda et al. 1995a, b, 1997a, b, 1998a–c, Ueda and Yamamura 1999a, b; Ueda et al. 1999c, d, 2000a, b). For example, **1** is effective as leaf opening factor for *Albizzia julibrissin* Durazz. at 10^{-5} M, but it was not effective for other genus, such as *Cassia*, *Phyllanthus*, and *Mimosa* even at 10^{-1} M. These findings support the idea that nyctinasty is controlled by genus-specific leaf closing and opening factors (Ueda et al. 2000b, Ueda and Yamamura 2000c). In the following sections, we demonstrate that these leaf movement factors operate as endogenous chemical factors controlling nyctinasty in the plant body.

Bioorganic Studies of Nyctinasty Using Functionalized Leaf Movement Factors as Molecular Probes: Fluorescence Studies on Nyctinasty

Most of the physiological studies on nyctinasty have been carried out in plants belonging to the genus *Albizzia* (for a review, see Lee 1990). We isolated **5** (a closing factor) and **10** (an opening factor) as a leaf opening factor and leaf closing factor, respectively. And it was also revealed that these are commonly involved among three *Albizzia* plants, such as *A. julibrissin*, *A. saman*, *A. lebbeck* (Ueda et al. 2000a), and are ineffective for plants belonging to other plant genera. We used **5** and **10** as a tool for tracing their molecular dynamism in vivo.

To identify target cells of leaf closing and opening factors, we synthesized FITC-labeled leaf closing factor (**11**) and rhodamine-labeled leaf opening factor

Leaf–closing Substances

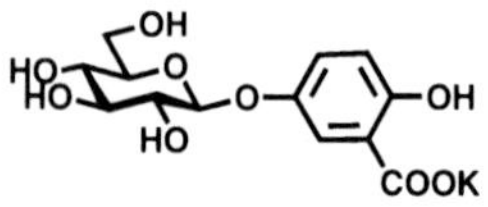

Pottasium 5-O-β-D-gluco-
pyranosyl gentisate (1)
(*Mimosa pudica* L.)

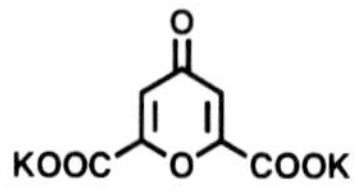

Potassium Chelidonate (2)
(*Cassia mimosoides* L.,
Cassia occidentalis L.)

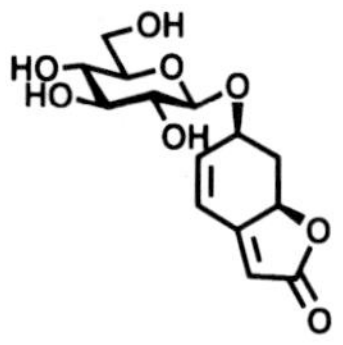

Phyllanthurinolactone (3)
(*Phyllanthus urinaria* L.)

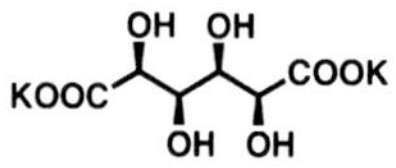

Potassium D-idarate (4)
(*Lespedeza cuneata* G. Don)

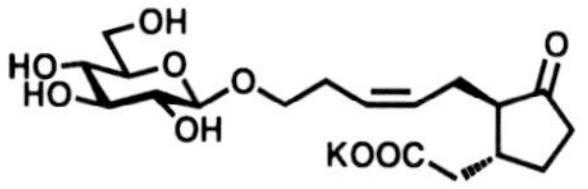

Potassium β-D-glucopyranosyl 12-
hydroxyjasmonate (5)
(*Samanea saman* & *Albizzia julibrissin*
Durazz)

Leaf–opening Substances

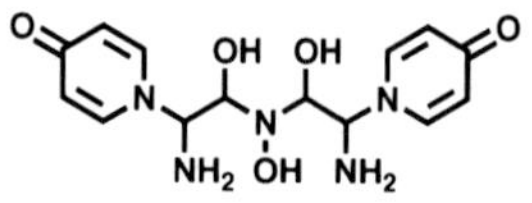

mimopudine (6)
(*Mimosa pudica* L.)

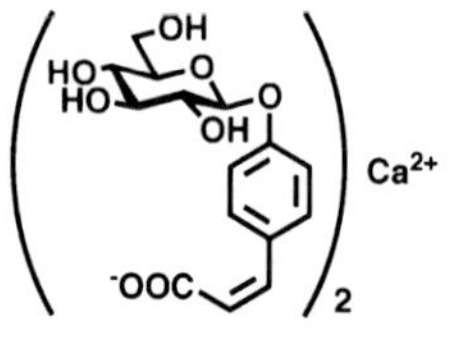

Calcium 4-O-β-D-gluco-
pyranosyl *cis-p*-coumarate (7)
(*Cassia mimosoides* L.)

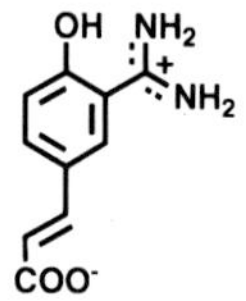

Phyllurine (8)
(*Phyllanthus urinaria* L.)

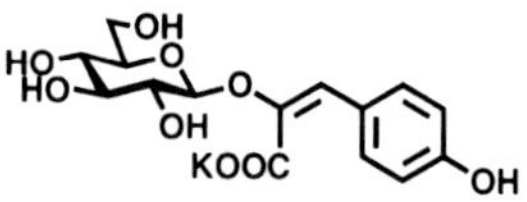

Potassium lespedezate (9)
(*Lespedeza cuneata* G. Don)

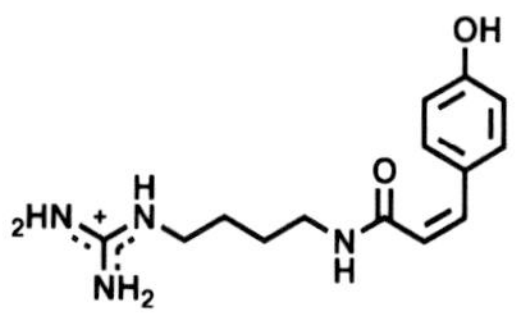

cis-p-coumaroylagmatine (10)
(*Samanea saman* & *Albizzia julibrissin*
Durazz)

Fig. 5.3 Leaf movement factors of five nyctinastic plants

5 Regulatory Mechanism of Plant Nyctinastic Movement

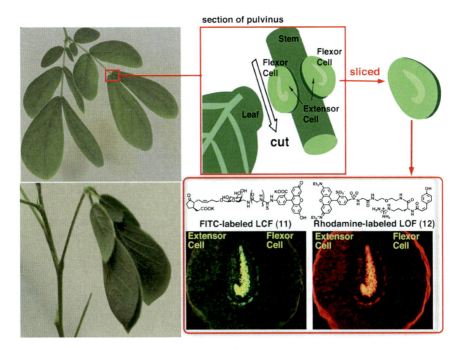

Fig. 5.4 Double fluorescence study of *Albizzia saman* using FITC-labeled leaf closing factor (**11**) and rhodamine-labeled leaf opening factor (**12**). The section containing motor cell was prepared by cutting the pulvini of *Albizzia saman* perpendicular to the vessel. Then, the section was stained by both **11** and **12**, and monitored by using appropriate filters, respectively. The fluorescence images of the section after staining were shown at the *lower right*. Staining of xylem cells appeared to be non-specific, because both **11** and **11′** stained xylem cells

(**12**) based on the results of structure–activity relationship studies (Fig. 5.4) (Nagano et al. 2003; Nakamura et al. 2006a, c). Probes **11** and **12** retain their activity to close and open *Albizzia* leaves, respectively, but they have no activity for other genera, similar to non-labeled natural products (**5** and **10**). A double fluorescence-labeling experiment using **11** and **12** was conducted to identify their target cell in the plant body (Nakamura et al. 2006c). Figure 5.4 shows the fluorescence image of the sections in pulvini of *A. saman* that were cut perpendicular to the vessel. Somehow unexpectedly, both of the probes bound to the motor cells located over the half-side of pulvini section, especially called extensor cell. (see Nakamura et al. 2006b). Strong fluorescence observed in the xylem and epidermis are due to the non-specific binding of the probe **11**, which is irrelevant to their bioactivity, which will be discussed later.

The motor cells in pulvinus of nyctinastic plants are two types; extensors, which locate at the upper (adaxial) side of a leaflet, and flexors, which locate at the lower (abaxial) side of a leaflet (Fig. 5.4). Leaflets move upward during closure and move downward during opening. Extensor cells increase their turgor pressure during opening and decrease their turgor during closing; while flexor cells

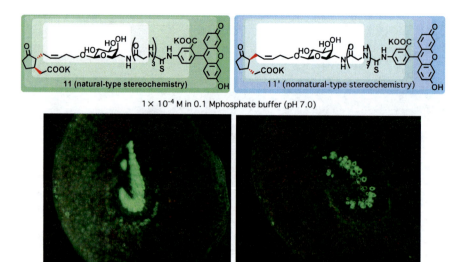

Fig. 5.5 An enantio-differential fluorescence study using a pair of probes **11** and **11′**, which was prepared from an enantiomeric pair of **5**, respectively. By the comparison between *right* and *left* fluorescence images, it was shown that stereospecific recognition of the probes was observed only in the extensor motor cell of *left* image

decrease their turgor pressure during opening and increase their turgor during closing. Figure 5.3 shows that **5** and **10** bind to the extensor cell in the pulvini to induce leaf opening and leaf closing, respectively. Thus, the induction of volume change in the extensor cell should be the trigger for leaf movement. It is likely that the binding of **5** and **10** to the extensor cell induce the changes in K$^+$ channel activity in the cell which results in changes in K$^+$ levels in the extensor cell, which in turn causes leaf closing and opening. Therefore, an important issue yet to be examined is whether these factors affect K$^+$ channel activity.

Moreover, an enantio-difference observed in the binding affinity of both enantiomers clearly demonstrated the involvement of some specific binding protein for **5** in the extensor cell. We synthesized probe **11′**, which has an enantiomeric aglycon of **11**, as a negative control against probe **11** because SAR study on **5** showed that the structure modification in the stereochemistry in aglycon part greatly diminished its leaf closing activity (Nakamura et al. 2006b). Comparison of the results using **11** and **11′** enables the differentiation between specific binding of the probe concerning bioactivity and non-specific binding of the probe due to the chemical adsorption. Bioactive probe **11** stained the vessel, epidermis, and extensor cell in the pulvini section (Fig. 5.4). In contrast, probe **11′**, which is inactive for opening leaves, stained the vessel and epidermis of the pulvini, but did not stain motor cells at all (Fig. 5.5). Thus, it was clearly shown that some specific binding protein that can differentiate the stereochemistry of the ligands is involved in the extensor cell. And it was also shown that strong fluorescence observed in the vessel and epidermis of the pulvini which is irrelevant to stereochemistry of the ligand is due to the non-specific bindings of the probes.

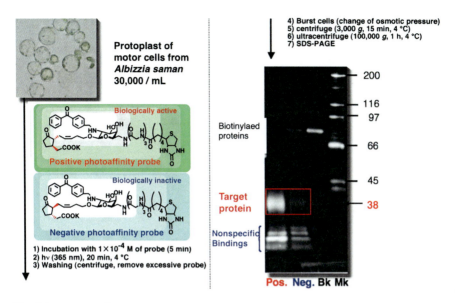

Fig. 5.6 Enantio-differential photoaffinity labeling experiments using positive and negative probes against protoplasts: SDS-PAGE analysis of membrane fractions prepared from photo-labeled protoplasts (From the *left*, membrane fraction treated by **positive probe**, membrane fraction treated by **negative probe**, untreated membrane fraction, and MW marker)

The specificity of the binding of probe **12** was confirmed by repelling with excess amount of non-labeled **10**. Thus, these results suggested that receptors for the closing factor **5** and opening factor **10,** which are involved in the changes in turgor in the extensor cell are both present in the extensor cells in *A. saman*.

In addition, the target cell of leaf movement factors was also confirmed to be a motor cell in other nyctinastic species, *C. mimosoides* L. and *P. urinaria* L., by using fluorescence-labeled leaf movement factors of corresponding plant (Sugimoto et al. 2001; Sato et al. 2005).

Cell-Shrinking in the Protoplast of Motor Cell in S. saman

It is also revealed that the **11** selectively binds to extensor motor cells in *S. saman* section (Nakamura et al. 2008). Moreover, photoaffinity labeling studies demonstrate that the **5** is targeted to the membrane protein in the motor protoplast cells (Fig. 5.6). These results strongly suggest that **5** directly affects the motor cells, leading to leaf closing activity.

To examine whether **5** has the capacity to induce rapid volume changes in extensor and/or flexor motor cell, the cell volume changes in protoplasts isolated from the extensor or flexor side of tertiary pulvini of *S. saman* are monitored

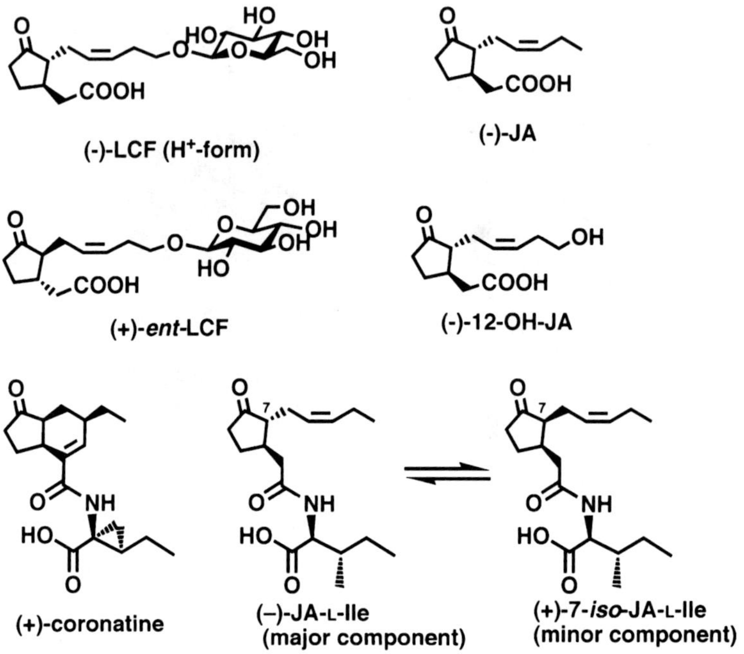

Fig. 5.7 Structures of (−)-JA, (−)-LCF (as H⁺-form), (+)-*ent*-LCF, (−)-12-OH-JA, (−)-JA-Ile (shown as a mixture of (−)-JA-Ile and (+)-7-*iso*-JA-Ile), and (+)-coronatine. JA-Ile is known to exist as a 95:5 ratio of (−)-JA-Ile and (+)-7-*iso*-JA-Ile, respectively

(Gorton and Satter 1984; Nakamura et al. 2011). **5** is effective at concentrations as low as 1 μM and an appreciable decrease in the volume of the protoplasts from extensor cells is established within 10–20 min, amounting to a total decrease of approximately 5% within 50 min. The shrinking of motor cell protoplasts upon **5** application would be the result of an immediate ion channel-mediated process. Moreover, the cell-shrinking by **5** is not observed in flexor cells and the dosage of other jasmonates such as *ent*-**5**, JA, 12-OH-JA, and coronatine does not induce the volume changes in extensor cells (Fig. 5.7). Leaf closure is cell and substrate specifically regulated, consistent with physiological data.

Potassium Fluxes in Motor Cell Protoplast in *S. saman*

Outward-directed K⁺ efflux from extensor motor cell is mediated by opening of K⁺-selective channels, which can be specifically blocked by TEA (Moran et al. 1990). For the **5**-promoted leaf closure, protoplast shrinking is also induced and inhibited by TEA indicating that **5**-induced and dark-promoted protoplast shrinkage shares a common signaling pathway. In extensor cell, K⁺ fluxes play important roles in cell-shrinking as well as leaf closure.

Cytosolic calcium ion has been the central focus in most studies of signaling in the pulvinus, with attempts to confirm it as part of the phosphatidylinositol signaling pathway. However, the effects of cytosolic calcium ion concentration on K⁺ fluxes are rather minor in the pulvinus of *S. saman* and is not required for 5-induced and dark-promoted protoplast shrinkage (Moshelion and Moran 2000).At least in part, these data strongly suggest that the molecular mechanism involves activation of selective ion channels.

Differences Between Jasmonic Acid Glycocide and Jasmonic Acid Signaling

Jasmonates are oxylipin-based plant hormones originating from poly-unsaturated fatty acids that act in response to developmental or environmental stimuli (Devoto and Turner 2003; Wasternack 2007; Howe and Jander 2008; Browse 2009; Wasternack and Kombrink 2010). Environmental cues for activating the jasmonate signaling pathway include wounding, insect attack or, UV light and as such jasmonate belongs to so-called stress hormones. Jasmonates are perceived intracellularly by a jasmonate receptor belonging to the F-box protein family that upon binding to the plant hormone targets a repressor of the jasmonate response pathway for degradation (Chini et al. 2007; Thines et al. 2007; Chung et al. 2008). The activated jasmonate form, most efficiently bound by the jasmonate receptor, was found to be a JA-Ile (Fonseca et al. 2009). JA-Ile is catalyzed by an enzyme encoded by the *JAR1* gene (Staswick et al. 2002; Swiatek et al. 2004). Recently it was demonstrated that coreceptor complex consisting of COI1, JAZ, and inositol pentakisphosphate strongly binds to JA-Ile and thus functions as jasmonate receptor (Sheard et al. 2010).

On the other hand, 5 is synthesized from 12-OH-JA with its tuber-inducing activity in *Solanaceous* species (Yoshihara et al. 1989). They belong to the other synthetic pathway compared to the derivatives of JA signaling. 12-OH-JA is inactive in mediating typical JA responses, such as tendril coiling (Blechert et al. 1999), inhibition of root growth and seed germination (Miersch et al. 2008), or activation of JA-responsive genes in barley, tomato, and Arabidopsis (Miersch et al. 1999, 2008; Gidda et al. 2003; Kienow et al. 2008; Nakamura et al. 2011). 5 is also inactive in JA-responsive genes in Arabidopsis and JA-induced emission of organic volatiles in *S. saman* and lima bean (Nakamura et al. 2011). By contrast, JA activates the typical JA-inducible biosynthesis of specific secondary metabolites in *S. saman* and lima bean (Nakamura et al. 2011). Under those reasons, 5 and 12-OH-JA have been considered as a by-product of switching off jasmonate signaling by the metabolic conversion of bioactive JA into inactive metabolites, and its broad and abundant occurrence in many different plant species suggests that this is a common mechanism (Miersch et al. 2008).

However, 5 has a unique bioactivity, such as leaf close activity. Differed from JA signaling, there would be original signal pathways for leaf closing attributing by 5. Identification of membrane-localized 5 binding protein would be a first step to find a key to the unrevealing of 5 signal pathways. Finally, the whole story of leaf closure by 5 with a different activity of other jasmonates would be given in molecular levels.

References

Acharya BR, Assmann SM (2009) Hormone interactions in stomatal function. Plant Mol Biol 69:451–462
Albrecht V, Weinl S et al (2003) The calcium sensor CBL1 integrates plant responses to abiotic stresses. Plant J 36:457–470
Bihler H, Eing C et al (2005) TPK1 is a vacuolar ion channel different from the slow-vacuolar cation channel. Plant Physiol 139:417–424
Blechert S, Bockelmann C, Fußlein M, Schrader T, Stelmach B, Niesel U, Weiler EW (1999) Structure-activity analyses reveal the existence of two separate groups of active octadecanoids in elicitation of the tendril-coiling response of *Bryonia dioica* Jacq. Planta 207:470–479
Briggs WR, Christie JM (2002) Phototropins 1 and 2: versatile plant blue-light receptors. Trends Plant Sci 7:204–210
Browse J (2009) Jasmonate passes muster: a receptor and targets for the defense hormone. Annu Rev Plant Biol 60:183–205
Chini A, Fonseca S, Fernandez G, Adie B, Chico JM, Lorenzo O, Garcıa-Casado G, Lopez-Vidriero I, Lozano FM, Ponce MR et al (2007) The JAZ family of repressors is the missing link in jasmonate signalling. Nature 448:666–671
Chung HS, Koo AJK, Gao X, Jayanty S, Thines B, Jones AD, Howe GA (2008) Regulation and function of Arabidopsis *JASMONATE ZIM*-domain genes in response to wounding and herbivory. Plant Physiol 146:952–964
Darwin C (1880) The power of movement in Plants. John Murray Inc.,
De Angeli A, Monachello D et al (2006) The nitrate/proton antiporter AtCLCa mediates nitrate accumulation in plant vacuoles. Nature 442:939–942
Devoto A, Turner JG (2003) Regulation of jasmonate-mediated plant responses in Arabidopsis. Ann Bot (Lond) 92:329–337
Diekmann W, Hedrich R, Raschke K, Robinson DG (1993) Osmocytosis and vacuolar fragmentation in guard cell protoplasts: their relevance to osmotically-induced volume changes in guard cells. J Exp Bot 44:1569–1577
Fonseca S, Chini A, Hamberg M, Adie B, Porzel A, Kramell R, Miersch O, Wasternack C, Solano R (2009) (+)-7-iso-Jasmonoyl-L-isoleucine is the endogenous bioactive jasmonate. Nat Chem Biol 5:344–350
Gao XQ, Li CG et al (2005) The dynamic changes of tonoplasts in guard cells are important for stomatal movement *Vicia faba*. Plant Physiol 139:1207–1216
Geiger D, Scherzer S et al (2010) Guard cell anion channel SLAC1 is regulated by CDPK protein kinases with distinct Ca^{2+} affinities. Proc Natl Acad Sci U S A 107:8023–8028
Geiger D, Scherzer S et al (2009) Activity of guard cell anion channel SLAC1 is controlled by drought-stress signaling kinase-phosphatase pair. Proc Natl Acad Sci U S A 106:21425–21430
Gidda SK, Miersch O, Levitin A, Schmidt J, Wasternack C, Varin L (2003) Biochemical and molecular characterization of a hydroxyjasmonate sulfotransferase from *Arabidopsis thaliana*. J Biol Chem 278:17895–17900

Gobert A, Isayenkov S et al (2007) The two-pore channel TPK1 gene encodes the vacuolar K+ conductance and plays a role in K+ homeostasis. Proc Natl Acad Sci U S A 104:10726–10731

Gorton HL, Satter RL (1984) Extensor and flexor protoplasts from *Samanea* Pulvini. Plant Physiol 76:680–684

Hafke JB, Hafke Y et al (2003) Vacuolar malate uptake is mediated by an anion-selective inward rectifier. Plant J 35:116–128

Hamamoto S, Marui J et al (2008a) Characterization of a tobacco TPK-type K+ channel as a novel tonoplast K+ channel using yeast tonoplasts. J Biol Chem 283:1911–1920

Hamamoto S, Yabe I et al (2008b) Electrophysiological properties of NtTPK1 expressed in yeast tonoplast. Biosci Biotechnol Biochem 72:2785–2787

Hedrich R, Neher E (1987) Cytoplasmic calcium regulates voltage dependent ion channels in plant vacuoles. Nature 329:833–835

Hidaka H, Inagaki M et al (1984) Isoquinolinesulfonamides, novel and potent inhibitors of cyclic nucleotide dependent protein kinase and protein kinase C. Biochemistry 23:5036–5041

Howe GA, Jander G (2008) Plant immunity to insect herbivores. Annu Rev Plant Biol 59:41–66

Inoue S, Kinoshita T et al (2005) Possible involvement of phototropins in leaf movement of kidney bean in response to blue light. Plant Physiol 138:1994–2004

Israelsson M, Siegel RS et al (2006) Guard cell ABA and CO_2 signaling network updates and Ca^{2+} sensor priming hypothesis. Curr Opin Plant Biol 9:654–663

Kienow L, Schneider K, Bartsch M, Stuible H-P, Weng H, Miersch O, Wasternack C, Kombrink E (2008) Jasmonates meet fatty acids: functional analysis of a new acyl-coenzyme a synthetase protein family from *Arabidopsis thaliana*. J Exp Bot 59:403–419

Kinoshita T, Shimazaki K (1999) Blue light activates the plasma membrane H(+)-ATPase by phosphorylation of the C-terminus in stomatal guard cells. EMBO J 18:5548–5558

Kinoshita T, Doi M et al (2001) Phot1 and phot2 mediate blue light regulation of stomatal opening. Nature 414:656–660

Kovermann P, Meyer S et al (2007) The Arabidopsis vacuolar malate channel is a member of the ALMT family. Plant J 52:1169–1180

Latz A, Becker D et al (2007) TPK1, a Ca(2+)-regulated Arabidopsis vacuole two-pore K(+) channel is activated by 14-3-3 proteins. Plant J 52:449–459

Lee Y (1990) Satter RL, Gorton HL, Vogelmann TC (eds) The Pulvinus: motor organ for leaf movement. American Society of Plant Physiologists, Rockville, pp 130–141

Lorenz A, Kaldenhoff R et al (2003) A major integral protein of the plant plasma membrane binds flavin. Protoplasma 221:19–30

MacRobbie EA (1998) Signal transduction and ion channels in guard cells. Philos Trans R Soc Lond B Biol Sci 353:1475–1488

Merlot S, Mustilli AC et al (2002) Use of infrared thermal imaging to isolate Arabidopsis mutants defective in stomatal regulation. Plant J 30:601–609

Miersch O, Kramell R, Parthier B, Wasternack C (1999) Structure–activity relations of substituted, deleted or stereospecifically altered jasmonic acid in gene expression of barleyleaves. Phytochemistry 50:353–361

Miersch O, Neumerkel J, Dippe M, Stenzel I, Wasternack C (2008) Hydroxylated jasmonates are commonly occurring metabolites of jasmonic acid and contribute to a partial switch-off in jasmonate signaling. New Phytol 117:114–127

Miyoshi E, Shizuri Y, Yamamura S (1987) Isolation of potassium chelidonate as a bioactive substance concerning with circadian rhythm in nyctinastic plants. Chem Lett: 511–514

Moran N, Ehrenstein G et al (1988) Potassium channels in motor cells of *Samanea saman*: a patch-clamp study. Plant Physiol 88:643–648

Moran N, Fox D, Satter RL (1990) Interaction of the depolarization-activated K+ channel of *Samanea saman* with inorganic ions: a patch-clamp study. Plant Physiol 94:424–431

Moran N (1996) Membrane-delimited phosphorylation enables the activation of the outward-rectifying K channels in motor cell protoplasts of *Samanea saman*. Plant Physiol 111: 1281–1292

Moshelion M, Moran N (2000) Potassium-efflux channels in extensor and flexor cells of the motor organ of *Samanea saman* are not identical. Effects of cytosolic calcium. Plant Physiol 124:911–919

Moshelion M, Becker D et al (2002) Diurnal and circadian regulation of putative potassium channels in a leaf moving organ. Plant Physiol 128:634–642

Mustilli AC, Merlot S et al (2002) Arabidopsis OST1 protein kinase mediates the regulation of stomatal aperture by abscisic acid and acts upstream of reactive oxygen species production. Plant Cell 14:3089–3099

Nagano H, Kato E, Yamamura S, Ueda M (2003) Fluorescence studies on nyctinasty which suggest the existence of genus-specific receptors for leaf-movement factor. Org Biomol Chem 1:3186–3192

Nakamura Y, Kiyota H, Kumagai T, Ueda M (2006a) Direct observation of the target cell for jasmonate-type leaf-closing factor: genus-specific binding of leaf-movement factors to the plant motor cell. Tetrahedron Lett 47:2893–2897

Nakamura Y, Miyatake R, Matsubara A, Kiyota H, Ueda M (2006b) Enantio-differential approach to identify the target cell for glucosyl jasmonate-type leaf-closing factor, by using fluorescence-labeled probe compounds. Tetrahedron 62:8805–8813

Nakamura Y, Matsubara A, Okada M, Kumagai T, Ueda M (2006c) Double fluorescence-labeling study on genus *Albizzia* using a set of fluorecence-labeled leaf-movement factors to identify the spatial distribution of their receptors. Chem Lett 35:744–745

Nakamura Y, Miyatake R, Ueda M (2008) Enantiodifferential approach for the detection of the target membrane protein of the Jasmonate Glycoside that controls the leaf movement of *Albizzia saman*. Angew Chem Int Ed 47:7289–7292

Nakamura Y, Mithofer A, Kombrink E, Boland W, Hamamoto S, Uozumi N, Tohma K, Ueda M (2011) 12-Hydroxyjasmonic acid glucoside is a COI1-JAZ-Independent activator of leaf-closing movement in *Samanea saman*. Plant Physiol 155:1226–1236

Pei ZM, Ward JM et al (1996) A novel chloride channel in *Vicia faba* guard cell vacuoles activated by the serine/threonine kinase, CDPK. EMBO J 15:6564–6574

Peiter E, Maathuis FJ et al (2005) The vacuolar Ca^{2+}-activated channel TPC1 regulates germination and stomatal movement. Nature 434:404–408

Raz V, Fluhr R (1993) Ethylene signal is transduced via protein phosphorylation events in plants. Plant Cell 5:523–530

Roelfsema MR, Hedrich R (2005) In the light of stomatal opening: new insights into 'the Watergate'. New Phytol 167:665–691

Sato A, Gambale F et al (2010) Modulation of the Arabidopsis KAT1 channel by an activator of protein kinase C in *Xenopus laevis* oocytes. FEBS J 277:2318–2328

Sato A, Sato Y et al (2009) Threonine at position 306 of the KAT1 potassium channel is essential for channel activity and is a target site for ABA-activated SnRK2/OST1/SnRK2.6 protein kinase. Biochem J 424:439–448

Sato H, Inada M, Sugimoto T, Kato N, Ueda M (2005) Direct observation of a target cell of leaf-closing factor by using novel fluorescence-labeled phyllanthurinolactone. Tetrahedron Lett 46:5537–5541

Satter RL, Geballe GT et al (1974) Potassium flux and leaf movement in *Samanea saman*. I. Rhythmic movement. J Gen Physiol 64:413–430

Satter RL, Galston AW (1981) Mechanisms of control of leaf movements. Annu Rev Plant Physiol 32:83–110

Satter RL, Moran N (1998) Ionic channels in plant cell membranes. Physiol Plant 72:816–820

Satter RL, Gorton HL, Vogelmann TC (eds) (1990) The pulvinus: motor organ for leaf movement. Current topics in plant physiology, vol 3. American Society of Plant Physiologists, Rockville

Schildcknecht H (1983) Turgorins, hormones of the endogeneous daily rhythms of higher organized plants—detection, isolation, structure, synthesis, and activity. Angew Chem Int Ed Engl 22:695–710

Sheard LB, Tan X, Mao H, Withers J, Ben-Nissan G, Hinds TR, Kobayashi Y, Hsu FF, Sharon M, Browse J et al (2010) Jasmonate perception by inositol-phosphate-potentiated COI1–JAZ co-receptor. Nature 468:400–405

Shigemori H, Sakai N, Miyoshi E, Shizuri Y, Yamamura S (1989) Potassium lespedezate and potassium isolespedezate, bioactive substances concerned with the circadian rhythm in nyctinastic plants. Tetrahedron Lett 30:3991–3994

Staswick PE, Tiryaki I, Rowe ML (2002) Jasmonate response locus *JAR1* and several related Arabidopsis genes encode enzymes of the firefly luciferase superfamily that show activity on jasmonic, salicylic, and indole-3-acetic acids in an assay for adenylation. Plant Cell 14:1405–1415

Sugimoto T, Wada Y, Yamamura S, Ueda M (2001) Fluorescence study on the nyctinasty of *Cassia mimosoides* L. using novel fluorescence-labeled probe compounds. Tetrahedron 57:9817–9825

Sugimoto T, Yamamura S, Ueda M (2002) Visualization of the precise structure recognition of leaf-opening substance by using biologically inactive probe compounds: fluorescence studies of nyctinasty in legumes 2. Chem Lett 11:1118–1119

Suh S, Moran N et al (2000) Blue light activates potassium-efflux channels in flexor cells from *Samanea saman* motor organs via two mechanisms. Plant Physiol 123:833–843

Swiatek A, van Dongen W, Esmans EL, van Onckelen H (2004) Metabolic fate of jasmonates in Tobacco Bright Yellow-2 cells. Plant Physiol 135:161–172

Thines B, Katsir L, Melotto M, Niu Y, Mandaokar A, Liu G, Nomura K, He SY, Howe GA, Browse J (2007) JAZ repressor proteins are targets of the SCF^{COI1} complex during jasmonate signalling. Nature 448:661–665

Ueda M, Niwa M, Yamamura S (1995a) Trigonelline, a leaf-closing factor of the nyctinastic plant, *Aeschynomene Indica*. Phytochemistry 39:817–819

Ueda M, Shigemori-Suzuki T, Yamamura S (1995b) Phyllanthurinolactone, a leaf-closing factor of nyctinastic plant, *Phyllanthus urinaria* L. Tetrahedron Lett 36:6267–6270

Ueda M, Ohnuki T, Yamamura S (1997a) The chemical control of leaf-movement in a nyctinastic plant, *Lespedeza cuneata* G. Don. Tetrahedron Lett 38:2497–2500

Ueda M, Tashiro C, Yamamura S (1997b) cis-p-Coumaroylagmatine, the genuine leaf-opening substance of a nyctinastic plant, *Albizzia julibrissin* Durazz. Tetrahedron Lett 38:3253–3256

Ueda M, Asano M, Yamamura S (1998a) Phyllurine, leaf-opening substance of a nyctinastic plant, *Phyllanthus urinaria* L. Tetrahedron Lett 39:9731–9734

Ueda M, Ohnuki T, Yamamura S (1998b) Chemical substances controlling the leaf-movement of a nyctinastic plant, *Cassia mimosoides* L. Phytochemistry 49:633–635

Ueda M, Sawai Y, Shibazaki Y, Tashiro C, Ohnuki T, Yamamura S (1998c) Leaf-opening substance of a nyctinastic plant, *Albizzia julibrissin* Durazz. Biosci Biotechnol Biochem 62:2133–2137

Ueda M, Yamamura S (1999a) Leaf-opening substance of *Mimosa pudica* L.; chemical studies on the other leaf-movement of mimosa. Tetrahedron Lett 40:353–356

Ueda M, Yamamura S (1999b) Leaf-closing substance of *Mimosa pudica* L.; chemical studies on another leaf-movement of mimosa II. Tetrahedron Lett 40:2981–2984

Ueda M, Asano M, Sawai Y, Yamamura S (1999a) The chemistry of leaf-movement in *Mimosa pudica* L. Tetrahedron 55:5781–5792

Ueda M, Sawai Y, Yamamura S (1999b) Syntheses and novel bioactivities of artificial leaf-opening substances of *Lespedeza cuneata* G. Don, designed for the bioorganic studies of nyctinasty. Tetrahedron 55:10925–10936

Ueda M, Okazaki M, Ueda K, Yamamura S (2000a) A leaf-closing substance of *Albizzia julibrissin Durazz*. Tetrahedron 56:8101–8105

Ueda M, Shigemori H, Sata N, Yamamura S (2000b) The diversity of chemical substances controlling the nyctinastic leaf-movement in plants. Phytochemistry 53:39–44

Ueda M, Yamamura S (2000) The chemistry and biology of the plant leaf-movements. Angew Chem Int Ed 39:1400–1414

Umezawa T, Nakashima K et al (2010) Molecular basis of the core regulatory network in ABA responses: sensing, signaling and transport. Plant Cell Physiol 51:1821–1839

Vavasseur A, Raghavendra AS (2005) Guard cell metabolism and CO_2 sensing. New Phytol 165:665–682

Wasternack C (2007) Jasmonates: an update on biosynthesis, signal transduction and action in plant stress response, growth and development. Ann Bot (Lond) 100:681–697

Wasternack C, Kombrink E (2010) Jasmonates: structural requirements for lipid- derived signals active in plant stress responses and development. ACS Chem Biol 5:63–77

Ward JM, Pei ZM et al (1995) Roles of ion channels in initiation of signal transduction in higher plants. Plant Cell 7:833–844

Ward JM, Schroeder JI (1994) Calcium-activated K+ channels and calcium-induced calcium release by slow vacuolar ion channels in guard cell vacuoles implicated in the control of stomatal closure. Plant Cell 6:669–683

Yoshida R, Hobo T et al (2002) ABA-activated SnRK2 protein kinase is required for dehydration stress signaling in Arabidopsis. Plant Cell Physiol 43:1473–1483

Yoshihara T, Omer ESA, Koshino H, Sakamura S, Kikuta Y, Koda Y (1989) Structure of a tuber-inducing stimulus from potato leaves (*Solanum tuberosum* L.). Agric Biol Chem 53:2835–2837

Yu L, Becker D et al (2006) Phosphorylation of SPICK2, an AKT2 channel homologue from *Samanea* motor cells. J Exp Bot 57:3583–3594

Yu L, Moshelion M et al (2001) Extracellular protons inhibit the activity of inward-rectifying potassium channels in the motor cells of *Samanea saman* pulvini. Plant Physiol 127:1310–1322

Chapter 6
Signal Transduction in Plant–Insect Interactions: From Membrane Potential Variations to Metabolomics

Simon Atsbaha Zebelo and Massimo E. Maffei

Abstract Upon herbivore attack plants react with a cascade of signals. Early events are represented by ion flux unbalances that eventually lead to plasma transmembrane potential (Vm) variations. These events are triggered by mechanical wounding implicated by chewing/piercing herbivores along with the injection of oral secretions (OS) containing plant response effectors and elicitors. Vm depolarization has been found to be a common event when plants interact with different biotrophs, and to vary depending on type and feeding habit of the biotroph. Here we show recent advances of internal and external signal transduction in plant-insect interactions by analyzing the differential impact of mechanical and herbivore damage on plants. Vm variations, calcium signaling, and ROS production precede the late events represented by gene expression, proteomics, and metabolomics. Transcriptomics allows to decipher genomic expression following Vm variations and signaling upon herbivory; proteomics helps to understand the biological function of expressed genes, whereas metabolomics gives feedbacks on the combined action of gene expression and protein synthesis, by showing the complexity of plant responses through synthesis of direct and indirect plant defense molecules. The practical application of modern methods starting from signal transduction to metabolic responses to insect herbivory are discussed and documented.

S. A. Zebelo · M. E. Maffei (✉)
Plant Physiology Unit, Department of Life Sciences and Systems Biology,
Innovation Centre, University of Turin, Via Quarello 11/A, 10135 Turin, Italy
e-mail: massimo.maffei@unito.it

Present Address:
S. A. Zebelo
Department of Entomology and Plant Pathology, Auburn University,
301 Funchess Hall, 36349 Auburn, Alabama, USA

6.1 Introduction

Electrophysiology is a developing discipline in the field of plant physiology. It is aimed at establishing the structure of information networks that exist within the plant, which are manifested as responses to biotic and abiotic environmental stimuli by means of electrochemical signals (Maffei et al. 2004; Baluska et al. 2004; Maffei and Bossi 2006; Howe and Jander 2008; Wu and Baldwin 2010). These signals seem to complement other plant signals such as hydraulic, mechanical, and hormonal, which are already well documented in plant science (Fromm and Lautner 2007; Yan et al. 2009). Electrical signals may serve for translation of environmental parameters and cues, obtained via sensory systems, into biological information and processes. In plants, most cells are electrically excitable and active, releasing and propagating electric signals, which may affect central physiological processes such as photosynthesis and respiration (Masi et al. 2009).

A few years ago, we reviewed the existing knowledge on membrane potential variation upon herbivore attack (Maffei and Bossi 2006). We now focus on early and late events following membrane potential variations and address our discussion to some interesting key questions in plant electrophysiology.

6.2 Characteristics of Electric Signals During Insect Herbivory

Plants rely on three known types of electrical signaling during biotic and abiotic stress [e.g., action potentials (APs), variation potentials (VPs), and system potentials (SPs)] to carry out critical life functions such as respiration, photosynthesis, water and nutrient uptake, and to respond to environmental stimuli and stress. Insect herbivory is known to cause both APs and VPs, but it is still unclear whether insect herbivory can causes SPs. SPs have been recently described as a novel electrical long distance apoplastic signal in plants induced by wounding. They act as the forerunners of slower traveling chemical signals. SPs serve as back up to APs and VPs and remain overlapped with APs and VPs in some instances (Zimmermann et al. 2009). Having this brief background about SPs, it is hard to exclude the occurrences of SPs during insect herbivory even though no evidence has been reported so far. Hereunder we describe how electric signal propagated as APs and VPs during herbivory on plants.

6.2.1 Action Potentials

APs in higher plants are theorized as the information carriers of intercellular and intracellular communication in the presence of environmental stressors. The generation of APs in cells of vegetative organs of higher plants is preceded by the appearance of a gradual bioelectric response. Excitation of APs in plant cells depends on Ca^{2+}, Cl^-, and K^+ ions. This gradual response consists in plasma membrane depolarization whose properties are similar to those of the receptor potential. When plasma membrane depolarization attains a certain critical threshold level, AP is generated according to the "all or none" rule (Pyatygin et al. 2008).

Like other wounding, insect herbivory is known to cause APs, which usually appears as a single pulse; in rare cases several repeated pulses are generated during herbivory, then APs propagate along the conducting bundles of the stem beyond the area of its generation. Herbivory-induced APs changes are followed by a fast electrical signal that travels through the entire plant from the point of origin of the perceived input with a speed up to 40 m s^{-1} (Volkov and Mwesigwa 2000). After reaching leaves, roots, ovary, etc., the propagating APs induce functional responses in these organs. The importance of APs as an early recognition between the host and pathogen or herbivore is the role of signal molecules that can affect V_m either directly or via receptors (Ebel and Mithöfer 1998). In the presence of leaf-feeding larvae of the Colorado potato beetle (*Leptinotarsa decemlineata*), the speed at which insect-induced APs moved downward through the stem was about 0.05 cm s^{-1} (Volkov and Haack 1995).

There are many more examples of the involvement of APs in plant signaling. The speeds of propagation of thermally-induced APs in green plants were found to be comparable to those occurring in various mammalian species. A single application of localized heat stress induced fast APs in *Aloe vera* (67 m s^{-1}) (Volkov et al. 2007). The Venus flytrap (*Dionaea muscipula*) possesses an active trapping mechanism to capture insects with one of the most rapid movements in the plant kingdom. The cumulative character of electrical stimuli points to the existence of electrical memory in the Venus flytrap (Volkov et al. 2008). Recently the electrical properties of the *Mimosa pudica* were investigated using electrostimulation of a petiole or pulvinus by the charged capacitor method. An equivalent electrical circuit was proposed to explain the obtained data (Volkov et al. 2010).

6.2.2 Variation Potentials

VPs consist of transient changes in the plasma transmembrane (Vm) potential (depolarization and subsequent slow repolarization), where the high persistence over time represents the main difference from APs. VPs are characterized by a continuous reduction in amplitude and velocity, which decreases with the distance from the site of occurrence of the stimulus (Oyarce and Gurovich 2011). Thus, the main difference with respect to APs is that VPs show longer, delayed repolarizations and a large range of variation (Maffei and Bossi 2006). VPs that vary with the intensity of the stimulus are nonself-perpetuating and appear to be a local response to either a hydraulic pressure wave or chemicals transmitted in the xylem/phloem. In VPs the ionic mechanism differs from that underlying APs; VPs are thought to involve a transient shutdown of a P-type H^+-ATPase in the plasma membrane (Stephens et al. 2006) and some authors also claim that the level of changes caused

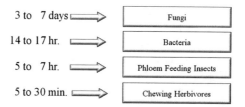

Fig. 6.1 VP variations are a common event occurring upon biotroph attacks. Here we show a typical time-response scheme obtained from V_m measurement of *A. thaliana* leaves attacked by two herbivores (*S. littoralis* and *M. persicae*) and infected by two plant pathogens (*P. syringae* and *B. cinerea*). As described in the text V_m response of the plant is proportional to the speed of damage brought by the biotroph. From Bricchi et al. unpublished results

by a stimulus via AP or VP might depend on the amount of Ca^{2+} and other ions entering or leaving the cytosol (Mcainsh et al. 2000; Shao et al. 2008).

While APs do not carry much information with respect to the nature or intensity of the triggered stimulus by the insect herbivory, VPs are modulated in amplitude as well as in their interdependent ion fluxes, from which the plant or the affected organ may be able to gain information about the nature and intensity of the injury, either brought by insect herbivory or other biotic and abiotic stresses. Within minutes from the Vm event, gene activations and subsequent metabolic changes occur.

6.3 VPs are a Common Events in Plant–Biotroph Interactions

Time-course measurements of Vm in *Arabidopsis* performed in our laboratory showed that after *Spodoptera littoralis* herbivory, a strong and rapid Vm depolarization (with respect to mechanical wounding) occurs after a few minutes from the herbivore wound, with recovery of the Vm between 5 and 6 h. When *Arabidopsis* was stressed by the aphid *Myzus persicae* almost the same extent of Vm depolarization was observed as in *S. littoralis* herbivory, but the timing of Vm depolarization peaked between 4 and 6 h, returning near to the basal V_m value between 16 and 24 h. A remarkable delay in Vm depolarization was observed when *Arabidopsis* leaves were infected by the avirulent strain of *Pseudomonas syringae*. Even in this case Vm depolarization was not significantly different from the values observed after *S. littoralis* and *M. persicae* herbivory, but the maximal Vm depolarization occurred between 16 and 18 h from inoculation. Finally, when *Arabidopsis* was infected by the fungus *Botrytis cinerea* a strong Vm depolarization occurred, but the timing of this event occurred after 3 days from infection.

These results indicate that *Arabidopsis* responds to biotic stress by inducing a strong and significant Vm depolarization and that the timing of this event depends on the rate and intensity of the wounding (Fig. 6.1) (Bricchi et al. unpublished results).

6.4 Herbivory-Induced VPs are Triggered by Calcium Ions

Electrical signals are the most important physical signals in the organisms and are capable of transmitting signals more quickly over long distances when compared with chemical signals (e.g., hormones). The electrical events that constitute signaling in the plant cell system depend on the distribution and concentration variation of ions on intercellular and extracellular membrane of the plant cell continuum. Upon herbivore attack, electric signaling starts where the leaf has been wounded and herbivore's elicitors are introduced. These events depend on ion unbalances plant Vm variations (reviewed by Maffei et al. 2007a). The candidate ion species for Vm variations in plant cells during biotic and abiotic stress are calcium (Ca^{2+}), protons (H^+), potassium (K^+), and chlorine (Cl^{1-}).

Ca^{2+} is one of the principal intracellular messengers (Arimura and Maffei 2010; Reddy et al. 2011). During electric signal generation and propagation, Ca^{2+} enters the cytoplasm through voltage-gated Ca^{2+} channels of the plasma membrane and is released from intracellular stores. The balance of Ca^{2+} concentration in cytoplasm is achieved via the presence of a large number of Ca^{2+} stores which can release Ca^{2+}, Ca^{2+}-specific channels/pumps, which regulate both Ca^{2+} influx and efflux in cells and subcellular compartments, and various Ca^{2+}-binding proteins, which bind to Ca^{2+} either to sequester it or to perform some other complex tasks (Mithöfer et al. 2009b). All these components involved in the regulation of the Ca^{2+} concentrations at its equilibrium level constitute the complex network of the Ca^{2+} homeostasis system. In the resting state, Ca^{2+} levels are high in the apoplast, mitochondria, vacuole, and ER (about 1 mM), and are very low in the cytosol (about 0.0001 mM) (Reddy et al. 2011). Within the framework of this signaling, Ca^{2+} ions bind to a regulatory protein calmodulin, thus acquiring the ability to activate a series of enzymes, e.g., protein kinases that phosphorylate various intracellular and membrane proteins and modulate their functional activity (see below).

After herbivore feeding, there is a dramatic cytosolic Ca^{2+} influx limited to a few cell layers lining the wounded zone (Maffei et al. 2004). This response is limited to herbivory or biotrophic activity, since neither single nor repeated mechanical wounding is able to induce significant changes in cytosolic Ca^{2+} ion influx (Bricchi et al. 2010). The Ca^{2+}ion influx occurs in a manner dependent on Ca^{2+} channel activity, according to studies consisting mostly of pharmacological analyses using, for instance, verapamil (a voltage-gated Ca^{2+} channel antagonist), EGTA (a Ca^{2+} chelator), and ruthenium red (an inhibitor of Ca^{2+} release from internal stores) (Maffei et al. 2006). Oral secretions (OS) from feeding insects may contain herbivore-specific compounds with elicitor-like properties able to induce Ca^{2+} variations (Mithöfer and Boland 2008).

Since ion fluxes through channels directly influence Vm, it seems reasonable to assume that molecules able to act on channel activity might be considered as important factors inducing electrical signals. A considerable efflux of K^+ from the cells where electric signaling is generated and propagated under injuring treatments might have a protective role related to modulation of cell metabolism, enzyme activities, turgor, etc. (Pyatygin et al. 2008). By considering all these circumstances, it is possible to assume that the Ca^{2+} variations in the cytoplasm and the efflux of K^+ during cell excitation may be of key significance for the formation of electric signaling effectors response.

6.4.1 Herbivory Versus Mechanical Wounding

Mechanical wounding of plant tissues is an inevitable consequence caused by insect herbivory. However, abiotic and biotic wounds are perceived distinctively by the plant, as it has been demonstrated in different plant-herbivore interactions (Maffei et al. 2007a; Mithöfer et al. 2009a; Bricchi et al. 2010). Insect wounding and chemical interactions (regurgitates, OS, and factors) play a vital role in recognition of the type of the damage by the plant (VanDoorn et al. 2010; Wu and Baldwin 2010; Bonaventure et al. 2011).

Two phenomena explain how plants discriminate insect herbivory from mechanical damage. The first is plant recognition of compounds present in insect OS. This view is supported by the identification of several insect-derived factors that elicit defense responses when applied to artificial wounds (Bonaventure et al. 2011). Plants may also differentiate mechanical wounding from herbivory through the use of as yet unknown mechanisms that gauges the quantity and quality of tissue damage. Control experiments have been performed using either manual punchers or robotic mechanical wound makers (Mecworm) mimicking damage which resemble herbivore wounding (Mithöfer et al. 2005; Bricchi et al. 2010). Caterpillar feeding involves the action of specialized mandibles that remove similarly sized pieces of leaf tissue in a highly choreographed and predictable manner, an action that is difficult to mimic with robotic devices. However, recent findings by Bricchi and co-workers offered a clear picture of plant-insect interaction, by dissecting events occurring during herbivory (HW), single (MD), and robotic (MW) mechanical damage (Bricchi et al. 2010). The results show that within 24 h after MD on Lima bean (*Phaseolus lunatus*) leaves, no significant Vm variations were observed, whereas a small but significant Vm depolarization occurred after 30 h. In contrast, HW caused by the Egyptian armyworm *S. littoralis* prompted a sudden and substantial Vm depolarization, reaching maximal values as early as 30 min after wounding. Thereafter, Vm depolarization decreased and returned to a steady value after 30 h, which was still significantly higher than the values measured after MD. MW caused a significantly stronger Vm depolarization over the entire measurement period, when compared to MD. However, MW Vm values remained significantly lower than those observed after HW. In order to assess the contribution of OS of the herbivorous larvae of *S. littoralis* on Vm depolarization, different dilutions of OS were applied on MW and MD leaves resulting in a strong Vm depolarization, with Vm values comparable to those observed after HW. No differences were observed between MD and MW responses to OS (Bricchi et al. 2010).

These results clearly show the correlation between Vm depolarization and the presence of OS, possibly containing effectors/elicitors able to alter the ion balance across the plasma membrane. Among the early events able to trigger such a significant Vm depolarization are variations in the cytosolic concentration of calcium ions [Ca^{2+}]$_{cyt}$ (Maffei et al. 2004, 2007a). Among the several techniques used to reveal calcium variation, confocal laser scanning microscopy is the preferred one (Mithöfer et al. 2009b). A significant increase in [Ca^{2+}]$_{cyt}$ was observed only after HW, whereas MD and MW did not affect this second messenger. Another signaling molecule is H$_2$O$_2$ which is generated within 2–3 h after leaf damage by HW and MW, whereas MD induced only half of the H$_2$O$_2$ levels compared to the other treatments. Both HW and MW also induced a marked accumulation of NO, but with distinct temporal patterns. NO production after MD followed the same trend but reached significantly lower values (Bricchi et al. 2010). The results confirm that chemical signals from the herbivores are responsible for the induction of the earliest signaling events and that these changes appear to be characteristic for the reaction to herbivory.

The experiments performed with repetitive mechanical wounding of lima bean leaves were also instrumental to understand what elicits secondary metabolite induction, such as the emission of volatile organic compounds. MW caused a pattern of volatile emission that was qualitatively similar to that induced by caterpillar attack, but quantitatively different (Arimura et al. 2000; Mithöfer et al. 2005; Bricchi et al. 2010). Altogether these findings clearly demonstrate that herbivore-like wounding is necessary but not sufficient, and that the mutual chemical interaction with folivory is required to trigger the full response. Thus, plants in order to avoid wasting defensive resources must differentiate insect (biotic) feeding from (abiotic) mechanical damage.

Researchers have also developed novel approaches to disentangle the effects of MD from those of OS. One approach in studying the role of insect saliva in plant-lepidopteran interactions is the obliteration of labial salivary glands. When such method was applied to *Helicoverpa zea* (corn ear worm) it provided compelling evidence that salivary secretions qualitatively affect plant defense responses to caterpillar feeding (Bede et al. 2006; Musser et al. 2006). This finding points to OS as the first biochemical line of offense of herbivore feeding.

6.5 Role of Herbivore's OS and Their Elicitors on Early Electric Signaling

It is important to stress the difference between OS (or regurgitate) and elicitors/effectors. OS is intended as the fluid that is spitted or vomited by the herbivore mouth during either feeding or as a consequence of a mechanic or biotic stress. OS contain several components, but not all of them are able to elicit a plant response. On the other hand, herbivore-associated elicitors (HAEs) act during insect herbivory and are diverse in structure, ranging from enzymes (e.g., glucose oxidase, beta glucosidase) to modified forms of lipids [e.g., fatty acid–amino acid conjugates (FACs), sulfur-containing fatty acids (caeliferins)] and from fragments of cell walls (e.g., pectins and oligogalacturonides) to peptides released from digested plant proteins (e.g., inceptins: proteolytic fragments of the chloroplastic ATP synthase gamma subunit) as recently reviewed (Bonaventure et al. 2011).

6.5.1 Herbivore-Associated Elicitors

Recently an interesting question has been posed by Tumlinson and Mori (Yoshinaga et al. 2008). Why lepidopteran larvae produce these potent elicitors? By using ^{14}C-labeled glutamine, glutamic acid, and linolenic acid in feeding studies of *Spodoptera litura* larvae, combined with tissue analyses, they found glutamine in the midgut cells to be a major source for biosynthesis of FACs. The role of FACs like volicitin as elicitors of induced responses in plants has been well documented (Alborn et al. 1997), but FACs were also found to play an active role in nitrogen assimilation in *S. litura*. In fact, glutamine-containing FACs in the gut lumen may function as a form of storage of glutamine, a key compound of nitrogen metabolism (Yoshinaga et al. 2008).

An additional level of complexity associated with the perception of FACs was revealed by the analysis of the metabolism of *N*-linolenoyl glutamic acid (18:3-Glu) on wounded *Nicotiana attenuata* leaf surfaces. A large fraction (~50–70%) of this abundant FAC in the *Manduca sexta* OS was metabolized within a few seconds upon contact with wounded leaf tissue. Linolenyl-Glu (LeaGln) was converted into an array of more and less polar derivatives and some of these forms were shown to be active elicitors and to induce defense responses. The array of derivatives detected opened the possibility that different modified FAC forms are responsible for eliciting distinct *N. attenuata* responses (VanDoorn et al. 2010).

It is interesting to note that most of the elicitors are not general elicitors of responses against insect herbivores in all plant species, rather they are usually restricted to particular plant-insect associations. For example, neither in Lima bean (*P. lunatus*) nor in cotton (*G. hirsutum*) induction of indirect defenses could be demonstrated after volicitin was added to mechanically damaged plant tissue (Spiteller et al. 2001; Maffei et al. 2004), even though some of these elicitors exhibited defense-related activities when added to certain mechanical wound of their associated plants (Felton and Korth 2000; Schmelz et al. 2006).

In addition to elicitors from OS, egg deposition and tarsal contact might also elicit plant responses (Hilker et al. 2005; Hilker and Meiners 2006). Induction of plant defensive responses by insect egg deposition is caused by the egg or egg-associated components of several insects, although the responsible chemistry has been identified only in bruchid beetles: long-chain α, γ-monounsaturated C_{22} diols and α, γ-mono- and di-unsaturated C_{24} diols, mono- or diesterified with 3-hydroxypropanoic acid (Kopke et al. 2010; Hilker and Meiners 2010). Egg deposition by the phytophagous sawfly *Diprion pini* L. (Hymenoptera, Diprionidae) is known to induce locally and systemically the emission of volatiles in Scots pine (*Pinus sylvestris* L.)

that attracts the egg parasitoid *Chrysonotomyia ruforum* Krausse (Hymenoptera, Eulophidae). The elicitor inducing the pine's response is known to be located in the oviduct secretion which the female sawfly applies to the eggs when inserting them into a slit in the pine needle using the sclerotized ovipositor valves. The elicitor in the oviduct secretion is only active when transferred to slit pine needles, since its application on undamaged needles did not induce the emission of attractive volatiles (Hilker et al. 2005). Similarly, it is possible that there are potent elicitors released by herbivorous arthropods during tarsal contact with a plant but they have not so far been found (Hilker and Meiners 2010).

Most of the studies on HAE have been done on chewing insects. In contrast to chewing insects, however, little is known about oral elicitors from sucking arthropods (spider mites and aphids). It has recently been proposed that the release of aphid elicitors (e.g., oligogalacturonides) due to cell wall digestion by gel saliva enzymes may induce Ca^{2+} influx (Torsten and van Bel 2008). Aphids, similar to plant pathogens, deliver effectors inside their hosts to manipulate host cell process enabling successful infestation of plants (Arimura et al. 2011). Recently, a functional genomics approach for the identification of candidate aphid effector proteins from the aphid species *M. persicae* (green peach aphid) based on common features of plant pathogen effectors has been developed (Bos et al. 2010). Data mining of salivary gland expressed sequence tags (ESTs) made it possible to identify 46 putative secreted proteins from *M. persicae*. Functional analyses of these proteins showed that, among them, Mp10 induced chlorosis and weakly induced cell death in *Nicotiana benthamiana*, and suppressed the oxidative burst induced by the bacterial PAMP flagellin 22 (flg22). In addition, using a medium throughput assay based on transient overexpression in *N. benthamiana*, two candidate effectors (Mp10 and Mp42) have been identified as reducing aphid performance, whereas MpC002 enhanced aphid performance (Bos et al. 2010). Overall, aphid-secreted salivary proteins share features with plant pathogen effectors and therefore may function as aphid effectors by perturbing host cellular processes (Arimura et al. 2011).

6.5.2 Alamethicin, HAE, and OS Exhibit Ion Channel Forming Activities

The effects of HAE and OS have been characterized by using an electrophysiological approach called planar (black) lipid bilayer technique. Black lipid bilayer membranes (BLM) are widely used to elucidate the molecular mechanisms of activity of various biologically active substances (Winterhalter 2000). On artificial BLM, the existence of ion channel-like facilitators has recently been demonstrated. Using this BLM technique, Boland and co-workers demonstrated that insect-derived elicitors and OS may directly interact with artificial lipid bilayers, generating channel-like activities that are highly conductive and selective for certain ions (Maischak et al. 2007). The addition of fresh samples of *Spodoptera exigua*-derived OS (1 μl/ml) to the *cis*-compartment of BLM generated a channel-like activity when a constant voltage (25 mV) was applied across the membrane. OS initiated cation flux from the *cis*- to the *trans*-compartment or the flow of anions in the opposite direction. The calculated conductance obtained was about 530 ± 65 pS. The dominating current was characterized by discrete opening steps of long lasting dwell time and various amplitudes. Strikingly, multiple conductance steps were detected that indicated the simultaneous as well as consecutive opening of more than one single channel. Interestingly, the effects of the OS component LeaGln on BLM currents were completely different. Long lasting open states were superimposed by rapid flickers, and opening and closing events (Maischak et al. 2007). Such types of currents suggest the presence of disturbed membrane structures caused by the added compound. This strongly supports the interpretation of data on intact lima bean plant cells, where the effects of LeaGln and derivatives have been attributed to their amphiphilic character and, hence, detergent-like properties, which could cause phase transitions in the membrane (Maffei et al. 2004).

Formation of channel-like pores in a plant membrane was induced within seconds after application of an aqueous solution containing regurgitate of the insect larvae *S. littoralis* (Lühring et al. 2007). Also elicitors with different functions like cellulolytic enzymes employed in cell wall digestion that had previously been presumed to act on plant cells via receptor binding sites proved to directly interact with membranes by formation of channel-like membrane pores (Klusener and Weiler 1999). Alamethicin is well known to produce pores in artificial membranes as well as in animal and plant cell membranes (Duclohier et al. 2003). To investigate the ability of alamethicin to induce pore formation in a plant membrane of *Chara*, the molecule was applied to *Chara* tonoplast at a concentration of 11 µM as a component of the patch pipette filling solution. Alamethicin was shown to form conductive pores depending on the polarity of the applied voltage in plasma membranes (Lühring et al. 2007).

6.6 Electric Signals Trigger Cascade of Events Leading to Gene Expression

Before gene expression, four steps characterize the transfer of signals between/ within plants: the release of the signal by the emitter plant and its transport, absorption, and perception by the receiver plant. Furthermore, two general types of signaling can be distinguished: within plants (internal signaling) and between plants (external signaling).

Internal signaling requires that signal transduction pathways transfer the perception of the signaling molecules to the nuclear genomic machinery by a comprehensive network of interacting pathways downstream of sensors/receptors (Bos et al. 2010). Plants have evolved a large array of interconnected cell signaling cascades, resulting in local resistance and long-distance signaling. Such responses initiate with the recognition of physical and chemical signals of the attacking biotrophs, activation of subsequent signal transduction cascades, and finally activation of genes involved in defense responses that consequently enhance feedback signaling and metabolic pathways (Arimura et al. 2005). Physicochemical processes, including interactions with odorant binding proteins and resulting in changes in Vms, can underlie signaling processes (Maffei and Bossi 2006; Maffei et al. 2007a; Heil et al. 2008).

As we discussed earlier, plant responses to herbivory seem to reflect an integrative "cross-talk" between Ca^{2+}-ions, reactive oxygen species (ROS), jasmonic acid (JA) and JA-conjugates, 12-oxophytodienoic acid (OPDA), salicylic acid (SA), ethylene, and still unknown members of the octadecanoid family (Mithöfer et al. 2009a). In distinct signaling processes, phytohormones play an important role in the transduction of signals. In internal signaling, four major phytohormones, SA (Vlot et al. 2008), JA (Memelink 2009; Diezel et al. 2011), ABA (Bodenhausen and Reymond 2007), and ethylene (Adie et al. 2007; Onkokesung et al. 2010) play important roles in the plant defense.

In *external signaling*, molecules released into the atmosphere are perceived by neighboring plants by still unknown mechanisms, even though Vm variations appear to be the first line of perception. This perception involves activation of separate defense genes in the receiver plants that are not activated when these plants are exposed to volatiles from artificially wounded leaves. Several VOCs are responsible for this gene activation and the expression of these genes requires calcium influx, protein phosphorylation/dephosphorylation, and involves the presence of ROS (Maffei et al. 2006, 2007b). In general, for a molecule to function as a plant–plant airborne signal in a natural setting, certain criteria have to be met. First, if the signal is constitutively released, independent of damage, then the signal from damaged plants must be released in significantly greater quantities for the receiver to distinguish it from the background signal. Alternatively, qualitative, rather than quantitative, changes in the signal could provide the information. Second, the signal must be received, and not only emitted, at physiologically active levels. The possible dilution that occurs represents the most onerous challenge for a potential plant–plant signal. Clearly, the greater the distance, over which a signal is to function, the greater the released amounts must be, even though higher sensitivity for specific molecules may play an important role. Once these criteria are achieved, a compound may be considered as a potential airborne signal (Preston et al. 2001). Recently, several plant species have been shown to respond to plant volatiles (Maffei 2010; for recent reviews on plant volatiles and their involvement on plant–insect interaction refer to: Baldwin 2010; Maffei et al. 2011).

Herbivore and mechanical wounding have been often profiled and compared, and the application of defense regulators has been assessed (Bricchi et al. 2010). A combination of approaches was employed to develop transcript profiles, including suppression subtractive hybridization (SSH), macroarray, northern blot, and cluster analysis. Comparative macroarray analyses revealed that most of the insect-induced transcripts were methyl jasmonate (MeJA) and ethylene regulated. The effects of mild insect infestation and the exogenous application of signaling compounds on larval feeding behavior were also monitored and bioassays were performed to measure dispersal percentage and growth of larvae on elicited plants. Larvae released on elicited plants had decreased larval performance, demonstrating the central role of induced plant defense against herbivory. Similarly, wounding and exogenous application of MeJA and ethylene also affected larval growth and feeding behavior. These results demonstrated that insect attack may upregulate large transcriptional changes and induce plant defense responses (Singh et al. 2008).

An interesting aspect in plant-to-plant communication is the ability of plants to deter other plants. Allelopathy studies the interactions among plants, fungi, algae, and bacteria with the organisms living in a certain ecosystem, interactions that are mediated by the secondary metabolites produced and exuded into the environment (Macias et al. 2007). Many allelochemicals are released by plants into the surrounding environment. When roots are directly exposed to these compounds, plants produce altered roots. Interestingly, in some cases these roots from seedlings exposed to some molecules [i.e., (+)-bornyl acetate] may be significantly longer than controls. The relationship between root length and the concentrations of some compounds indicates that plants can specifically respond to the molecular configurations (Horiuchi et al. 2007).

Thus, plants have evolved sophisticated systems to cope with herbivore challenges. When plants perceive herbivore-derived physical and chemical cues, such as elicitors in insects' OS and compounds in oviposition fluids, plants dramatically reshape their transcriptomes, proteomes, and metabolomes (Wu and Baldwin 2010). Moreover, insect herbivory leads to induced resistance to subsequent infestations in plants. This is due in part to feeding-induced expression of genes that can lead to reduced palatability and/or digestibility of the plant material. Based on structural similarities the known resistance genes (especially in plant pathogen interactions) have been classified in five classes: (1) a class of cytoplasmic receptor-like proteins with a nucleotide binding site and a leucine-rich repeat (NBS-LRR), (2) Cf-X that are putative transmembrane receptors carrying extracellular LRR motifs, (3) Xa21 or FLS2 that possess an intracellular serine/threonine kinase domain in addition to an extracellular LRR motif, (4) Pto that are encoding for a cytoplasmic serine/threonine kinase, and (5) RPW8 that carries a signal anchor at the N-terminus and a coiled-coil motif (Dangl and Jones 2001).

Changes in gene expression underlying inducible responses to herbivory are known to be complex and multifaceted, and studies of responses to herbivory and mechanical wounding suggest a similar pattern of multiple independent, but networked defense response pathways (Mithöfer et al. 2009a). More specifically, changes in gene expression after herbivory do not always reflect that of hormone or wounding treatment and, in addition, herbivores with different feeding styles induce both unique and overlapping changes (Maffei et al. 2011). Distinct differentially expressed genes from plants that were either infested by insects or mechanically wounded have been identified and time-course analysis revealed diverse timing of peak transcript accumulation. Some highly insect-inducible genes were also found to be wound inducible (Lawrence et al. 2006).

Among the several genes involved in plant-insect interactions, mitogen-activated protein kinase (MAPK) cascades are important pathways downstream of sensors/receptors that regulate cellular responses. There is evidence of MAPKs playing a role in the signaling of (a) biotic stresses, pathogens, and plant hormones (Maffei et al. 2007a). In *N. attenuata*, activation of MAPK cascade results in rapid phosphorylation of two critical wound- and SA-induced protein kinases, WIPK and SIPK, respectively (Wu and Baldwin 2010). As we discussed earlier, when insects attack their host plant, the plant's wound response is reconfigured at transcriptional, phytohormonal, and defensive levels due to the introduction of OS into wounds during feeding. OS-containing FACs dramatically amplify wound-induced MAPK. Recent findings show that after applying OS to wounds created in one portion of a leaf, SIPK is activated in both wounded and specific unwounded regions of the leaf but not in phylotactically connected adjacent leaves. This fact led to the interesting hypothesis that some herbivores (e.g., *M. sexta*) attack elicits a mobile signal that travels to nonwounded regions of the attacked leaf where it activates MAPK signaling and, thus, downstream responses; subsequently, a different signal is transported by the vascular system to systemic leaves to initiate defense responses without activating MAPKs in systemic leaves (Wu et al. 2007). Our results on Vm long-distance signaling-dependent depolarization fully support this hypothesis (Atsbaha et al. unpublished data).

Calcium-dependent protein kinases (CDPKs) are regularly involved in signal transduction of a variety of biotic and abiotic stresses, their involvement as active protein cascades in herbivore/wound responses has been documented (Arimura and Maffei 2010; Arimura et al. 2011; Reddy et al. 2011). To investigate the roles CPKs play in a herbivore response-signaling pathway, the characteristics of *Arabidopsis* CPK mutants damaged by a feeding generalist herbivore, *S. littoralis*, have been screened. Following insect attack, the *cpk*3 and *cpk*13 mutants showed lower transcript levels of plant defensin gene *PDF*1.2 compared to wild-type plants. The CPK cascade was not directly linked to the herbivory-induced signaling pathways that were mediated by defense-related phytohormones such as JA and ethylene. CPK3 was also suggested to be involved in a negative feedback regulation of the cytosolic Ca^{2+} levels after herbivory and wounding damage. In vitro kinase assays of CPK3 protein with a suite of substrates demonstrated that the protein phosphorylates transcription factors (TFs) (including ERF1, HsfB2a, and CZF1/ZFAR1) in the presence of Ca^{2+}. CPK13 strongly phosphorylated only HsfB2a, irrespective of the presence of Ca^{2+}. Furthermore, in vivo agroinfiltration assays showed that CPK3-or CPK13-derived phosphorylation of a heat shock factor (HsfB2a) promotes *PDF*1.2 transcriptional activation in the defense response. Thus, in *Arabidopsis* two CPKs (CPK3 and CPK13) are involved in the herbivory-induced signaling network via HsfB2a-mediated regulation of the defense-related transcriptional machinery. However, this cascade is not involved in the phytohormone-related signaling pathways, but rather directly impacts TFs for defense responses (Kanchiswamy et al. 2010).

JA was shown to be a major signal controlling the upregulation of defense genes in response to insect feeding. When *Arabidopsis thaliana* mutants coil-1, ein2-1, and sid2-1 impaired in JA, ethylene, and SA signaling pathways were challenged with the specialist small cabbage white (*Pieris rapae*) and the generalist Egyptian cotton worm (*S. littoralis*), larval growth was affected by the JA-dependent defenses, but *S. littoralis* gained much more weight on coil-1 than *P. rapae*. Moreover, ethylene and SA mutants had an altered transcript profile after *S. littoralis* herbivory, but not after *P. rapae* herbivory. In contrast, both insects yielded similar transcript signatures in the abscisic acid (ABA)-biosynthetic mutants aba2-1 and aba3-1, and ABA controlled transcript levels both negatively and positively in insect-attacked plants. This study revealed a new role for ABA in defense against insects in *Arabidopsis* and identified some components important for plant resistance to herbivory (Bodenhausen and Reymond 2007). In other studies, the time course and the amount of induced JA and OPDA levels in corn seedlings were strikingly different after wounding, application of caterpillar regurgitate, or treatment with the green leaf volatile *cis*-3-hexenyl acetate (Z-3-6:AC). Exposure to Z-3-6:AC induced accumulation of transcripts encoded by three putative 12-oxophytodienoate 10,11-reductase genes (ZmOPR1/2, ZmOPR5, and ZmOPR8). Although changes in ZmOPR5 RNAs were detected only after exposure to Z-3-6:AC, ZmOPR1/2 RNAs, and ZmOPR8 RNAs also were abundant after treatment with crude regurgitant elicitor or mechanical damage (Engelberth et al. 2007). Transgenic plant lines silenced in signaling and foliar defense traits have been evaluated in a field study for resistance against attack by naturally occurring herbivores. *N. attenuata* plants silenced in early JA biosynthesis, JA perception, proteinase inhibitors, and nicotine direct defenses and lignin biosynthesis were infested more frequently than wild-type plants. This implies that that influence a plant's apparency, stem hardness, and pith direct defenses all contribute to resistance against herbivores (Diezel et al. 2011).

Cytochrome P450 monooxygenases (P450s) are integral in defining the relationships between plants and insects. Secondary metabolites produced in plants for protection against insects and other organisms are synthesized via pathways that include P450s in many different families and subfamilies. Survival of insects in the presence of toxic secondary metabolites depends on their metabolism by more limited groups of P450s (Schuler 2011). Inducible plant defenses against herbivores are controlled by a transient burst of JA and its conversion to the active hormone (3R,7S)-jasmonoyl-L-isoleucine, JA-Ile (Woldemariam et al. 2011). JA-Ile signals through the COI1-JAZ co-receptor complex to control key aspects of plant growth, development, and immune function. Despite detailed knowledge of the JA-Ile biosynthetic pathway, little is known about the genetic basis of JA-Ile catabolism and inactivation. Recently, the identification of a wound-and jasmonate-responsive gene from *Arabidopsis* that encodes a cytochrome P450 (CYP94B3) involved in JA-Ile turnover has been reported. CYP94B3-overexpressing plants displayed phenotypes indicative of JA-Ile deficiency, including defects in male fertility, resistance to JA-induced growth inhibition, and susceptibility to insect attack. Increased accumulation of JA-Ile in wounded CYP94B3 leaves was associated with enhanced expression of JA-responsive genes. These results demonstrate that CYP94B3 exerts negative feedback control on JA-Ile levels and performs a key role in attenuation of jasmonate responses (Koo et al. 2011).

The discussion above points to the fact that JAs trigger an important transcriptional reprogramming of plant cells to modulate both basal development and stress responses. In spite of the importance of transcriptional regulation, only one TF, the *A. thaliana* basic helix-loop-helix MYC2, has been described so far as a direct target of JAZ repressors. Recently, by means of yeast two hybrid screening and tandem affinity purification strategies, two previously unknown targets of JAZ repressors, the TFs MYC3 and MYC4, phylogenetically closely related to MYC2 have been identified. MYC3 and MYC4 interact in vitro and in vivo with JAZ repressors and also form homo-and heterodimers with MYC2 and among themselves. They both are nuclear proteins that bind DNA with sequence specificity similar to that of MYC2. Loss-of-function mutations in any of these two TFs impair full responsiveness to JA and enhance the JA insensitivity of myc2 mutants. Moreover, the triple mutant myc2, myc3, myc4 is as impaired as coi1-1 in the activation of several, but not all, JA-mediated responses such as the defense against bacterial pathogens and insect herbivory. MYC3 and MYC4 are activators of JA-regulated programs that act additively with MYC2 to regulate specifically different subsets of the JA-dependent transcriptional response (Fernandez-Calvo et al. 2011).

High-throughput quantitative real-time PCR technology can be used to analyze the response of susceptible and resistant plants to insect infestation. Many other TF genes belonging to several TF families are responsive to insect infestation. In some cases, more TF genes were responsive in the resistant interaction than in the susceptible interaction. Recent results suggest that some TFs are among the early insect-responsive genes in resistant plants, and may play important roles in resistance to multiple aphid species (Gao et al. 2010). Since acquisition of stress tolerance requires orchestration of a multitude of biochemical and physiological changes, research during the last 2 decades has been focused to evaluate whether different stresses may cause signal-specific changes in cellular messenger levels. The modulation of diverse physiological processes is important for stress adaptation and many Ca^{2+} and Ca^{2+}/calmodulin (CaM)-binding TFs have been identified in plants (Reddy et al. 2011). Functional analyses of some of these TFs indicate that they play key roles in stress signaling pathways. Recent progress in this area with emphasis on the roles of Ca^{2+} and Ca^{2+}/CaM-regulated transcription in stress responses have been reviewed (Reddy et al. 2011).

Serial analysis of gene expression (SAGE) is a technique that allows for the absolute quantification of mRNA abundance by quantifying the relative frequencies of individual short (13 nt) transcripts signatures tags (Velculescu et al. 1995). An upgraded development of the technique (SuperSAGE) doubled the generation of nucleotide tags which substantially improved the annotation of tags when aligned to sequences in public nucleotide databases (Matsumura et al. 2005). Combining these techniques with next-generation sequencing (NGS) allows for the detection and analysis of very low abundant transcripts. By using SuperSAGE in combination with NGS for the quantification of the early changes (within 30 min) occurring in the transcriptome of plants after a single event of elicitation, genes encoding for potential regulatory components of the elicitor-mediated responses have been detected by looking for low abundant transcripts that were rapidly and

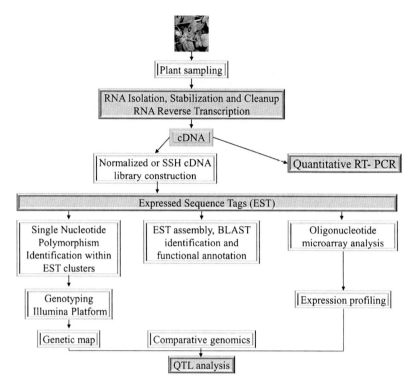

Fig. 6.2 A typical workflow for genomic studies. See description in the text. Suppression subtractive hybridization (*SSH*)

transiently induced after elicitation. For example, the analysis of the early changes in the transcriptome of *N. attenuata* after FAC elicitation using SuperSAGE/NGS has identified regulatory genes involved in insect-specific mediated responses in plants. Moreover, it has provided a foundation for the identification of additional novel regulators associated with this process (Gilardoni et al. 2010).

The coordination of defense gene transcription is often coupled with significant adjustments in the levels of expression of primary metabolic and structural genes to relocate resources, repair damage, and/or induce senescence. This complicates the process of finding suitable 'housekeeping' or reference genes to use in measurements of gene expression by real-time reverse transcription (RT-PCR) in response to herbivore attack. Recently it has been proposed a method using RNA quantification in combination with an external spike of commercially available mRNA as normalization factors in studies involving herbivory, multiple stress treatments or species where stable reference genes are unknown (Rehrig et al. 2011).

The seeking of developing a genetic model system to analyze insect-plant interactions still remains one of the main challenges (Whiteman et al. 2011). In order to give a general view of procedures used in gene expression analyses, Fig. 6.2 shows a typical workflow for genomic studies.

6.7 Proteomic Responses to Herbivory

While mRNA is a message, the actual biological function of a gene, catalysis of reactions, maintenance of structural and chemical homeostasis, and detection of the environment, is done by proteins (Sergeant and Renaut 2010). The number of proteins in a biological system by far exceeds the number of coding sequences making the identification and sequence determination of proteins a task more difficult than the sequencing of polynucleotides. Analyses of protein expression patterns are useful in integrating physiology and growth or reproductive effects of stressed host plant and the changes in the transcriptome described above translate into changes in the proteome and metabolome, that eventually account for the phenotype of increased resistance. However, it is becoming increasingly clear that proteomic changes cannot be directly predicted from changes in the transcriptome. Furthermore, in a given cellular microenvironment, both proteins and transcripts interact with other molecules in specific ways, and these interactions determine the regulation, expression, activity, and stability of specific mRNA and protein molecules (Giri et al. 2006).

Differential analysis of protein expressions on the interaction between insect herbivores and plants is usually studied via 2D electrophoresis combined with high-throughput mass spectrometry (MS) analyses. On the basis of the abundance of expression proteins in the plant fed by the insect, the intensities of significantly changed spots in protein spots are detected (see Fig. 6.3).

The understanding of the role of plant proteins in defense against herbivores lags behind that of proteins involved in defense against pathogens. However, recent microarray and proteomic approaches have revealed that a broader array of proteins may be involved with defense against herbivores than previously appreciated. Because arthropods possess a diverse range of feeding habits and styles, including chewing as well as phloem- or xylem-feeding species, arthropod-inducible proteins (AIPs) may be regulated by multiple signaling hormones, including phytohormones (Kant and Baldwin 2007; Zhu-Salzman et al. 2008). A typical workflow for proteomic studies is depicted in Fig. 6.3.

Although the proteomic changes elicited by herbivore attack are still poorly known, a growing body of evidence is emerging from the literature.

In a recent study, *S. exigua* caterpillar-specific post-translational modification of *A. thaliana* soluble leaf proteins was investigated by LC-ESI-MS/MS. It has to be noted that caterpillar labial saliva contains numerous oxidoreductases and effectors that are secreted onto plant tissues during feeding and believed to be responsible for undermining plant-induced defenses. One of these proteins is glucose oxidase, an enzyme producing H_2O_2 that can lead to activation of signaling pathways through cysteine oxidation (Bede et al. 2006). Among proteins modified in a caterpillar-specific manner, some were associated with photosynthesis. Oxidative modifications, such as caterpillar-specific denitrosylation of Rubisco activase and chaperonin, cysteine oxidation of Rubisco, DNA-repair enzyme, and chaperonin and caterpillar-specific 4-oxo-2-nonenal modification of the DNA-repair enzyme were also

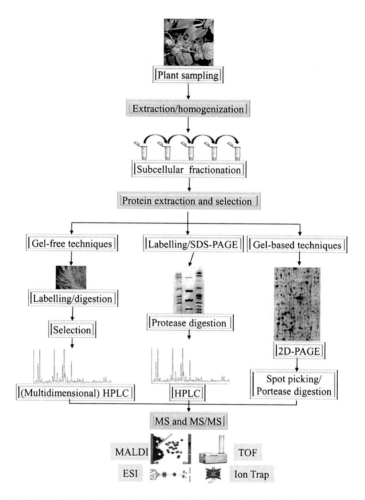

Fig. 6.3 A typical workflow for proteomic studies. See description in the text

observed (Thivierge et al. 2010). Furthermore, by using 2D electrophoresis of differentially expressed proteins, combined with MALDI-TOF MS and MALDI-TOF/TOF MS, several other *S. exigua*-feeding-responsive proteins were identified, all of which were involved in metabolic regulation, binding functions or cofactor requirement of protein, cell rescue, and defense and virulence. About 50% of these were involved in metabolism, including transketolase, *S*-adenosylmethionine synthase 3, 2,3-biphosphoglycerate-independent phosphoglycerate mutase, beta-ureidopropionase, GDP-D-mannose 3′,5′-epimerase, and fatty acid synthase (Zhang et al. 2010).

When *Arabidopsis* leaves were attacked by specialist *Plutella xylostella*, among the responsive proteins some were associated with the Calvin-Benson cycle in the chloroplast and TCA cycle in the mitochondria, confirming that carbon metabolism-related proteins may play crucial roles in induced defense response in plants under insect infestation. About 50% of proteins were in the chloroplast, which shows the chloroplast has a key role in the insect feeding response for plant (Liu et al. 2010). Furthermore, when physiological factors affecting feeding behavior by *P. xylostella* were studied on herbivore-susceptible and herbivore-resistant *Arabidopsis*, the proteomic results, together with detection of increased production of hydrogen peroxide in resistant recombinant plants, showed a correlation between *P. xylostella* resistance and the production of increased levels of ROS, in particular H_2O_2, and this was expressed prior to herbivory. Consistent with the occurrence of greater oxidative stress in the resistant recombinant plants is the observation of greater abundance in susceptible recombinant plants of polypeptides of the photosynthetic oxygen-evolving complex, which are known to be damaged under oxidative stress. Thus, enhanced production of ROS may be a major pre-existing mechanism of *P. xylostella* resistance in *Arabidopsis*, but definitive corroboration of this requires much further work (Collins et al. 2010).

The protein SGT1 (Suppressor of G-2 allele of SKP1) is required for immune responses to pathogens in humans and plants and it is thought to confer at least partial resistance to pathogens through its interaction with heat shock protein HSP90 (Shirasu 2009). In plants, SGT1 and HSP90 mediate the stability of NB-LRR type R proteins (Boter et al. 2007). The roles of SGT1 in induced resistance to leaf-chewing herbivores have been recently discovered in the interaction between *N. attenuata* and *M. sexta* (Meldau et al. 2011). By using virus-induced gene silencing (VIGS) to knock down the transcript levels of NaSGT1 the role of NaSGT1 in modulating herbivory-induced responses in *N. attenuata* was analyzed. NaSGT1-silenced plants show decreased amounts of defensive metabolites after herbivory, and *M. sexta* larvae gain more weight on these plants than on empty vector plants. Furthermore, it was shown that NaSGT1 is required for the normal regulation of MeJA-induced transcriptional responses in *N. attenuata* (Meldau et al. 2011).

Differentially expressed proteins were identified in *Citrus clementina* leaves after infestation by the two-spotted spider mite *Tetranychus urticae*. Significant variations were observed for several protein spots after spider mite infestation as well as after MeJA treatments. While the majority constituted photosynthesis- and metabolism-related proteins, some others were oxidative stress-associated enzymes, including phospholipid glutathione peroxidase, a salt stress-associated protein, ascorbate peroxidase, Mn-superoxide dismutase, and defense-related proteins, such as the pathogenesis-related acidic chitinase, the protease inhibitor miraculin-like protein, and a lectin-like protein (Maserti et al. 2011). In another study, a proteomic approach was applied to gain insight into the physiological processes, including glucosinolate metabolism, in response to MeJA in *Arabidopsis*. Functional classification analysis of differentially expressed proteins showed that photosynthesis and carbohydrate anabolism was repressed after MeJA treatment, while carbohydrate catabolism was upregulated. Additionally, proteins related to the JA biosynthesis pathway, stress and defense, and secondary metabolism were upregulated. Among the differentially expressed proteins, many were involved in oxidative tolerance. Thus, MeJA elicits a defense response at the proteome level through a mechanism of redirecting growth-related metabolism to defense-related metabolism (Chen et al. 2011).

To access phloem sap, aphids have developed a furtive strategy, their stylets progressing toward sieve tubes mainly through the apoplasmic compartment. Aphid feeding requires that they overcome a number of plant responses, ranging from sieve tube occlusion and activation of phytohormone-signaling pathways to expression of anti-insect molecules (Giordanengo et al. 2010). By investigating the responses of wheat to infestation by aphids (*Sitobion avenae*), the majority of proteins altered by aphid infestation were involved in metabolic processes and, even in this case, photosynthesis. Other proteins identified were involved in signal transduction, stress and defense, antioxidant activity, regulatory processes, and hormone responses. Responses to aphid attack at the proteome level were broadly similar to basal nonspecific defense and stress responses in wheat, with evidence of down-regulation of insect-specific defense mechanisms, in agreement with the observed lack of aphid resistance in commercial wheat lines (Ferry et al. 2011).

By studying expression profiles of proteins in rice leaf sheaths in responses to infestation by the brown planthopper (*Nilaparvata lugens*), significant changes were found for JA synthesis proteins, oxidative stress response proteins, beta-glucanases, protein kinases, clathrin protein, glycine cleavage system protein, photosynthesis proteins and aquaporins. Proteomic and transcript responses that were related to wounding, oxidative and pathogen stress overlapped considerably between insect-resistant and susceptible rice lines (Wei et al. 2009).

Key posttranscriptional changes were studied by identifying proteins and metabolites that were increased in abundance in both resistant and susceptible soybeans following *Heterodera glycines* (the soybean cyst nematode, SCN) attack. Identified proteins were grouped into six functional categories. Furthermore, metabolite analysis by gas chromatography (GC)-MS identified several metabolites, among which some were altered by one or more treatment (Afzal et al. 2009).

From the above results it is becoming evident that understanding how the plant responses to insects at the proteomic level will provide tools for a better management of insect pest in the field.

6.8 Electrical Signal Ultimate Target: The Induction of Metabolic Responses

Metabolomics is a level downstream from transcriptomics and proteomics and has been widely advertised as a functional genomics and systems biology tool (Macel et al. 2010). Combining the technological advances in metabolomics, transcriptomics, and genomics facilitated the detection of genes that contribute to diversification in plant secondary metabolism, as it has been recently reviewed (Kroymann 2011).

Plants may produce hundreds of thousands metabolites, of which only a marginal part has been identified to date. Many of these so-called secondary metabolites have a defensive function, while others are required for growth, development, or reproduction of the plant. Owing to their bioactive properties, secondary plant metabolites are a rich source of compounds, such as insecticides, fungicides, and plant-derived medicines (Jansen et al. 2009). Several aspects of metabolomics have been studied. For example, environmental metabolomics characterizes the interactions of organisms with their environment by studying organism–environment interactions and assessing organism function and health at the molecular level. These interactions are studied from individuals to populations, which can be related to the traditional fields of ecophysiology and ecology, and from instantaneous effects to those over evolutionary time scales (Bundy et al. 2009). Metabolites involved in resistance to herbivory have also been identified using an ecometabolomic approach, showing the potential of metabolomics to identify bioactive compounds involved in plant defense (Kuzina et al. 2009). Metabolomic approaches have been often demonstrated to be useful in discovering unexpected bioactive compounds involved in ecological interactions between plants and their herbivores and higher trophic levels (Jansen et al. 2009; Maffei et al. 2011).

Methods to pursue genuine metabolomics have not yet been developed due to the extensive chemical diversity of plant primary and secondary metabolites, although chemometric and bioinformatic advances promise to enhance our global understanding of plant metabolism. Among the separation-based approaches are GC or liquid chromatography (LC) combined with MS. Fourier transform-ion cyclotron resonance (FT-ICR)-MS is better suited for rapid, high-throughput applications and is currently the most sensitive method available. Unlike MS-based analyses, nuclear magnetic resonance (NMR) spectroscopy provides a large amount of information regarding molecular structure, and novel software innovations have facilitated the unequivocal identification and absolute quantification of compounds within composite samples (Hagel and Facchini 2008). Figure 6.4 shows a typical workflow diagram illustrating methodologies and technical platforms used for different analytical approaches in metabolomics.

High-throughput molecular techniques are now widely used to unravel the complex interaction between insects and plants. The nature of these interactions may range from positive, for example interactions with pollinators, to negative, such as interactions with pathogens and herbivores (van Dam 2009). The production of secondary metabolites from herbivore-damaged plants can be nearly 2.5-fold higher than that of intact plants, and this is particularly true for volatile compounds (Dudareva et al. 2006; Dudareva and Pichersky 2008). Among the several metabolites produced, plants respond to insect herbivory with the emission of volatiles that attract the enemies of the herbivores, such as insect predators and parasitoids. Moreover, plant volatiles may be exploited by any organism in the environment and this results in many more infochemically mediated interactions (Snoeren et al. 2007). The insect feeding-induced emission of volatiles has been demonstrated for several higher plant species, among others the model plant *A. thaliana*, maize (*Zea mays*), lima bean (*P. lunatus*), *N. attenuata*, *Medicago*

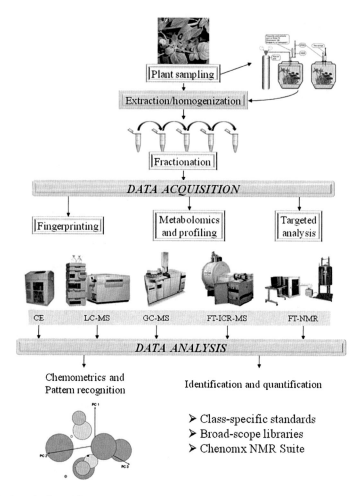

Fig. 6.4 A typical workflow for metabolomic studies. See description in the text

truncatula, and spruce (*Pinus glabra*), as well as for lower plants like ferns (Maffei 2010; Maffei et al. 2011, and references cited therein). In general, plant volatiles can carry various types of information: (1) for herbivores to localize their host plants, (2) for indirect defense employing a third trophic level by attracting natural enemies of the plant's offender, and for (3) neighboring plants and (4) distant parts of the same plant, respectively, to adjust their defensive phenotype accordingly (Heil and Silva Bueno 2007).

Many plant species contain toxins that control the maximum per capita daily biomass intake by herbivores (McLean and Duncan 2006). Mixtures of secondary compounds can also have a synergistic ecological effect, causing greater negative impacts on herbivores than when compared with equal amounts of single compounds (Berenbaum and Zangerl 1996). Several lines of evidence show that cardenolides continue to be detrimental to both above and belowground herbivores (Rasmann et al. 2011). Using several species of milkweed (*Asclepias* spp.), the variation among plant species in the induction of toxic cardenolides was explained by latitude, with higher inducibility evolving more frequently at lower latitudes. A greater cardenolide investment by a species was accompanied by an increase in an estimate of toxicity (measured as chemical polarity). Furthermore, analyses of root and shoot cardenolides showed concordant patterns, making milkweed species from lower latitudes better defended with higher inducibility, greater diversity, and added toxicity of cardenolides (Rasmann and Agrawal 2011).

When a generalist herbivore feeds in the absence of plant toxins, adaptive foraging generally increases the probability of coexistence of plant species populations, because the herbivore switches more of its effort to whichever plant species is more common and accessible. In contrast, toxin-determined selective herbivory can drive plant succession toward dominance by the more toxic species, as previously documented above in boreal forests and prairies. When the toxin concentrations in different plant species are similar, but species have different toxins with nonadditive effects, herbivores tend to diversify foraging efforts to avoid high intakes of any one toxin. This diversification leads the herbivore to focus more feeding on the less common plant species. Thus, uncommon plants may experience dispensatory mortality from herbivory, reducing local species diversity. The dispensatory effect of herbivory may inhibit the invasion of other plant species that are more palatable or have different toxins (Feng et al. 2009). There is also the possibility that the key determinant of leaf age feeding preferences in folivores is not the concentration of either nutrients or secondary metabolites, but is the ratio of the two, making the toxin: nutrient ratio is a potentially important determinant of feeding preferences (Lambdon and Hassall 2005).

Insects experience a wide array of chemical pressures from plant allelochemicals and pesticides and have developed several effective counterstrategies to cope with such toxins. Plants fed by Colorado potato beetles showed a significantly greater production of glycoalkaloids than in control plants and manually defoliated plants for both skin and inner tissue of tubers (Dinkins et al. 2008). Cotton (*Gossypium* spp.) produces gossypol, a sesquiterpene that occurs naturally in seed and other parts of the cotton plant, and a variety of other gossypol-like terpenoids that exhibit toxicity to a wide range of herbivores. Root feeding by *Meloidogyne incognita* has little influence on direct and indirect defenses of *G. hirsutum* against insect herbivory, whereas local and systemic induction of volatiles occurred with *Heliocoverpa zea* damage to leaves, and it increased in levels when root herbivory was added (Olson et al. 2008). Moreover (+)-gossypol is as inhibitory to *H. zea* larvae as racemic or (−)-gossypol, and thus, cotton plants containing predominantly the (+)-enantiomer in foliage may maintain significant defense against insect herbivory (Stipanovic et al. 2006).

Conium maculatum contains high concentrations of piperidine alkaloids that act as chemical defenses against herbivores. The caterpillar *Agonopterix alstroemeriana* can cause severe damage, resulting in some cases in complete defoliation of the toxic plant. Total alkaloid production in *C. maculatum* was positively correlated with *A. alstroemeriana* herbivory levels. Individual plants with lower concentrations of alkaloids experienced more damage by *A. alstroemeriana*, indicative of a preference on the part of the insect for plants with less chemical defense (Castells et al. 2005).

Defensive sulfur compounds, particularly dimethyl disulfide (DMDS), are highly toxic for nonadapted species. The toxicity of DMDS in these insects is due to disruption of the cytochrome oxidase system of their mitochondria. Larvae of the specialist *Acrolepiopsis assectella* are less susceptible to DMDS than adults of the Leguminosae specialist *Callosobruchus maculatus*. Activity of glutathione S-transferase (GST), a key enzyme in the detoxification system (Dixon et al. 2010; Cummins et al. 2011), was increased in *C. maculatus* adults and larvae after exposure to DMDS, whereas no effect on GST activity were found in *A. assectella*. This finding implies that induced GST is involved in *C. maculatus* tolerance to DMDS (Dugravot et al. 2004).

6.9 Concluding Remarks

Evolutionary arm race for survival between plants and insect herbivory started million years ago and continues today. Despite years of intense research, our knowledge on the significance of electric variations in plants upon stress is still poor. Nevertheless, much work has been done to promote plant resistance and tolerance against insect herbivory by promoting early detection following biotic attack. In this context, electrophysiology has become an invaluable tool for fast measurement, by alerting investigators much before the evident symptoms of herbivory are visible. The deciphering of Vm variations and their connection to all subsequent steps in the signaling cascade plant responses is a challenge that will lead to the development of new and promising tools for early detection.

References

Adie B, Chico JM, Rubio-Somoza I, Solano R (2007) Modulation of plant defenses by ethylene. J Plant Growth Regul 26:160–177

Afzal AJ, Natarajan A, Saini N, Iqbal MJ, Geisler M, El Shemy HA, Mungur R et al (2009) The nematode resistance allele at the rhg1 locus alters the proteome and primary metabolism of soybean roots. Plant Physiol 151:1264–1280

Alborn HT, Turlings TCJ, Jones TH, Stenhagen G, Loughrin JH, Tumlinson JH (1997) An elicitor of plant volatiles from beet armyworm oral secretion. Science 276:873–879

Arimura G, Maffei ME (2010) Calcium and secondary CPK signaling in plants in response to herbivore attack. Biochem Biophys Res Commun 400:455–460

Arimura G, Ozawa R, Shimoda T, Nishioka T, Boland W, Takabayashi J (2000) Herbivory-induced volatiles elicit defence genes in lima bean leaves. Nature 406:512–515

Arimura G, Kost C, Boland W (2005) Herbivore-induced, indirect plant defences. Biochim Biophys Acta, Mol Cell Biol Lipids 1734:91–111

Arimura GI, Ozawa R, Maffei ME (2011) Recent advances in plant early signaling in response to herbivory. Int J Mol Sci 12:3723–3739

Baldwin IT (2010) Plant volatiles. Curr Biol 20:R392–R397

Baluska F, Mancuso S, Volkmann D, Barlow P (2004) Root apices as plant command centres: the unique 'brain-like' status of the root apex transition zone. Biologia 59:7–19

Bede JC, Musser RO, Felton GW, Korth KL (2006) Caterpillar herbivory and salivary enzymes decrease transcript levels of *Medicago truncatula* genes encoding early enzymes in terpenoid biosynthesis. Plant Mol Biol 60:519–531

Berenbaum MR, Zangerl AR (1996) Phytochemical diversity: adaptation or random variation? In: Romeo JT, Saunders IA, Barbosa P (eds) Phytochemical diversity and redundancy in ecological interactions. Plenum Press, New York

Bodenhausen N, Reymond P (2007) Signaling pathways controlling induced resistance to insect herbivores in *Arabidopsis*. Mol Plant Microbe Interact 20:1406–1420

Bonaventure G, VanDoorn A, Baldwin IT (2011) Herbivore-associated elicitors: FAC signaling and metabolism. Trends Plant Sci 16:294–299

Bos JIB, Prince D, Pitino M, Maffei ME, Win J, Hogenhout SA (2010) A functional genomics approach identifies candidate effectors from the aphid species *Myzus persicae* (green peach aphid). PLoS Genet 6:e1001216

Boter M, Amigues B, Peart J, Breuer C, Kadota Y, Casais C, Moore G et al (2007) Structural and functional analysis of SGT1 reveals that its interaction with HSP90 is required for the accumulation of Rx, an R protein involved in plant immunity. Plant Cell 19:3791–3804

Bricchi I, Leitner M, Foti M, Mithofer A, Boland W, Maffei ME (2010) Robotic mechanical wounding (mecworm) versus herbivore-induced responses: early signaling and volatile emission in lima bean (*Phaseolus lunatus* L.). Planta 232:719–729

Bundy JG, Davey MP, Viant MR (2009) Environmental metabolomics: a critical review and future perspectives. Metabolomics 5:3–21

Castells E, Berhow MA, Vaughn SF, Berenbaum MR (2005) Geographic variation in alkaloid production in *Conium maculatum* populations experiencing differential herbivory by *Agonopterix alstroemeriana*. J Chem Ecol 31:1693–1709

Chen YZ, Pang QY, Dai SJ, Wang Y, Chen SX, Yan XF (2011) Proteomic identification of differentially expressed proteins in *Arabidopsis* in response to methyl jasmonate. J Plant Physiol 168:995–1008

Collins RM, Afzal M, Ward DA, Prescott MC, Sait SM, Rees HH, Tomsett A (2010) Differential proteomic analysis of *Arabidopsis thaliana* genotypes exhibiting resistance or susceptibility to the insect herbivore, *Plutella xylostella*. PLoS ONE 5

Cummins I, Dixon DP, Freitag-Pohl S, Skipsey M, Edwards R (2011) Multiple roles for plant glutathione transferases in xenobiotic detoxification. Drug Metab Rev 43:266–280

Dangl JL, Jones JDG (2001) Plant pathogens and integrated defence responses to infection. Nature 411:826–833

Diezel C, Kessler D, Baldwin IT (2011) Pithy protection: *Nicotiana attenuata*'s jasmonic acid-mediated defenses are required to resist stem-boring weevil larvae. Plant Physiol 155:1936–1946

Dinkins CLP, Peterson RKD, Gibson JE, Hu Q, Weaver DK (2008) Glycoalkaloid responses of potato to Colorado potato beetle defoliation. Food Chem Toxicol 46:2832–2836

Dixon DP, Skipsey M, Edwards R (2010) Roles for glutathione transferases in plant secondary metabolism. Phytochemistry 71:338–350

Duclohier H, Alder G, Kociolek K, Leplawy MT (2003) Channel properties of template assembled alamethicin tetramers. J Pept Sci 9:776–783

Dudareva N, Pichersky E (2008) Metabolic engineering of plant volatiles. Curr Opin Biotechnol 19:181–189

Dudareva N, Negre F, Nagegowda DA, Orlova I (2006) Plant volatiles: recent advances and future perspectives. Crit Rev Plant Sci 25:417–440

Dugravot S, Thibout E, Abo-Ghalia A, Huignard J (2004) How a specialist and a non-specialist insect cope with dimethyl disulfide produced by *Allium porrum*. Entomol Exp Appl 113: 173–179

Ebel J, Mithöfer A (1998) Early events in the elicitation of plant defence. Planta 206:335–348

Engelberth J, Seidl-Adams I, Schultz JC, Tumlinson JH (2007) Insect elicitors and exposure to green leafy volatiles differentially upregulate major octadecanoids and transcripts of 12-oxo phytodienoic acid reductases in *Zea mays*. Mol Plant Microbe Interact 20:707–716

Felton GW, Korth KL (2000) Trade-offs between pathogen and herbivore resistance. Curr Opin Plant Biol 3:309–314

Feng ZL, Liu RS, DeAngelis DL, Bryant JP, Kielland K, Stuart Chapin F, Swihart R (2009) Plant toxicity, adaptive herbivory, and plant community dynamics. Ecosystems 12:534–547

Fernandez-Calvo P, Chini A, Fernandez-Barbero G, Chico JM, Gimenez-Ibanez S, Geerinck J, Eeckhout D et al (2011) The *Arabidopsis* bHLH transcription factors MYC3 and MYC4 are targets of JAZ repressors and act additively with MYC2 in the activation of jasmonate responses. Plant Cell 23:701–715

Ferry N, Stavroulakis S, Guan WZ, Davison GM, Bell HA, Weaver RJ, Down RE et al (2011) Molecular interactions between wheat and cereal aphid (*Sitobion avenae*): analysis of changes to the wheat proteome. Proteomics 11:1985–2002

Fromm J, Lautner S (2007) Electrical signals and their physiological significance in plants. Plant Cell Environ 30:249–257

Gao LL, Kamphuis LG, Kakar K, Edwards OR, Udvardi MK, Singh KB (2010) Identification of potential early regulators of aphid resistance in *Medicago truncatula* via transcription factor expression profiling. New Phytol 186:980–994

Gilardoni PA, Schuck S, Jungling R, Rotter B, Baldwin IT, Bonaventure G (2010) SuperSAGE analysis of the *Nicotiana attenuata* transcriptome after fatty acid-amino acid elicitation (FAC): identification of early mediators of insect responses. BMC Pediatrics 10:66

Giordanengo P, Brunissen L, Rusterucci C, Vincent C, van Bel A, Dinant S, Girousse C et al (2010) Compatible plant-aphid interactions: how aphids manipulate plant responses. C R Biol 333:516–523

Giri AP, Wunsche H, Mitra S, Zavala JA, Muck A, Svatos A, Baldwin IT (2006) Molecular interactions between the specialist herbivore *Manduca sexta* (Lepidoptera, Sphingidae) and its natural host *Nicotiana attenuata*. VII. Changes in the plant's proteome. Plant Physiol 142:1621–1641

Hagel JM, Facchini PJ (2008) Plant metabolomics: analytical platforms and integration with functional genomics. Phytochem Rev 7:479–497

Heil M, Silva Bueno JC (2007) Within-plant signaling by volatiles leads to induction and priming of an indirect plant defense in nature. Proc Natl Acad Sci U S A 104:5467–5472

Heil M, Lion U, Boland W (2008) Defense-inducing volatiles: in search of the active motif. J Chem Ecol 34:601–604

Hilker M, Meiners T (2006) Early herbivore alert: insect eggs induce plant defense. J Chem Ecol 32:1379–1397

Hilker M, Meiners T (2010) How do plants "notice" attack by herbivorous arthropods? Biol Rev 85:267–280

Hilker M, Stein C, Schroder R, Varama M, Mumm R (2005) Insect egg deposition induces defence responses in *Pinus sylvestris*: characterisation of the elicitor. J Exp Biol 208:1849–1854

Horiuchi J-I, Muroi A, Takabayashi J, Nishioka T (2007) Exposing *Arabidopsis* seedlings to borneol and bornyl acetate affects root growth: specificity due to the chemical and optical structures of the compounds. J Plant Interact 2:101–104

Howe GA, Jander G (2008) Plant immunity to insect herbivores. Annu Rev Plant Biol 59:41–66

Jansen JJ, Allwood JW, Marsden-Edwards E, van der Putten WH, Goodacre R, van Dam NM (2009) Metabolomic analysis of the interaction between plants and herbivores. Metabolomics 5:150–161

Kanchiswamy CN, Muroi A, Maffei ME, Yoshioka H, Sawasaki T, Arimura G (2010) Ca^{2+}-dependent protein kinases and their substrate HsfB2a are differently involved in the heat response signaling pathway in *Arabidopsis*. Plant Biotechnol 27:469–473

Kant MR, Baldwin IT (2007) The ecogenetics and ecogenomics of plant-herbivore interactions: rapid progress on a slippery road. Curr Opin Genet Dev 17:519–524

Klusener B, Weiler EW (1999) Pore-forming properties of elicitors of plant defense reactions and cellulolytic enzymes. FEBS Lett 459:263–266

Koo AJK, Cooke TF, Howe GA (2011) Cytochrome P450 CYP94B3 mediates catabolism and inactivation of the plant hormone jasmonoyl-L-isoleucine. Proc Natl Acad Sci U S A 108: 9298–9303

Kopke D, Beyaert I, Gershenzon J, Hilker M, Schmidt A (2010) Species-specific responses of pine sesquiterpene synthases to sawfly oviposition. Phytochemistry 71:909–917

Kroymann J (2011) Natural diversity and adaptation in plant secondary metabolism. Curr Opin Plant Biol 14:246–251

Kuzina V, Ekstrom CT, Andersen SB, Nielsen JK, Olsen CE, Bak S (2009) Identification of defense compounds in *Barbarea vulgaris* against the herbivore *Phyllotreta nemorum* by an ecometabolomic approach. Plant Physiol 151:1977–1990

Lambdon PW, Hassall M (2005) How should toxic secondary metabolites be distributed between the leaves of a fast-growing plant to minimize the impact of herbivory? Funct Ecol 19: 299–305

Lawrence SD, Dervinis C, Novak N, Davis JM (2006) Wound and insect herbivory responsive genes in poplar. Biotechnol Lett 28:1493–1501

Liu LL, Zhang J, Zhang YF, Li YC, Xi JH, Li SY (2010) Proteomic analysis of differentially expressed proteins of *Arabidopsis thaliana* response to specialist herbivore *Plutella xylostella*. Chem Res Chin Univ 26:958–963

Lühring H, Nguyen VD, Schmidt L, Roese US (2007) Caterpillar regurgitant induces pore formation in plant membranes. FEBS Lett 581:5361–5370

Macel M, van Dam NM, Keurentjes JJB (2010) Metabolomics: the chemistry between ecology and genetics. Mol Ecol Res 10:583–593

Macias FA, Molinillo JMG, Varela RM, Galindo JCG (2007) Allelopathy—a natural alternative for weed control. Pest Manage Sci 63:327–348

Maffei ME (2010) Sites of synthesis, biochemistry and functional role of plant volatiles. S Afr J Bot 76:612–631

Maffei M, Bossi S (2006) Electrophysiology and plant responses to biotic stress. In: Volkov A (ed) Plant electrophysiology—theory and methods. Springer, Berlin

Maffei M, Bossi S, Spiteller D, Mithöfer A, Boland W (2004) Effects of feeding *Spodoptera littoralis* on lima bean leaves. I. Membrane potentials, intracellular calcium variations, oral secretions, and regurgitate components. Plant Physiol 134:1752–1762

Maffei ME, Mithofer A, Arimura GI, Uchtenhagen H, Bossi S, Bertea CM, Cucuzza LS et al (2006) Effects of feeding *Spodoptera littoralis* on lima bean leaves. III. Membrane depolarization and involvement of hydrogen peroxide. Plant Physiol 140:1022–1035

Maffei ME, Mithofer A, Boland W (2007a) Before gene expression: early events in plant-insect interaction. Trends Plant Sci 12:310–316

Maffei ME, Mithofer A, Boland W (2007b) Insects feeding on plants: rapid signals and responses preceding the induction of phytochemical release. Phytochemistry 68:2946–2959

Maffei ME, Gertsch J, Appendino G (2011) Plant volatiles: production, function and pharmacology. Nat Prod Rep 28:1359–1380

Maischak H, Grigoriev PA, Vogel H, Boland W, Mithofer A (2007) Oral secretions from herbivorous lepidopteran larvae exhibit ion channel-forming activities. FEBS Lett 581: 898–904

Maserti BE, Del Carratore R, la Croce CM, Podda A, Migheli Q, Froelicher Y, Luro F et al (2011) Comparative analysis of proteome changes induced by the two spotted spider mite *Tetranychus urticae* and methyl jasmonate in citrus leaves. J Plant Physiol 168:392–402

Masi E, Ciszak M, Stefano G, Renna L, Azzarello E, Pandolfi C, Mugnai S et al (2009) Spatiotemporal dynamics of the electrical network activity in the root apex. Proc Natl Acad Sci U S A 106:4048–4053

Matsumura H, Ito A, Saitoh H, Winter P, Kahl G, Reuter M, Kruger DH et al (2005) Supersage. Cell Microbiol 7:11–18

Mcainsh MR, Gray JE, Hetherington AM, Leckie CP, Ng C (2000) Ca^{2+} signalling in stomatal guard cells. Biochem Soc Trans 28:476–481

McLean S, Duncan AJ (2006) Pharmacological perspectives on the detoxification of plant secondary metabolites: implications for ingestive behavior of herbivores. J Chem Ecol 32:1213–1228

Meldau S, Baldwin IT, Wu JQ (2011) SGT1 regulates wounding- and herbivory-induced jasmonic acid accumulation and *Nicotiana attenuata*'s resistance to the specialist lepidopteran herbivore *Manduca sexta*. New Phytol 189:1143–1156

Memelink J (2009) Regulation of gene expression by jasmonate hormones. Phytochemistry 70:1560–1570

Mithöfer A, Boland W (2008) Recognition of herbivory-associated molecular patterns. Plant Physiol 146:825–831

Mithöfer A, Wanner G, Boland W (2005) Effects of feeding *Spodoptera littoralis* on lima bean leaves. II. Continuous mechanical wounding resembling insect feeding is sufficient to elicit herbivory-related volatile emission. Plant Physiol 137:1160–1168

Mithöfer A, Boland W, Maffei ME (2009a) Chemical ecology of plant-insect interactions. In: Parker J (ed) Molecular aspects of plant disease resistance. Wiley-Blackwell, Chirchester

Mithöfer A, Mazars C, Maffei M (2009b) Probing spatio-temporal intracellular calcium variations in plants. In: Pfannschmidt T (ed) Plant signal transduction. Humana Press Inc., Totowa

Musser RO, Farmer E, Peiffer M, Williams SA, Felton GW (2006) Ablation of caterpillar labial salivary glands: technique for determining the role of saliva in insect-plant interactions. J Chem Ecol 32:981–992

Olson DM, Davis RF, Wackers FL, Rains GC, Potter T (2008) Plant-herbivore-carnivore interactions in cotton, *Gossypium hirsutum*: linking belowground and aboveground. J Chem Ecol 34:1341–1348

Onkokesung N, Galis I, von Dahl CC, Matsuoka K, Saluz HP, Baldwin IT (2010) Jasmonic acid and ethylene modulate local responses to wounding and simulated herbivory in *Nicotiana attenuata* leaves. Plant Physiol 153:785–798

Oyarce P, Gurovich L (2011) Evidence for the transmission of information through electric potentials in injured avocado trees. J Plant Physiol 168:103–108

Preston CA, Laue G, Baldwin IT (2001) Methyl jasmonate is blowing in the wind, but can it act as a plant–plant airborne signal? Biochem Syst Ecol 29:1007–1023

Pyatygin S, Opritov V, Vodeneev V (2008) Signaling role of action potential in higher plants. Russ J Plant Physiol 55:285–291

Rasmann S, Agrawal AA (2011) Latitudinal patterns in plant defense: evolution of cardenolides, their toxicity and induction following herbivory. Ecol Lett 14:476–483

Rasmann S, Erwin AC, Halitschke R, Agrawal AA (2011) Direct and indirect root defences of milkweed (*Asclepias syriaca*): trophic cascades, trade-offs and novel methods for studying subterranean herbivory. J Ecol 99:16–25

Reddy ASN, Ali GS, Celesnik H, Day IS (2011) Coping with stresses: roles of calcium- and calcium/calmodulin-regulated gene expression. Plant Cell 23:2010–2032

Rehrig EM, Appel HM, Schultz JC (2011) Measuring 'normalcy' in plant gene expression after herbivore attack. Mol Ecol Res 11:294–304

Schmelz EA, Carroll MJ, LeClere S, Phypps SM, Meredith J, Chourey PS, Alborn HT et al (2006) Fragments of ATP synthase mediate plant perception of insect attack. Proc Natl Acad Sci U S A 103:8894–8899

Schuler MA (2011) P450s in plant-insect interactions. Biochim Biophys Acta 1814:36–45

Sergeant K, Renaut J (2010) Plant biotic stress and proteomics. Curr Proteomics 7:275–297

Shao HB, Song WY, Chu LY (2008) Advances of calcium signals involved in plant anti-drought. C R Biol 331:587–596

Shirasu K (2009) The HSP90-SGT1 chaperone complex for NLR immune sensors. Annu Rev Plant Biol 60:139–164

Singh A, Singh IK, Verma PK (2008) Differential transcript accumulation in *Cicer arietinum* L. in response to a chewing insect *Helicoverpa armigera* and defence regulators correlate with reduced insect performance. J Exp Bot 59:2379–2392

Snoeren TAL, De Jong PW, Dicke M (2007) Ecogenomic approach to the role of herbivore-induced plant volatiles in community ecology. J Chem Ecol 95:17–26

Spiteller D, Pohnert G, Boland W (2001) Absolute configuration of volicitin, an elicitor of plant volatile biosynthesis from lepidopteran larvae. Tetrahedron Lett 42:1483–1485

Stephens NR, Cleland RE, Van Volkenburgh E (2006) Shade-induced action potentials in *Helianthus anuus* L. originate primarily from the epicotyl. Plant Signaling Behav 1:15–22

Stipanovic RD, Lopez JD, Dowd MK, Puckhaber LS, Duke SE (2006) Effect of racemic and (+)- and (-)-gossypol on the survival and development of *Helicoverpa zea* larvae. J Chem Ecol 32:959–968

Thivierge K, Prado A, Driscoll BT, Bonneil E, Thibault P, Bede JC (2010) Caterpillar- and salivary-specific modification of plant proteins. J Proteome Res 9:5887–5895

Torsten W, van Bel AJ (2008) Induction as well as suppression: how aphid saliva may exert opposite effects on plant defense. Plant Signaling Behav 3:427–430

van Dam NM (2009) How plants cope with biotic interactions. Plant Biol 11:1–5

VanDoorn A, Kallenbach M, Borquez AA, Baldwin IT, Bonaventure G (2010) Rapid modification of the insect elicitor N-linolenoyl-glutamate via a lipoxygenase-mediated mechanism on *Nicotiana attenuata* leaves. BMC Plant Biol 10:164

Velculescu VE, Zhang L, Vogelstein B, Kinzler KW (1995) Serial analysis of gene-expression. Science 270:484–487

Vlot AC, Klessig DF, Park SW (2008) Systemic acquired resistance: the elusive signal(s). Curr Opin Plant Biol 11:436–442

Volkov AG, Haack RA (1995) Insect-induced bioelectrochemical signals in potato plants. Bioelectrochem Bioenerg 37:55–60

Volkov AG, Mwesigwa J (2000) Interfacial electrical phenomena in green plants: action potentials. In: Volkov AG (ed) Liquid interfaces in chemical, biological, and pharmaceutical applications. Dekker, New York

Volkov AG, Lang RD, Volkova-Gugeshashvili MI (2007) Electrical signaling in *Aloe vera* induced by localized thermal stress. Bioelectrochemistry 71:192–197

Volkov AG, Adesina T, Markin VS, Jovanov E (2008) Kinetics and mechanism of *Dionaea muscipula* trap closing. Plant Physiol 146:694–702

Volkov AG, Foster JC, Markin VS (2010) Signal transduction in *Mimosa pudica*: biologically closed electrical circuits. Plant Cell Environ 33:816–827

Wei Z, Hu W, Lin QS, Cheng XY, Tong MJ, Zhu LL, Chen RZ et al (2009) Understanding rice plant resistance to the brown planthopper (*Nilaparvata lugens*): a proteomic approach. Proteomics 9:2798–2808

Whiteman NK, Groen SC, Chevasco D, Bear A, Beckwith N, Gregory TR, Denoux C et al (2011) Mining the plant-herbivore interface with a leafmining *Drosophila* of *Arabidopsis*. Mol Ecol 20:995–1014

Winterhalter M (2000) Black lipid membranes. Curr Opin Colloid Interface Sci 5:250–255

Woldemariam MG, Baldwin IT, Galis I (2011) Transcriptional regulation of plant inducible defenses against herbivores: a mini-review. J Plant Interact 6:113–119

Wu JQ, Baldwin IT (2010) New insights into plant responses to the attack from insect herbivores. Annu Rev Genet 44:1–24

Wu J, Hettenhausen C, Meldau S, Baldwin IT (2007) Herbivory rapidly activates MAPK signaling in attacked and unattacked leaf regions but not between leaves of *Nicotiana attenuata*. Plant Cell 19:1096–1122

Yan XF, Wang ZY, Huang L, Wang C, Hou RF, Xu ZL, Qiao XJ (2009) Research progress on electrical signals in higher plants. Prog Nat Sci 19:531–541

Yoshinaga N, Aboshi T, Abe H, Nishida R, Alborn HT, Tumlinson JH, Mori N (2008) Active role of fatty acid amino acid conjugates in nitrogen metabolism in *Spodoptera litura* larvae. Proc Natl Acad Sci U S A 105:18058–18063

Zhang JH, Sun LW, Liu LL, Lian J, An SL, Wang X, Zhang J et al (2010) Proteomic analysis of interactions between the generalist herbivore *Spodoptera exigua* (Lepidoptera: Noctuidae) and *Arabidopsis thaliana*. Plant Mol Biol Rep 28:324–333

Zhu-Salzman K, Luthe DS, Felton GW (2008) Arthropod-inducible proteins: broad spectrum defenses against multiple herbivores. Plant Physiol 146:852–858

Zimmermann MR, Maischak H, Mithoefer A, Boland W, Felle HH (2009) System potentials, a novel electrical long-distance apoplastic signal in plants, induced by wounding. Plant Physiol 149:1593–1600

Chapter 7
Phytosensors and Phytoactuators

Alexander G. Volkov and Vladislav S. Markin

Abstract Plants continuously sense a wide variety of perturbations and produce various responses known as tropisms in plants. It is essential for all plants to have survival sensory mechanisms and actuators responsible for a specific plant response process. Plants are ideal adaptive structures with smart sensing capabilities based on different types of tropisms, such as chemiotropism, geotropism, heliotropism, hydrotropism, magnetotropism, phototropism, thermotropism, electrotropism, thigmotropism, and host tropism. Plants can sense mechanical, electrical and electromagnetic stimuli, gravity, temperature, direction of light, insect attack, chemicals and pollutants, pathogens, water balance, etc. Here we show how plants sense different environmental stresses and stimuli and how phytoactuators response to them. Plants generate various types of intracellular and intercellular electrical signals in response to these environmental changes. This field has both theoretical and practical significance because these phytosensors and phytoactuators employ new principles of stimuli reception and signal transduction and play a very important role in the life of plants.

7.1 Introduction

Plants are ideal adaptive structures with smart sensing capabilities (Taya 2003) based on different types of tropisms, such as chemiotropism, geotropism, heliotropism, hydrotropism, magnetotropism, phototropism, thermotropism, electrotropism, thigmotropism, and host tropism. A phytoactuator is a part of a plant

---

A. G. Volkov (✉)
Department of Chemistry, Oakwood University, Huntsville, AL 35896, USA
e-mail: agvolkov@yahoo.com

V. S. Markin
Department of Neurology, University of Texas Southwestern Medical Center at Dallas, Dallas, TX 75390-8813, USA

responsible for moving or controlling a specific plant response process. It is operated by a source of electrochemical energy or hydraulic pressure and converts that energy into motion. A phytosensor is defined as a device that can either detect, record, and transmit information related to a physiological change/process in a plant. It can also use plant tissue to monitor the presence of various chemicals in a substance. In most successful phytosensors, the principle behind the determination of a chemical or biological molecule is the specific interaction of such an analyte molecule with the plant tissue present in the phytosensor probe device. Even though a variety of biological materials and transduction methods have been investigated in the development of novel phytosensors, the most successful commercial systems include immobilized enzymes and electrochemical transducers. As an alternative to enzyme-based phytosensors, a plant-tissue based sensor was first developed by immobilizing slices of yellow squash tissue as a CO_2 gas sensor (Kuriyama and Rechnitz 1981). Since then, a variety of plant tissues has been incorporated into various electrochemical transducers to detect and quantify a range of biologically important analytes including drugs, hormones, toxicants, neurotransmitters and amino acids (He and Rechnitz 1995; Liawruangrath et al. 2001; Sidwell and Rechnitz 1986; Wijesuriya and Rechnitz 1993). Plant tissues have received considerable attentions in recent years as alternative biocatalyst for replacing isolated enzymes to construct phytosensors due to their high activity, stability, and low cost (Chen and Tan 1995; Mei et al. 2007; Quin et al. 2000; Zhu et al. 2004). However, they often encounter problems of long response time, low sensitivity, and complex sensor assembly (Quin et al. 2000). A detailed discussion on phytosensors utilized in chemical or biological analysis is beyond the scope of this chapter. Based on fast bioelectrochemical signaling events in plants, here we discuss the evidence supporting the foundation for utilizing the entire green plant as a fast phytosensor for monitoring the environmental perturbations in close vicinity of a living plant.

Nerve cells in animals and phloem cells in plants share one fundamental property: they possess excitable membranes through which electrical excitations can propagate in the form of action potentials (Bertholon 1783; Bose 1907, 1926, 1928; Burdon-Sanderson 1873; Davies 2006; Fromm and Bauer 1994; Fromm and Spanswick 1993; Ksenzhek and Volkov 1998; Volkov 2000, 2006a, b). Plants generate bioelectrochemical signals that resemble nerve impulses, which are present in plants at all evolutionary levels. Prior to the morphological differentiation of nervous tissues, the inducement of nonexcitability after excitation and the summation of subthreshold irritations were developed in the vegetative and animal kingdoms in protoplasmatic structures (Goldsworthy 1983).

The cells, tissues, and organs of plants transmit electrochemical impulses over short and long distances. It is conceivable that action potentials are the mediators for intercellular and intracellular communication in response to environmental irritants (Sinukhin and Britikov 1967; Volkov 2000, 2006a, b). Action potential is a momentary change in electrical potential on the surface of a cell that takes place when it is stimulated, especially by the transmission of an impulse.

Initially, plants respond to irritants at the site of stimulation; however, excitation waves can be conducted along the membranes throughout the entire plant. Bioelectrical impulses travel from the root to the stem and vice versa (Volkov and Haack 1995a, b). Chemical treatment, intensity of the irritation, mechanical wounding, previous excitations, temperature, and other irritants influence the speed of propagation (Volkov et al. 2000, 2001a, b, 2007b; Volkov and Mwesigwa 2001a, b).

Conductive bundles of vegetative organisms sustain the flow of material and trigger the conduction of bioelectrical impulses. This feature supports the harmonization of biological processes involved in the fundamental activity of vegetative organisms.

The conduction of bioelectrochemical excitation is a rapid method of long distance signal transmission between plant tissues and organs. Plants quickly respond to changes in luminous intensity, osmotic pressure, temperature, cutting, mechanical stimulation, water availability, wounding, and chemical compounds such as herbicides, plant growth stimulants, salts, and water. Once initiated, electrical impulses can propagate to adjacent excitable cells. The change in transmembrane potential creates a wave of depolarization or action potential, which affects the adjoining resting membrane.

The phloem is a sophisticated tissue in the vascular system of plants. Representing a continuum of plasma membranes, the phloem is a potential pathway for transmission of electrical signals. It consists of two types of conducting cells: the characteristic sieve-tube elements, and the companion cells. Sieve-tube elements are elongated cells that have end walls perforated by numerous minute pores through which dissolved materials can pass. Sieve-tube elements are connected in a vertical series known as sieve tubes. The companion cells have nuclei and they are adjacent to the sieve-tube elements. It is hypothesized that they control the process of conduction in the sieve tubes. Thus, when the phloem is stimulated at any point, the action potential can propagate over the entire cell membrane and along the phloem with constant voltage.

Electrical potentials have been measured at the tissue and whole plant level by using the experimental set-up shown in Fig. 7.1. Measurements were taken inside a Faraday cage mounted on a vibration-stabilized table. Nonpolarizable reversible Ag/AgCl electrodes were used to measure the electrical signals. The temperature was held constant since these electrodes are sensitive to the temperature. When electrochemical signals are measured, it is extremely important to take into consideration the *sampling rate* which determines how often the measurement device samples an incoming analog signal. According to the sampling theorem, the original analog signal must be adequately sampled in order to be properly represented by the sampled signal. If the sampling rate is too slow, the rapid changes in the original signal between any two consecutive samples cannot be accurately recorded. As a result, higher frequency components of the original signal will be misrepresented as lower frequencies. In signal processing, this problem is known

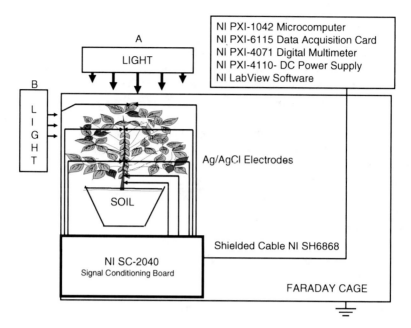

Fig. 7.1 Experimental set-up for measuring electrical signals in green plants

as *aliasing*. According to the Nyquist Criterion, the sampling frequency must be at least twice the bandwidth of the signal to avoid aliasing. Undersampling may result in the mispresentation of the measured signal.

The cells of many biological organs generate an electric potential that may result in the flow of an electric current. Electrical impulses may arise spontaneously or they may result from stimulation. Once initiated, they can propagate to the adjacent excitable cells. The change in transmembrane potential creates a wave of depolarization, or action potential that affects the adjoining membrane.

Plants constantly communicate with the external world in order to maintain homeostasis. Internal biological processes and their concomitant responses to the environment are closely associated with the phenomenon of excitability in plant cells. The extreme sensitivity of the protoplasm to chemical effects is the foundation for excitation. The excitable cells, tissues and organs alter their internal condition and external reactions under the influence of environmental factors, referred to as irritants; this excitability can be monitored. Plants generate different types of extracellular electrical responses in connection to environmental stress. Recent findings have indicated that plants may use a common defense system to respond to various abiotic and biotic stresses, such as heat, cold, drought, flooding, osmotic shock, wounding, high light intensity, UV-radiation, ozone, and pathogens. Using cDNA microarrays, a large number of genes have been found to be coordinately regulated and overlap under different stresses.

7.2 Host Tropism: Insect-Induced Electrochemical Signals in Plants

Understanding plant–insect interactions is important for ecology and for the development of novel crop protection strategies (Maffei et al. 2007). Volkov and Haack (1995a, b) were the first to create the unique opportunity to investigate the role of electrical signals induced by insects in long-distance communication in plants. Action and resting potentials were measured in potato plants (*Solanum tuberosum* L.) in the presence of leaf-feeding larvae of the Colorado potato beetle (*Leptinotarsa decemlineata* (Say); Coleoptera: Chrysomelidae). When the larvae were allowed to consume upper leaves of the potato plants and action potentials with amplitudes of 40 ± 10 mV were recorded every 2 ± 0.5 h during a two-day test period. (Volkov and Haack 1995a, b). The resting potential decreased from 30 mV to a steady state level of 0 ± 5 mV. The action potential propagates from plant leaves with Colorado potato beetles down the stem, and to the potato tuber. The speed of propagation of the action potential does not depend on the location of a measuring electrode in the stem of the plant or tuber, or the distance between the measuring and reference electrodes. Therefore, plants can be used as phytosensors for insect attacks.

7.3 Phototropism and Heliotropism: Molecular Recognition of the Direction of Light by Plants

Plants gain energy from two sources: quantum and thermophysical processes. Photosynthesis is an example of a quantum process, whereas transpiration is a thermophysical one. Plants have evolved sophisticated systems to sense the environment.

Plants can be used as photosensors for the direction of light (Volkov et al. 2004a, b, 2005; Volkov 2006b). Light is important for plant development by influencing nearly all aspects of the life cycle from germination to flowering. Plants perceive light ranging from ultraviolet to far-red light by specific photoreceptors. Natural radiation simultaneously activates more than one photoreceptor in higher plants. These receptors initiate distinct signaling pathways leading to wavelength-specific light responses. Nastic motion causes plants to angle their stems so that their leaves face light sources. The three classes of plant photoreceptors that have been identified at the molecular level are phototropins, cryptochromes, and phytochromes.

Phototropin is a blue light (360–500 nm) flavoprotein photoreceptor responsible for phototropism and chloroplast orientation. The phototropins, such as phot1 and phot2, are a family of flavoproteins that function as the primary photoreceptors in plant phototropism and in intracellular chloroplast movements. Phot1 contains two 12 kD flavin mononucleotide binding domains. LOV1 (light, oxygen, and voltage) and LOV2 are found within its N-terminal region and a C-terminal serine/ threonine protein kinase domain. The protein conformation changes in light-activated phototropin. Phot1 and phot2 bind FMN and undergo light-dependent autophosphorylation. Phot2 is localized in the plasma membrane (Casal 2000; Short and Briggs 1994).

Cryptochromes (cry1 and cry2) are flavoproteins in the family of photoreceptors responsible for photomorphogenesis (Ahmad et al. 1998; Cashmore et al. 1999). They perceive (UV-A) light as well as blue light (360–500 nm). Although cryptochromes and phototropin share many similarities, they have different transduction pathways. Cry1 plays a significant role in the synthesis of anthocyanin and in the entrainment of circadian rhythms. Cry2 plays a part in the photoperiodic flowering and cotyledon expansion. Cryptochromes were predominantly found in the nucleus. Stomatal opening is also stimulated by blue light and UV irradiation. Zeaxanthin has been proposed to be the blue/green light photoreceptor (Frechilla et al. 2000).

Phytochrome (phy) is a protein photoreceptor that regulates many aspects of plant development. Plant phytochromes are also light-modulated protein kinases that process dual ATP-dependent autophosphorylation and protein phosphotransferase activities (Quail 1997).

Phototropism is one of the best-known plant tropic responses. A positive phototropic response is characterized by a bending or turning toward the source of light (Fig. 7.2). When plants bend or turn away from the source of light, the phototropic response is considered negative. A phototropic response is a sequence of the four following processes: reception of the directional light signal, signal transduction, transformation of the signal to a physiological response, and the production of directional growth response (Volkov et al. 2004a, b, 2005; Volkov 2006b).

The soybean plant was irradiated inside the Faraday cage in the direction A (Fig. 7.1) with white light for two days with a 12:12 h light:dark photoperiod prior to the conduction of experiments. Action potentials are not generated when the lights are turned off and on. Changing the direction of irradiation from direction A to direction B generates action potentials in soybean approximately after 1–2 min. The generation of action potentials depend on the wavelength of irradiation light (Fig. 7.3). Irradiation at wavelengths 400–500 nm induces fast action potentials in soybean; conversely, the irradiation of soybean in the direction B at wavelengths between 500 and 630 nm fails to generate action potentials (Volkov et al. 2004a, b, 2005; Volkov 2006b). Irradiation between 500 and 700 nm does not induce phototropism. Irradiation of soybean by blue light induces positive phototropism (Volkov 2006b; Volkov et al. 2004a, b, 2005).

The plasma membrane in a phloem facilitates the passage of electrical excitations in the form of action potentials. The action potential has a stereotyped form and an essentially fixed amplitude—an "all or none" response to a stimulus. Each impulse is followed by the absolute refractory period. The fiber cannot transmit a second impulse during the refractory period. The integral organism of a plant can

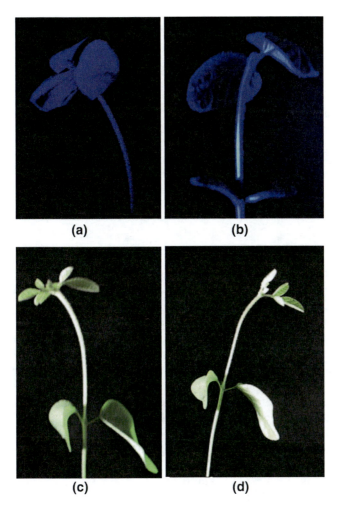

Fig. 7.2 Positive phototropism in soybean at 450 nm (**a**, **b**) and *white light* (**c**, **d**). Soybean was irradiated from *left side* (**a**, **c**) or from *right side* (**b**, **d**). Photographs were taken 30 min after beginning of the irradiation

be maintained and developed in a continuously varying environment only if all cells, tissues, and organs function in concordance.

These propagating excitations are modeled theoretically as traveling wave solutions of certain parameter-dependent nonlinear reaction–diffusion equations coupled with some nonlinear ordinary differential equations. These traveling wave solutions can be classified as single and multiple loop pulses, fronts and backs of periodic waves with different wave speed. This classification is matched by the classification of the electrochemical responses observed in plants. The experimental observations also show that under the influence of various pathogens, the

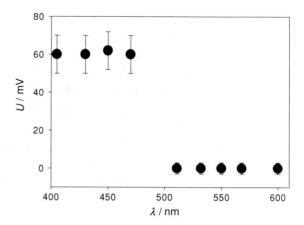

Fig. 7.3 Action spectrum: dependence of action potential amplitude in soybean on wavelength of irradiation. Distance between electrodes was 5 cm. The soil was preliminary treated by water every day. Volume of soil was 0.5 L. Frequency of scanning was 50,000 samples/s

shapes and speeds of the electrochemical responses undergo changes. From the theoretical perspective, the changes in the shapes and wave speeds of the traveling waves can be accounted by appropriate changes in parameters in the corresponding nonlinear differential equations (Volkov and Mwesigwa 2001b).

Fromm and Spanswick (1993) found that the electric stimulation of the plant is followed by ion shifts, which is most striking in the phloem cells. The amount of cytoplasmic calcium increased slightly while the content of K^+ and Cl^- was diminished after stimulation. Such evidence leads to the conclusion that Ca^{2+} influx, as well as K^+ and Cl^- efflux, is involved in the propagation of action potentials. In animal axons the excitation is underlined by the K^+ and Na^+ transmembrane transport; conversely, in phloem cells the K^+, Ca^+ and more than likely H^+ channels are involved in this process.

Babourina et al. (2002) have found that blue light induces significant changes in activity of H^+ and Ca^{2+} transporters within the first 10 min of exposure to blue light, peaking between 3 and 5 min. Blue light induced the opening of potassium and anion channels in plant cells.

Some voltage-gated ion channels work as plasma membrane nanopotentiostats. A blocker of K^+ ion channels, such as tetraethylammonium chloride (TEACl), stops the propagation of action potentials in soybean induced by blue light (Fig. 7.4) and inhibits phototropism (Volkov 2006a, b; Volkov et al. 2004a, b, 2005). Soybean plants lose the sensitivity to the direction of light if we compare Figs. 7.2 and 7.4. Voltage-gated ionic channels control the plasma membrane potential and the movement of ions across membranes; thereby, regulating various biological functions. These biological nanodevices play vital roles in signal transduction in higher plants phototropism.

The duration of the action potential is not influenced by the location of the measuring electrode in the stem or leaves of the plant or on the distance between the measuring and reference electrodes. Action potentials take an active part in the expedient character of response reactions of plants as a reply to external changes. These impulses transfer a signal about the changes of conditions in a conducting

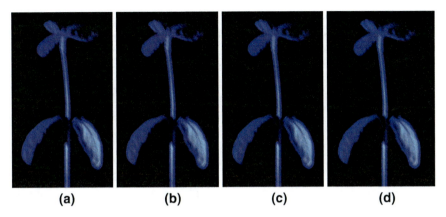

Fig. 7.4 Soybean plant after irradiation at 450 nm in the direction B (Fig. 7.1). Photographs were taken 1 min (**a**), 30 min (**b**), 40 min (**c**), and 70 min (**d**) after beginning of the irradiation. Soil around soybean was treated by 10 mM TEACl 24 h before photos were taken

Fig. 7.5 Mechanism of biosignaling in green plants

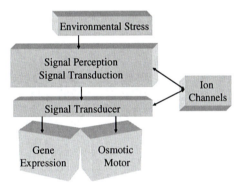

bundle of a plant from the root system to the point of growth and from the point of growth to the root system. Solitary waves due to impulses generated by changes in environmental conditions can be carriers of information in soybeans.

Green plants sense many parameters in order to adapt to the environment. Figure 7.5 illustrates the mechanism of bio-signaling in green plants.

7.4 Thigmotropism: Mechanosensation in Plants

The concept of mechanosensitive ion channels was primarily developed based on studies of specialized mechanosensory neurons (Markin and Sachs 2004). Due to the applied mechanical force on the cell membrane, these mechanically-gated channels, capable of converting mechanical stress into electrical or biochemical signals, are activated (Hamill and Martinac 2001; Volkov et al. 1998). As a result, they act as molecular transducers and play a vital role in regulating various physiological processes responsible for growth and development in all forms of life, as well as monitoring the surrounding environmental challenges for survival; for example, turgor control in plants. Although the current understanding of the structure and function of mechanosensitive ion channels found in living organisms is limited, significant progress has recently been made in the area of evolutionary origins of mechanosensitive ion channels.

The patch-clamp technique has provided the tool for identification of two basic types of mechanosensitive ion channels found in living cells: stretch-activated and stretch-inactivated ion channels (Hamill et al. 1981). Cosgrove and Hedrich (1991) studied the mechanosensory channels in the plasma membrane of guard cells of *Vicia faba L* using the patch-clamp technique, and identified three coexisting stretched-activated calcium, potassium and chloride ion channels. It has been found that such mechanosensitive ion channels play a vital role in the physiological function of the plant by controlling the ion transport across the plasma membrane, and hence influencing volume and turgor regulation of guard cells.

Plant response to mechanical stimulation has long been known. Perhaps all plants can react in response to the mechanical stimuli, only certain plants with rapid and highly noticeable touch-stimulus response have received much attention; for example, the trap closure of the Venus flytrap. Mechanosensation is considered to have evolved as one of the oldest sensory mechanisms in living organisms (Kloda and Martinac 2002; Martinac 2004; Martinac and Kloda 2003).

The Venus flytrap is a marvel of plant electrical, mechanical, and biochemical engineering. The rapid closure of the Venus flytrap upper leaf in about 0.3 s is one of the fastest movements in the plant kingdom. When a prey touches the trigger hairs, these mechanosensors trigger a receptor potential, which generate an electrical action potential. Two stimuli generate two action potentials, which close the trap at room temperature in a fraction of a second. Propagation of action potentials and the trap closing can be blocked by uncouplers, inhibitors of voltage gated channels, and aquaporins (Volkov et al. 2008c). We found that the electrical stimulus between a midrib and a lobe can close the Venus flytrap without mechanical stimulation (Volkov et al. 2007a, 2011b). It was also shown that the Venus flytrap has a short term electrical memory; using the charge stimulating method we demonstrated that Venus flytrap can accumulate small subthreshold charges, and when the threshold value is reached, the trap closes (Volkov et al. 2008a, b, 2009a).

In contrast to chemical signals such as hormones, electrical signals are able to transmit information rapidly over long distances. Biologically closed electrical circuits performing these functions operate over large distances in biological tissues. The activation of such circuit can lead to various physiological and biophysical responses (Volkov et al. 2008b, 2010a, b, c, d, e, f). It is often convenient to represent the real electrical and electrochemical properties of biointerfaces with idealized equivalent electrical circuit models consisting of discrete electrical components. We investigated the biologically closed electrical circuits in *Mimosa*

7 Phytosensors and Phytoactuators

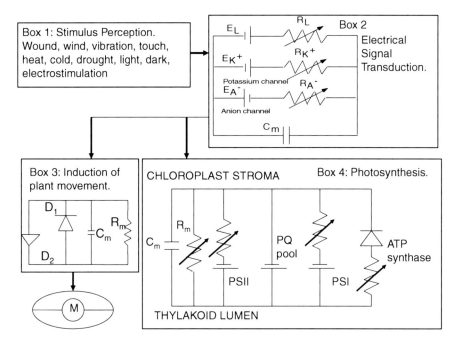

Fig. 7.6 Biologically closed electrical circuits in *Mimosa pudica*. Abbreviations: *C*—capacitance; *D*—diode as a model of an ion channel; *E*—electromotive force; *M*—osmotic motor; R_m—membrane resistance

pudica L. and in the upper leaf of the Venus flytrap (see Chap.1) and proposed the equivalent electrical circuits (Fig. 7.6). Stimulus perception (Box 1) can generate electrical signals such as action potentials. Box 2 shows the equivalent electrical circuit for a signal transduction (Volkov et al. 2010b). Electrical signals induce osmotic flow (Box 3) and starts the osmotic motor M. Box 4 shows the equivalent electrical circuit for photosynthesis in the *Mimosa pudica* (Volkov 1989).

Signal transduction in *Mimosa pudica* L. has been attracting the attention of researchers since the sixteenth century (Gardiner 1888; Haberlandt 1890; Hooke 1667; Pfeffer 1905; Ricca 1916; Ritter 1811). It is a sensitive plant in which the leaves and the petiole move in response to intensity of light, mechanical or electrical stimuli, drought, and hot or cold stimuli (Bose 1907, 1913, 1918, 1928; Hooke 1667). It was found by Haberlandt (1890) that long distance signal transduction in *Mimosa pudica* occurs through the phloem.

Mechanical movement of *Mimosa pudica* after electrical signal transduction is a defense mechanism (Bose 1918) and can be the simplest criterion of plant behavior and intelligence. Hooke (1667) stated long ago, "that this may be so, it seems with great probability to be argued from the strange phenomena of sensitive plants, wherein Nature seems to perform several animal actions with the same schematism or organization that is common to all vegetables, as may appear by some no less instructive then curious observations that were made by divers eminent members of the Royal Society on some of these kind of plants…".

Fig. 7.7 The structure of *Mimosa pudica*

Mimosa pudica L. is a thigmonastic or seismonastic plant in which the leaves close and the petiole hangs down in response to certain stressors. The unique anatomy of the *Mimosa pudica* contributes to the response mechanism of the plant (Fig. 7.6). The plant contains long slender branches called petioles, which can fall due to mechanical, thermal or electrical stimuli. The petioles contain smaller pinnae, arranged on the midrib of the pinna. The pinnules are the smallest leaflets while the entire leaf contains the petioles, pinnae, and pinnules. A pulvinus is a joint-like thickening at the base of a plant leaf or leaflet that facilitates thigmonastic movements (Fig. 7.7). Primary, secondary, and tertiary pulvini are responsible for the movement of the petiole, pinna, and leaflets, respectively (Shimmen 2006).

Thigmonastic or seismonastic movements in *Mimosa pudica* appear to be regulated by electrical, hydrodynamical, and chemical transduction (Allen 1969; Satter 1990; Stoeckel and Takeda 1993; Temmei et al. 2005). The pulvinus of *Mimosa pudica* shows elastic properties, and we found that electrically or mechanically induced movements of the petiole were accompanied by a change of pulvinus shape (Volkov et al. 2010b). As the petiole falls, the volume of the lower part of the pulvinus decreases and the volume of the upper part increases due to the redistribution of water between the upper and lower parts of the pulvinus (Fig. 7.8). This hydroelastic process is reversible. During the relaxation of the petiole, the volume of the lower part of the pulvinus increases and the volume of the upper part decreases. Redistribution of ions between the upper and lower parts of a pulvinus causes fast transport of water through aquaporins and causes a fast change in the volume of the motor cells (Fig. 7.9). The biologically closed electrochemical circuits in electrically and mechanically anisotropic pulvini of *Mimosa pudica* were analyzed using the charge stimulating method for electrostimulation at different voltages. Changing the polarity of electrodes leads to

7 Phytosensors and Phytoactuators

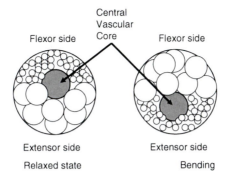

Fig. 7.8 Schematic diagram of a pulvinus cross section in relaxed and bended states

Fig. 7.9 Sequence of photos of a pulvinus of *Mimosa pudica* after mechanical stimulation (**a**) and during a relaxation after mechanical stimulation (**b**)

a strong rectification effect in a pulvinus and to different kinetics of a capacitor discharge if the applied initial voltage is 0.5 V or higher.

The exact mechanism of seismonastic movements in *Mimosa pudica* L. is unknown. There are three main hypotheses: chemical, muscular and osmotic motor mechanisms. All of these mechanisms of plant movements have experimental support, but each hypothesis does not take into account experimental data supporting the other two mechanisms.

Muscular hypothesis: Gardiner (1888) found that mechanical properties of the pulvinus are similar to animal muscles. The seismonastic movement of a petiole may be more efficient than typical animal muscle movements (Balmer and Franks 1975). The phosphorylation level of actin in the pulvinus affects the dynamic reorganization of actin filaments and causes seismonastic movement (Kameyama et al. 2000; Kanzawa et al. 2006; Pal et al. 1990; Yamashiro et al. 2001). If the flexor of the pulvinus is cut away, the extensor will still work, reacting to stimuli (Dutrochet 1837).

Chemical hypothesis: Pfeffer (1905) and Ricca (1916) suggested that an unknown chemical compound is responsible for the seismonastic movements in *Mimosa pudica*. According to Ricca (1916) this hydrophilic compound, so called a Ricca factor, moves through the xylem vessels to induce mechanical responses to stimuli. Schildknecht and Bender (1983) isolated and characterized the Ricca factor from *Mimosa pudica* L., which is turgorin 4-0-(β-D-gluco-pyranosyl-6'-sulfate) gallic acid. The binding site of this turgorin is located on the plasma membrane in the pulvinus (Varin et al. 1997). The mechanism of a turgorin action can be similar to acetylcholine effects in animal nerves (Schildknecht and Meier-Ausgenstein 1990).

Osmotic motor hypothesis: Mechanical stimuli induce action potential (Bose 1918), which can activate K^+ and Cl^- voltage gated ion channels (Samejima and Sibaoka 1982). Ion transport in the pulvinus induces osmotic movement of water and sudden turgor loss in the lower pulvinar cells (Weintraub 1951). The mechanism of leaf movement exhibited by *Mimosa pudica* is different from the movement of guard cells in the stomata (Asprey and Palmer 1955).

The exact mechanism of seismonastic movements in *Mimosa pudica* L. should include at least elements of all three hypotheses (Volkov et al. 2010a, b, c, d, e).

The pulvinus is comprised of three main parts: a central vascular core and two layers of flexor and extensor cells (Fig. 7.8). The action potential activates voltage gated channels and induces the redistribution of K^+, Cl^-, H^+ and Ca^{2+} ions between extensor and flexor layers (Toriyama 1955; Toriyama and Jaffe 1972) which in turn leads to the osmotic movement of water, causing the bending of the pulvinus and movement of the petiole in *Mimosa pudica* (Fig. 7.10). The potassium concentration in the apoplast of extensor cells of the *Mimosa pudica's* pulvinus increases from 30–70 to 100 mM. Thus the motor cells shrink and resultantly, the cells from the flexor site also take up K^+ ions from the apoplast and begin to swell (Kumon and Tsurumi 1984). The differential volume changes of flexor and extensor sites result from the transport of ions accompanied by osmotic transport of water. There is a high gradient of osmotic pressure between extensor and flexor cells of about 1 MPa in *Samanea* pulvini (Gardiner 1888; Gorton 1987).

Fig. 7.10 The mechanism of the seismonastic movements in *Mimosa pudica*

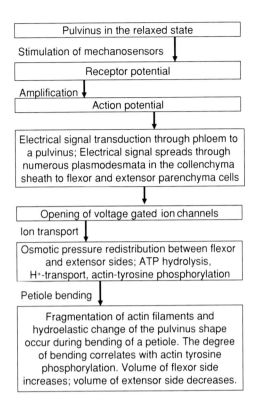

The identification and characterization of bioelectrochemical mechanisms for electrical signal transduction in plants would mark a significant step forward in understanding this under- explored area of plant physiology. Although plant mechanical and chemical sensing and corresponding responses are well known, membrane electrical potential changes in plant cells and the possible involvement of electrophysiology in transduction mediation of these sense-response patterns represents a new dimension of plant tissue and whole organism integrative communication. A short time ago we discovered the bioelectrochemical mechanisms of electrical signal transduction in biologically closed electrical circuits in the pinnae and petioles of *Mimosa pudica* and the plant's responses to electrostimulation (Volkov et al. 2010a, b, c, d, e). The studies of the mechanisms of concerted movements in plants from electrical signal transduction to cascades of cellular events will have a potentially broad impact on both fundamental sciences and engineering (Markin et al. 2008).

Mechanical movements in *Mimosa pudica* can be induced by electrostimulation if very high applied voltages 200–400 V were briefly applied between the soil and the primary pulvinus to measure the contractile characteristics of a petiole (Balmer and Franks 1975). Jonas (1970) used a 0.5 µF capacitor charged by 50, 100 and 150 V for electrostimulation and found oscillations of leaves and fast petiole movement in *Mimosa pudica* after the application of an electrical shock. The petioles bend downward and the pinnae close after the application of 9 V to *Mimosa pudica* (Yao et al. 2008). These high voltages are non-physiological and have a side effect—plant electrolysis. We analyzed, both experimentally and theoretically, the mechanism of mechanical movements in *Mimosa pudica* induced by low voltage electrostimulation of the petiole and pinna (Volkov et al. 2010a, b, c, d, e, f, 2011a). The charge stimulating method is a very efficient tool for the study of bioelectrochemistry of cells, clusters of cells, or for electrostimulation of whole plants and evaluation of biologically closed electrical circuits.

Some plants move their leaves upon sudden shaking or touch as seismonastic and thigmonastic movements use osmotic motors, powered by H^+-ATPases (Fleurat-Lessard et al. 1997; Fleurat-Lessard and Roblin 1982; Liubimova et al. 1964; Liubimova-Engel'gardt et al. 1978). The osmotic motor often resides in specialized leaf organs, midribs or pulvini, at the base of the leaves and leaflets. The osmotic motor transfers water through water channels or aquaporins.

In terms of electrophysiology, these responses in *Mimosa pudica* can be considered in three stages: (1) stimulus perception, (2) signal transmission and (3) induction of response (Fig. 7.10). Action potentials involve effluxes of K^+ and Cl^- and a temporary change of turgor, produced by osmotic motors. Like the action potential, a critical threshold depolarization triggers Ca^{2+} influx, opening of Ca^{2+}-sensitive Cl^- channels and K^+ channels; effluxes last over a short period of time and result in turgor regulation.

7.4.1 Mechanics of Petiole Movement

After mechanical stimulation of a petiole or a pulvinus, a petiole falls in a few seconds and relaxes to the initial state in 10–12 min. The maximal angle between a petiole and a stem varies from 25 to 100° in different plants. Electrostimulation of a pulvinus by a 47 µF capacitor charged to 1.5 V (+on upper part, −on lower part of a pulvinus) leads to a petiole bending similar to the effect of mechanical stimulation. The lower part of the pulvinus has a higher volume and curvature when a petiole is in a relaxed state. After mechanical or electrical stimulation of a pulvinus, the volume and curvature in the upper part of the pulvinus increases and a petiole hangs down. A pulvinus changes its shape during the hydro mechanical movement of a petiole.

Figure 7.11 shows the kinetics of a petiole bending, triggered by electrical stimulation. This bending is synchronized with the increased volume of the upper part of a pulvinus (Fig. 7.11) and the decreased volume of the lower part of a pulvinus (Fig. 7.11).

The mechanism of a petiole movement can be explained as follows. The pulvinus is a flexible hinge located at the base of the stalk of the leaf. It has a very anisotropic structure. The motor cells are organized in such a way that they allow changes with changing turgor only in length, but not in circumference (Mayer

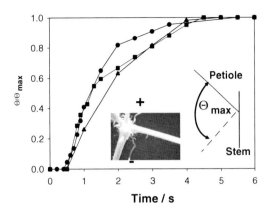

Fig. 7.11 Kinetics of a petiole bending triggered by electrical stimulation of *Mimosa pudica* by 47 μF charged capacitor with initial voltage of 1.5 V

et al. 1985). As a result, the antagonistic changes in the length of the flexor and extensor cells produce the petiole movements. This mechanism is similar to the work of a bimetallic strip that converts a temperature change into a mechanical displacement. The different expansions force the flat strip to bend one way if heated, and in the opposite direction if cooled below its normal temperature. In a similar way, the pulvinus uses a hydroelastic mechanism (Markin et al. 2008) that converts a difference of turgor pressure in extensor and flexor cells to the bending and ensuing rotation of the petiole. Bending of the pulvinus also manifests itself by a change in the curvature of the extensor and flexor sides. Movement of the petiole and the change in curvature closely follow each other. During the falling of a petiole, the volume of the lower part of the pulvinus decreases and the volume of the upper part increases in a few seconds. During relaxation of a petiole to its initial state, the volume of the extensor side of a pulvinus increases and the volume of the flexor side decreases in 20 min. This seems to occur due to the redistribution of water between the upper and lower parts of a pulvinus.

Through the usage of nuclear magnetic resonance, the movement of water from the lower half of the pulvinus to the upper half of the pulvinus following a mechanical stimulus was observed (Tamiya et al. 1988). This observation gives direct evidence for the theory of fast movement of water from the lower half to the upper half of the pulvinus. Movement of water from the upper half of the pulvinus to the lower half of the pulvinus during petiole relaxation is slow (Fig. 7.9). The details of the mechanism for the concerted action of the flexor and extensor sides of the pulvinus are not yet known, although osmotic pumps are definitely involved in water exchange in these two layers of the pulvinus.

The hydroelastic mechanism of *Mimosa pudica* movement parallels with the mechanism that closes the Venus fly trap. The hydroelastic model is based on the assumption that the leaf possesses curvature elasticity and consists of outer and inner hydraulic layers where different hydrostatic pressure can build up. The open state contains high elastic energy accumulated due to a hydrostatic pressure difference between the outer and inner layers of the leaf. Stimuli induce water flow from one hydraulic layer to another. This very fast process also involves water exchange between two layers of cells in the lobes of the trap with a consequent change of leaf curvature. The closing of the Venus flytrap was described by the hydroelastic curvature (HEC) mechanism (Markin et al. 2008).

Cl^- and H^+ ions also participate in osmotic changes in a pulvinus using anion voltage gated anion channels (Abe 1981) and proton pumps. There is an opinion that ATPase activity is strongly involved in the thigmonastic movement of *Mimosa pudica* because a high density of H^+-ATPases in the phloem and pulvini was found (Fleurat-Lessard et al. 1997; Fleurat-Lessard and Roblin 1982; Lyubimova et al. Liubimova et al. 1964; Liubimova-Engel'gardt et al. 1978).

7.5 Photoperiodism and Time Sensing: Biological Clock

The biological clock regulates a wide range of physiological and developmental processes in plants. In plants, circadian rhythms are linked to the light–dark cycle. Many of the circadian rhythmic responses to day and night continue in constant light or dark, at least for a period of time. The biological clock in a plant is an endogenous oscillator with a period of approximately 24 h. The circadian clock in plants is sensitive to light, which resets the phase of the rhythm. Molecular mechanism underlying circadian clock function is poorly understood, although it is now widely accepted for both plants and animals that it is based on circadian oscillators. The circadian clock was discovered in 1729 by De Mairan in his first attempt to resolve experimentally the origin of rhythm in the leaf movements of *Mimosa pudica* (De Mairan 1729). This rhythm continued even when the *Mimosa pudica* plant was maintained under continuous darkness. We also investigated the electrical activity of *Mimosa pudica* in the day light, at night, and in darkness the following day (Volkov et al. 2011a). The response in darkness the following day was similar to a typical daily response.

7.5.1 Circadian Rhythms in Electrical Circuits of Clivia miniata

We investigated electrical responses of *Clivia miniata* to electrical stimulation during the day in daylight, darkness at night, and the following day in darkness with different timing and voltages (Volkov et al. 2011a, c). A monocotyledon *C. miniata* is a vascular plant with dark green, strap shaped leaves with somewhat swollen leaf-bases which arise from fleshy roots (Lindley 1854; Regel 1864). The name *Clivia* was given to this ornamental genus of plants by John Lindley to compliment Lady Clive, the Duchess of Northumberland, who first cultivated this plant in England. *C. miniata* is a model for the study of circadian rhythms in

Fig. 7.12 Electrical equivalent schemes of a capacitor discharge in plant tissue. Abbreviations: C_1—charged capacitor from voltage source U_0; C_2—capacitance of plant tissue; R—resistance, D—a diode as a model of voltage gated ion channels

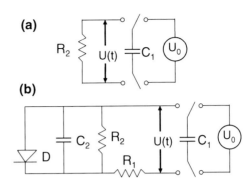

plants. Circadian variation of the *C. miniata*, *Aloe vera* and *Mimosa pudica* are sensitive to electrical stimulation (Volkov et al. 2011a, c, d). The biologically closed electrochemical circuits in the leaves of *C. miniata* (Kaffir lily), which regulate its physiology, were analyzed in vivo using the charge stimulation method. The electrostimulation was provided with different voltages and electrical charges. Resistance between Ag/AgCl electrodes in the leaf of *C. miniata* was higher at night than during the day or the following day in the darkness. The biologically closed electrical circuits with voltage gated ion channels in *C. miniata* are activated the next day, even in darkness. *C. miniata* memorizes daytime and night time. At continuous light, *C. miniata* recognizes night time and increases the input resistance to the night time value even under light. These results show that the circadian clock can be maintained endogenously and has electrochemical oscillators, which can activate voltage gated ion channels in biologically closed electrochemical circuits. The activation of voltage gated channels depends on the applied voltage, electrical charge and speed of transmission of electrical energy from the electrostimulator to the *C. miniata* leaves. We investigated the equivalent electrical circuits in *C. miniata* and its circadian variation to explain the experimental data (Volkov et al. 2011a, c).

As it is well known, if a capacitor of capacitance C with initial voltage U_0 is discharged through a resistor R (Fig. 7.12a), the voltage decreases with time t as

$$U(t) = U_0 \times e^{-t/\tau} \qquad (7.1)$$

Where

$$\tau = RC \qquad (7.2)$$

denotes the time constant. Eq. 7.1 in the logarithmic form reads:

$$\ln(U(t)/U_0) = -t/\tau Z \qquad (7.3)$$

The time constant, τ, can be determined from the slope of this linear function. The circuit time constant τ governs the discharging process. As the capacitance or resistance increases, the time of the capacitor discharge increases according to Eq. 7.1. The resistance of the linear circuit can be easily found from Eq. 7.1.

However, if the function (7.3) is not linear, then one can find the so called input resistance at any moment of the time (Volkov et al. 2010c):

$$R_{input} = -\frac{U}{C\, dU/dt} \tag{7.4}$$

We studied electrical discharge of 10 µF capacitor between two Ag/AgCl electrodes in the leaf of *C. miniata* parallel to the conductive bundles. The results obtained in the daylight during the first day are presented in Fig. 7.13a, b. The difference between the two experiments is the polarity of the electrodes: the positive pole is closer to the base of a leaf in Fig. 7.13a and the positive pole is closer to the apex in Fig. 7.3b. The dependence of resistivity on polarity can be explained by the opening of voltage gated ion channels, and can be modeled as diodes in Fig. 7.12b. Opening and closing of voltage gated channels results in the effect of electrical rectification. Similar rectification effects were found in *Aloe vera*, the Venus flytrap and *Mimosa pudica*. While using a silicon rectifier Schotky diode NTE583 as a model of a voltage gated channel, we reproduced experimental dependencies of the capacitor discharge in plant tissue (Volkov et al. 2009b).

The same electrical discharge in darkness during the nighttime is shown in Fig. 7.13c. The kinetics of the night discharge is significantly slower (Fig. 7.13c). This means that leaf resistance strongly increases at night (Fig. 7.13c).

The biological clock in *C. miniata* recognizes the daytime, even in darkness. The discharge during the following day in darkness is very similar to the first day. Input resistance in the initial moment of the capacitor discharge was the same as during the day light. During the third day when the lights are on, the results are the same as shown in Fig. 7.13, which were reproduced on different leaves of *C. miniata* plants.

In all three examples, kinetics of a capacitor discharge depends on polarity of electrodes in a leaf of *C. miniata* due to electrical anisotropy of the leaf. Dependence of the discharge kinetics on polarity of electrodes shows the rectification effect. This can be caused by the opening or closing of voltage gated ion channels and decreasing or increasing of the resistance in plant tissue, correspondingly (Volkov et al. 2009b, 2010b, 2011a, b, c).

Figure 7.14 illustrates the plant memory of a "sunset". Normally, we switch off the lights at 5:00 pm, however in this experiment we did not switch off the light at that time. Any time from the morning and during a day until 4:00 pm, the time dependencies of a 10 µF capacitor discharge coincide. At 5:00 pm resistance in leaves starts to increase and at 7:00 pm it reaches the same parameters as at night even at continuous light. Figure 7.14b presents the normalized voltage U/U_0 in semi-logarithmic coordinates. As one can see from Fig. 7.14, the capacitor discharge is fast during a daytime, but speed of the capacitor discharge decreases after 4:00 pm even under continuous light and reaches minimal value at 7:00 pm as in the dark during nighttime. Darwin (1880) found that leaves in *Clivia* move

7 Phytosensors and Phytoactuators 193

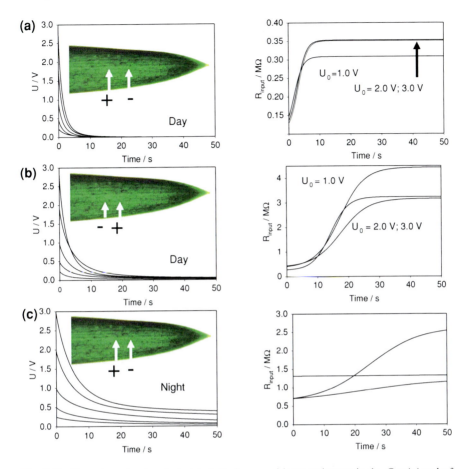

Fig. 7.13 Time dependencies of electrical discharge and input resistance in the *C. miniata* leaf between two Ag/AgCl electrodes connected to the 10 μF charged capacitor during a day (**a**, **b**) and at night (**c**). U is the capacitor voltage and U_0 is the initial voltage in volts

periodically: "A long glass filament was fixed to a leaf, and the angle formed by it with the horizon was measured occasionally during three successive days. It fell each morning until between 3 and 4 p.m., and rose at night. The smallest angle at any time above the horizon was 48°, and the largest 50°; so that it rose only 2° at night; but as this was observed each day, and as similar observations were nightly made on another leaf of a distinct plant, there can be no doubt that the leaves move periodically. The position of the apex when it stood highest was 0.8 of an inch above its lowest point." The periodical movement of leaves in *Clivia* has the electrophysiological component as it is shown in Fig. 7.14. Two hours before the light turning, the speed of electrical discharge decreases due to increasing of a resistance in leaves, which is probably related to the closing of voltage gated channels. This result is very impressive: the biological clock in *C. miniata*

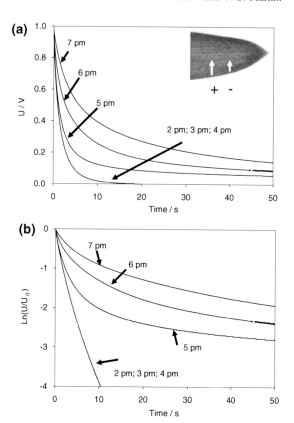

Fig. 7.14 Time dependence of the capacitor discharge in the *C. miniata* leaf afternoon and at evening during continuous light. U is the capacitor voltage and U_0 is the initial voltage in volts

recognizes the approaching of darkness after 4 p.m. even under constant light from 6.00 a.m. to 6.00 p.m.

Circadian oscillators are components of the biological clocks that regulate the activities of plants in relation to environmental cycles and provide an internal temporal framework. The circadian rhythm regulates a wide range of electrophysiological and developmental processes in plants such as floral development, flowers opening in the morning and closing towards evening, seeds germination, leaf growth and movement, stem elongation chloroplast movements, photosynthetic capacity, stomata conductance, cell division (Thomas and Vince-Prue 1997).

Plant tissues have biologically closed electrochemical circuits that are involved in these regulations. Voltage gated ion channels are devices in the engineering sense: they have signal inputs, power supplies, and signal outputs. We found periodic activation and deactivation of these circuits in the leaves of *C. miniata* controlled by internal clock, rather than environmental clues. We tested characteristics of capacitor discharge during the day, night, and next day in the darkness. Response of circuits in *C. miniata* during the day was significantly faster than at night (Figs. 7.3, 7.4 and 7.8a, c) and these variations were steadily repeated day after day.

While in darkness the following day, the plant remembers the time and the rate of discharge drastically increases approaching the rate of the first day. However, the input resistance is the same as during day light in the beginning of the capacitor discharge (Fig. 7.13a, b) and increases to night values during the discharge process (Fig. 7.3c). That means that the internal clock does change electrical conductance, but alone, without environmental clues (light), cannot ideally generate the same values of day and night conductance. The plant needs additional environmental information, and then the properties of electrical circuits will be restored to the same conditions.

These results demonstrate that the circadian clock can be maintained endogenously, probably involving electrochemical oscillators, which can activate or deactivate ion channels in biologically closed electrochemical circuits. This circadian rhythm can be related to the differences found in the membrane potentials during the day and nighttime, which were found in different plants (Kim et al. 1992, 1993; Racusen and Satter 1975; Scott and Gulline 1975; Thomas and Vince-Prue 1997). The expression of many ion transporters in plants is regulated by the circadian rhythm (Lebaudy et al. 2008).

7.5.2 Circadian Rhythms in Electrical Circuits of Aloe vera and Mimosa pudica

Mimosa pudica is a nyctinastic plant that closes its leaves in the evening; the pinnules fold together and the whole leaf droops downward temporarily until sunrise. The leaves open in the morning due to a circadian rhythm, which is regulated by a biological clock with a cycle of about 24 h.

During photonastic movement in *Mimosa pudica*, leaves recover their daytime position. During a scotonastic period, the primary pulvini straighten up and pairs of pinnules fold together about the tertiary pulvini. The closing of pinnae depends upon the presence of phytochrome in the far-red absorbing form (Fondeville et al. 1966).

Isolated pulvinar protoplasts are responsive to light signals in vitro (Cote 1995; Kim et al. 1992, 1993). In the dark period, the closed inward-directed K^+ channels of extensor cells are opened within 3 min by blue light. Conversely, the inward-directed K^+ channels of flexor cells, which are open in the darkness, are closed by blue light. In the light period, however, the situation is more complex. Premature darkness alone is sufficient to close the open channels of extensor protoplasts, but both darkness and a preceding pulse of red light are required to open the closed channels in the flexor protoplasts (Kim et al. 1992, 1993).

Aloe vera (L.) is a member of the Asphodelaceae (Liliaceae) family with crassulacean acid metabolism (CAM). In *Aloe vera*, stomata are open at night and closed during the day (Denius and Homann 1972). CO_2 acquired by *Aloe vera* at night is temporarily stored as malic and other organic acids, and is decarboxylated the following day to provide CO_2 for fixation in the Benson-Calvin cycle behind closed stomata. *Aloe vera* is a model for the study of plant electrophysiology with crassulacean acid metabolism.

The circadian rhythm can be related to the difference in the membrane potentials during the day and night time, which was found in pulvini of different plants (Kim et al. 1992, 1993; Kumon and Tsurumi 1984; Racusen and Satter 1975; Scott and Gulline 1975). During the day in darkness, there is still a rectification effect in the pulvinus (Fig. 7.11b), however, in the presence of light during the daytime, this rectification effect and the resistance decrease in the pulvinus are two times higher. These results demonstrate that the circadian clock can be maintained endogenously, probably involving electrochemical oscillators, which can activate or deactivate ion channels in biologically closed electrochemical circuits.

7.6 Plants as Phytosensors for Monitoring Atmospheric Electrochemistry: Acid Rain

The existence of ions in the atmosphere is the fundamental reason for atmospheric electricity. The voltage between the earth's surface and the ionosphere is approximately 40 kV, with an electrical current of approximately 2,000 A, and a current density about 5 pA/m^2 (Feynman et al. 1963; Wåhlin 1986). The earth is an electrode immersed in a weak gaseous electrolyte, the naturally ionized atmosphere. The earth's surface has a negative charge. The electrostatic field strength at the earth's surface is around 110–220 V/m, and depends on time of a day. It is approximately 110 V/m, however, at 7 p.m. GMT the electrostatic field strength is about 220–250 V/m (Wåhlin 1986). Oceans, lakes, and rivers cover a significant part of the earth, and their surface is also charged negatively against the atmosphere (Wåhlin 1986). Electrical polarity in soybean, potato, tomato, and cacti coincides with the electrical field of the electric double layer of earth: negative in roots and positive at the top of the plants (Lemström 1904; Volkov 2000; Volkov and Mwesigwa 2001a, b). Atmospheric change of the electrostatic field strength at 7 p.m. GMT does not induce action potentials or change in the electrical properties of soybean or potato plants.

Acid rain is the most serious environmental problem and has impact on agriculture, forestry, and human health. Chemical reactions involving aerosol particles in the atmosphere are derived from the interaction of gaseous species with the liquid water. These reactions are associated with aerosol particles and dissolved electrolytes. For example, the generation of HONO from nitrogen oxides takes place at the air/water interface of seawater aerosols or in clouds. Clouds convert between 50 and 80% of SO_2 to H_2SO_4. This process contributes to the formation of acid rain. Acid rain exerts a variety of influences on the environment by greatly increasing the solubility of different compounds, thus directly or indirectly affecting many forms of life.

Acid rain has pH below 5.6. Sulfuric acid and nitric acid are the two predominant acids in acid rain. Approximately, 70% of the acid content in acid rain is sulfuric acid, with nitric acid contributing to the rest 30%. Spraying the soybean plant with an aqueous solution of H_2SO_4 in the pH region from 5.0 to 5.6 does not induce action potentials. However, action potentials were generated in soybean either by spraying the leaves of the plant (1 mL) or deposition of 10 μL drops of aqueous solution of H_2SO_4 or HNO_3 in the pH region from 0 to 4.9 on leaves. The duration of single action potentials after spraying the plant with HNO_3 and H_2SO_4 was 0.2 s and 0.02 s, respectively. We have demonstrated that plant sensors can be used for immediately detection of acid rains (Shvetsova et al. 2002).

7.7 Chemiotropism: Electrical Signals Induced by Pesticides and Uncouplers

Plants can be used as sensors for pesticides (Labady et al. 2002; Mwesigwa et al. 2000; Shvetsova et al. 2001; Volkov et al. 2001; Volkov and Mwesigwa 2001a, b). In photosynthetic pathways, the radiation from the sun excites photosynthetic pigments. This excitement compels the pigments to donate electrons to the electron acceptors in the electron transport chain. Pheophytin is the first molecule to receive this energized electron. Proton gradients across the thylakoid membrane are established as a result of the charge transfer in the electron transport chain associated with sequential redox reaction. The energy produced from this gradient is used to drive ATP synthesis. It is at this stage where uncouplers have the ability to separate the flow of electrons in the electron transport chain and the H^+ pump in ATP synthesis. Uncouplers are preventing the energy transfer from electron transport chain to ATPase. They are thought to uncouple oxidative phosphorylation.

Uncouplers are generally weak acids. These chemicals are often used to inhibit photosynthetic water oxidation due to their ability to become oxidized by the manganese cluster of the O_2-evolving complex of photosystem II (PSII) and chloroplast. Most protonophoric uncouplers, widely used in photosynthesis research, are oxidized by the manganese cluster of the PSII O_2-evolving complex in chloroplasts and inhibit photosynthetic water oxidation. Oxidized uncouplers can be reduced by the membrane pool of plastoquinone, leading to formation of an artificial cyclic electron transfer chain around PSII involving uncouplers as redox carriers. Protonophores such as carbonylcyanide m-chlorophenylhydrazone (CCCP), 2,3,4,5,6-pentachlorophenol (PCP), and 4,5,6,7-tetrachloro-2-trifluorom-ethylbenzimidazole (TTFB) inhibit the Hill reaction with $K_3Fe(CN)_6$ in chloroplast and cyanobacterial membranes. Inhibition of the Hill reaction by uncouplers reaches maximum when the pH corresponds to the pK values of these compounds.

Uncouplers promote autooxidation of the high-potential form of cytochrome b559 and partially convert it to lower potential forms. Protonophores uncouple electron transport, accelerate the deactivation of the S-2 and S-3 states on the donor side, and facilitate the oxidation of cytochrome b559 on the acceptor side of PSII.

Once oxidized, uncouplers can then be reduced by plastoquinone, thereby facilitating the formation of artificial cyclic electron transport chain around PSII involving uncouplers as redox carriers. Autooxidation of high potential cytochrome b559 is enhanced by the presence of uncouplers. Cytochrome b559 is also converted to low potential forms in the presence of chemical uncouplers.

Protonophores are able to: (1) uncouple electron transport and H^+ pump in ATP synthesis, (2) accelerate the deactivation of the S-2 and S-3 states on the donor side, and (3) facilitate oxidation of cytochrome b559 on the acceptor side of photosystem II Volkov et al. 1997. Although the interaction of proton-conducting ionophores with photosynthetic electron transport has been extensively studied during the past decade, the mode of action of protonophores remained uncertain. Electrochemical measurements in real time are required for a better understanding of the molecular mechanism of action of protonophores.

Pesticides PCP, 2,4-Dinitrophenol (DNP), CCCP, and crbonylcyanide-4-trifluoromethoxyphenylhydrazone (FCCP) act as insecticides and fungicides. PCP is the primary source of dioxins found in the environment. This pollutant is a defoliant and herbicide. PCP is utilized in termite control, wood preservation, seed treatment, and snail control. The pesticide DNP is used to manufacture dye and wood preservative. DNP is often found in pesticide runoff water. The electrochemical effects of CCCP, PCP, DNP, and FCCP have been evaluated on soybean plants.

CCCP decreased the variation potentials of soybean from 80–90 to 0 mV after 20 h. CCCP induced fast action potentials in soybean with amplitude of 60 mV. The maximum speed of propagation was 25 m/s. Exudation is a manifestation of the positive root pressure in the xylem. After treatment with CCCP, the exudation from cut stems of the soybean remains the same. Therefore, the addition of CCCP did not cause a change in the pressure, although it may influence the zeta potential due to depolarization (Shvetsova et al. 2001).

The addition of aqueous solution of PCP also causes the variation potential in soybeans to stabilize at 0 mV after 48 h. Rapid action potentials are induced. These action potentials last for 2 ms, and have amplitudes of 60 mV. The speed of propagation is 12 ms^{-1}; after 48 h, the speed increased to 30 ms^{-1}.

DNP induces fast action potentials and decreases the variation potential to zero in soybeans. The addition of aqueous DNP to the soil induces fast action potentials in soybeans. After treatment with an aqueous solution of DNP, the variation potential, measured between two Ag/AgCl electrodes in a stem of soybean, slowly decreases from 80 to 90 mV (negative in a root, positive on the top of the soybean) to 0 during a 48 h time period. The duration of single action potentials, 24 h after treatment by DNP, changes from 3 to 0.02 s. The amplitude of action potentials is about 60 mV. The maximum speed of action potential propagation is 1 m/s. After two days, the variation potential stabilized at 0. Fast action potentials were generated in a soybean, with amplitude of about 60 mV, 0.02 s duration time, and a speed of 2 m/s. Fromm and Spanswick (1993) studied the inhibiting effects of DNP on the excitability of willow by recording the resting potential in the phloem cells. In willow, 10^{-4} M DNP rapidly depolarized the membrane potential by about 50 mV.

The FCCP also induced action potentials in soybean. The maximum speed of these action potentials within 20 h after the treatment was 10 ms^{-1}. After 100 h, the action potentials were still being produced. The amplitude of 60 mV remained constant. The duration was 0.3 ms, and the speed of propagation was 40 ms^{-1}.

7.8 Gravitropism in Plants

Gravitropism is the ability of plant organs for directional growth or movement as a result of the gravitational force of the earth. Naturally, the powerful gravitational force dictates most of other environmental stimuli and influences the growth and development of a plant. Primary roots grow downward reaching for water and mineral ions and shoots grow upward facilitating the efficient photosynthesis. These two differential growth patterns, toward and away from the earth's center of gravity, are known as positive and negative gravitropism, respectively (Boonsirichai et al. 2002; Sack 1991).

Since the early postulation on the importance of the root tip, the cap, as the essential element for graviperception to generate the physiological signal, developments have been made in the understanding of physiological and molecular processes fundamental to the root gravitropism (Blancaflor and Masson 2003; Blancafloret al. 1998; Chen et al. 1999; Darwin 1880; Morita and Tasaka 1996; Stankovic 2006; Tasaka et al. 1999). Understanding the details of gravitropism at the molecular level is essential since the perceived gravitropic stimulus triggers several important physiological processes including signal generation, intracellular and intercellular signal transduction, followed by growth control leading to bending and reorientation of the responding organ. The intelligent behavior of plant root apices as *command centers* has been critically discussed and reviewed based on the new information obtained from electrophysiology and cell/molecular biology of higher plants (Baluska et al. 2004, 2005).

In literature, the mechanism of gravitropism is explained by two long-surviving hypotheses: the Cholodney–Went hypothesis and the starch-statolith hypothesis. The Cholodny-Went hypothesis suggests that the gravitropic curvature is due to lateral transport of auxin across the gravistimulated plant organs resulting asymmetric growth. According to the starch-statolith hypothesis, the gravity perception in roots occurs in the root cap due to sedimentation of starch-filled amyloplasts within the cells in the columella, the central region of the root cap. Laser ablation experiments to remove the innermost columella cells with the highest amyloplast-sedimentation velocities in the root cap of *Arabidopsis* primary cells have resulted in the strongest inhibitory effect in response to gravity.

It is suggested that the sedimenting amyloplasts can disrupt the local actin filaments in the plant cytoskeleton causing the activation of mechanosensitive ion channels in the plasma and/or intracellular membrane, followed by rapid increases in cytoplasmic ion levels. Despite the ongoing studies, fully understanding of the mechanism by which the physiological or biochemical signal is generated from the physical signal due to amyloplast sedimentation still remains unanswered.

Even though the existence of electrical potentials induced by gravitropism in higher plants has long been known (Bose 1907, 1913, 1918, 1928), the concept has been rarely studied until recently. For example, fast extracellular gravielectric potentials with maximum amplitude of 17 mV have been observed in soybean hypocotyls due to directional change in the gravity vector (Tanada and Vinten-Johansen 1980). A transient of rapid surface potential with about 10 mV has resulted in gravistimulated bean epicotyls approximately after 30–120 s following gravistimulation (Shigematsu et al. 1994). Recently, the current understanding of the electrophysiology of plant gravitropism was reviewed by Stankovic 2006. A detailed understanding of plant's response to gravity should provide valuable insights into future research programs in space plant biology and horticulture.

7.9 Conclusion

Green plants interfaced with a computer through data acquisition systems can be used as fast biosensors for monitoring the environment, detecting effects of pollutants, pesticides, defoliants, predicting and monitoring climate changes, and in agriculture, directing and fast controlling of conditions influencing the harvest. The use of new computerized methods provides opportunities for detection of fast action potentials in green plants in real time.

Plants are continuously exposed to a wide variety of perturbations including variation of temperature and/or light, mechanical forces, gravity, air and soil pollution, drought, deficiency or surplus of nutrients, attacks by insects and pathogens, etc., and hence, it is essential for all plants to have survival sensory mechanisms against such perturbations. Plants have evolved sophisticated systems to sense environmental stimuli for adaptation, and to sense signals from other cells for coordinated action. Consequently, plants generate various types of intracellular and intercellular electrical signals in response to these environmental changes.

In this particular case their main research interest is to study the bioelectrochemical mechanisms of acquisition of external stimuli by plants, its transduction into electrical signals, memorizing and/or transferring these signals, and actuation of mechanical and chemical devices for defense or attack.

This field has both theoretical and practical significance because discovering these mechanisms would greatly advance our knowledge of natural sensors, principles of their functioning and integration into general system of defense and attack. These systems play a very important role in the life of plants, but their nature is still very poorly understood. Our studies could advance this important scientific field and open new perspectives in technical application of these biological principles.

Nerve cells in animals and phloem cells in plants share one fundamental property: they possess excitable membranes through which electrical excitations can propagate in the form of action potentials. The cells, tissues, and organs of plants transmit electrochemical impulses over short and long distances. It is conceivable that action potentials are the mediators for intercellular and intracellular communication in response to environmental irritants. Action potential is a momentary change in electric potential on the surface of a cell that takes place when it is stimulated. Initially, plants respond to irritants at the site of stimulation; however, excitation waves can be conducted along the membranes throughout the entire plant. Bioelectrical impulses travel from the root to the stem and vice versa. Chemical treatment, intensity of the irritation, mechanical wounding, previous excitations, temperature, and other irritants influence the speed of propagation.

Acknowledgement This work was supported by the grant CBET-1064160 from the National Science Foundation.

References

Abe T (1981) Chloride ion efflux during an action potential in the main pulvinus of *Mimosa pudica*. Bot Mag Tokyo 94:379–383

Ahmad M, Jarillo JA, Smirnova O, Cashmore AR (1998) Cryptochrome blue-light photoreceptors implicated in phototropism. Nature 392:720–723

Allen RD (1969) Mechanism of the seismonastic reaction in *Mimosa pudica*. Plant Physiol 44:1101–1107

Asprey GF, Palmer JH (1955) A new interpretation of the mechanics of pulvinar movement. Nature 175:1122–1123

Babourina O, Newman I, Shabala S (2002) Blue light-induced kinetics of H^+ and Ca^{2+} fluxes in etiolated wild-type and phototropin-mutant *Arabidopsis* seedlings. Proc Natl Acad Sci USA 99:2433–2438

Balmer RT, Franks JG (1975) Contractile characteristics of *Mimosa pudica* L. Plant Physiol 56:464–467

Baluska F, Mancuso S, Volkmann D, Barlow P (2004) Root apices as plant command centers: the unique 'brain-like" status of the root apex transition zone. Biologia (Bratisl) (Suppl) 13:1-13

Baluska F, Volkmann D, Menzel D (2005) Plant synapses: actin-based domains for cell-to-cell communication. Trends Plant Sci 10:106–111

Bertholon M (1783) De l'electricite des vegetaux: ouvrage dans lequel on traite de l'electricite de l'atmosphere sur les plantes, de ses effets sur leconomie des vegetaux, de leurs vertus medico. P.F. Didot Jeune, Paris

Blancaflor EB, Fasano JM, Gilroy S (1998) Mapping the functional roles of cap cells in the response of *Arabidopsis* primary roots to gravity. Plant Physiol 116:213–222

Blancaflor EB, Masson PH (2003) Plant Gravitropism. unraveling the ups and downs of a complex process. Plant Physiol 133:1677–1690

Boonsirichai K, Guan C, Chae R, Masson PH (2002) Root gravitropism: an experimental tool to investigate basic cellular and molecular processes underlying mechanosensing and signal transmission in plants. Ann Rev Plant Biol 53:421–447

Bose JC (1907) Comparative electro-physiology, a physico-physiological study. Longmans Green and Co., London

Bose JC (1913) Researches on irritability of plants. Longmans, London

Bose JC (1918) Life movements in plants. B.R. Publishing Corp, Delhi
Bose JC (1926) The Nervous mechanism of plants. Longmans Green and Co., London
Bose JC (1928) The Motor Mechanism of Plants. Longmans Green, London
Burdon-Sanderson J (1873) Note on the electrical phenomena which accompany stimulation of the leaf of *Dionaea muscipula*. Philos Proc R Soc Lond 21:495–496
Casal JJ (2000) Phytochromes, cryptochromes, phototropin: photoreceptor interactions in plants. Photochem Photobiol 71:1–11
Cashmore AR, Jarillo JA, Wu YJ, Liu D (1999) Cryptochromes: blue light receptors for plants and animals. Science 284:760–765
Chen AM, Rosen ES, Masson PH (1999) Update: gravitropism in higher plants. Plant Physiol 120:343–350
Chen Y, Tan TC (1995) Dopamine-sensing efficacy and characteristics of pretreated palnt tissue powder sensors. Sens Actuators B 28:39–48
Cosgrove DJ, Hedrich R (1991) Stretch-activated chloride, potassium, and calcium channels coexisting in plasma membranes of guard cells of *Vicia faba* L. Planta 186:143–153
Cote GG (1995) Signal transduction in leaf movements. Plant Physiol 109:729–734
Darwin C (1880) The power of movements in plants. John Murray, London
Davies E (2006) Electrical signals in plants: Facts and hypothesis. In: Volkov AG (ed) Plant Electrophysiology. Springer, Berlin, pp 407–422
Denius HR, Homann PH (1972) The relation between photosynthesis, respiration, and crassulacean acid methabolism in the leaf slices of *Aloe arborescent* Mill. Plant Physiol 49:873–880
De Mairan M (1729) Observation botanique. Histoire de l'Academie Royale de Sciences, Paris, pp. 35–36
Dutrochet MH (1837) Memoires pour server a l'histoire anatomique et physiologique des vegetaux et des animaux, Bruxelles
Feynman RP, Leighton RB, Sands M (1963) The Feynman lectures on physics. Addison-Wesley, Reading
Fleurat-Lessard P, Bouche-Pillion S, Leloup C, Bonnemain J (1997) Distribution and activity of the plasma membrane H^+-ATPase related to motor cell function in *Mimosa pudica* L. Plant Physiol 114:827–834
Fleurat-Lessard P, Roblin G (1982) Comparative histology of the petiole and the main pulvinus in *Mimosa pudica* L. Ann Bot 50:83–92
Fondeville JC, Bortwick HA, Hendricks SB (1966) Leaflet movement of *Mimosa pudica* L Indicative of phytochrome action. Planta 69:357–364
Frechilla S, Talbott LD, Bogomolni RA, Zeiger E (2000) Reversal of blue light-stimulated stomatal opening by green light. Plant Physiol 122:99–106
Fromm J, Bauer T (1994) Action potentials in maize sieve tubes change phloem translocation. J Exp Bot 45:463–469
Fromm J, Spanswick R (1993) Characteristics of action potentials in willow (*Salix viminalis* L.). J Exp Bot 44:1119–1125
Gardiner W (1888) On the power of contractility exhibited by the protoplasm of certain plant cells. Ann Bot os 1:362–367
Goldsworthy A (1983) The evolution of plant action potentials. J Theor Biol 103:645–648
Gorton HL (1987) Water relation in pulvini from *Samanea saman*. 2. effects of excision of motor tissues. Plant Physiol 83:945–950
Haberlandt G (1890) Das reisleitende Gewebesystem der Sinnpflanse. W. Engelmann, Leipzig
Hamill OP, Martinac B (2001) Molecular basis of mechanotransduction in living cells. Physiol Rev 81:685–740
Hamill OP, Marty A, Neher E, Sackmann B, Sigworth FJ (1981) Improved patch-clamp techniques for high-resolution current recording from cells and cell-free membrane patches. Pfluegers Arch Eur J Physiol 391:85–100
He X, Rechnitz GA (1995) Plant tissue-based fiber-optic pyruvate sensor. Anal Chim Acta 316:57–63

Hooke R (1667) Micrographia. The Royal Society, London
Jonas H (1970) Oscillations and movements of *Mimosa* leave due to electric shock. J Interdisc Cycle Res 1:335–348
Kameyama K, Kishi Y, Yoshomura M, Kanzawa N, Tsuchia T (2000) Thyrosine phosphorylation in plant bending. Nature 407:37
Kanzawa N, Hoshino Y, Chiba M, Hoshino D, Kobayashi H, Kamasawa N, Kishi Y, Osumi M, Sameshima M, Tsuchia T (2006) Change in the actin cytoskeleton during seismonastic movement of *Mimosa pudica*. Plant Cell Physiol 47:531–539
Kim HY, Coté GG, Crain RC (1992) Effect of light on the membrane potential of protoplasts from *Samanea saman* pulvini. involvement of K^+ channels and the H^+-ATPase. Plant Physiol 99:1532–1539
Kim HY, Coté GG, Crain RC (1993) Potasium channels in Samanea saman protoplasts controlled by phytochrome and the biological clock. Science 260:960–962
Kloda A, Martinac B (2002) Common evolutionary origins of mechanosensitive ion channels in archaea, bacteria and cell-walled eukarya. Archaea 1:35–44
Ksenzhek OS, Volkov AG (1998) Plant energetics. Academic, San Diego
Kumon K, Tsurumi S (1984) Ion efflux from pulvinar cells during slow downward movement of the petiole of *Mimosa pudica* L. induced by photostimulation. J Plant Physiol 115:439–443
Kuriyama S, Rechnitz GA (1981) Plant tissue-based bioselective membrane electrode for glutamate. Anal Chim Acta 131:91 96
Labady A, Thomas D'J, Shvetsova T, Volkov AG (2002) Plant electrophysiology: excitation waves and effects of CCCP on electrical signaling in soybean. Bioelectrochem 57:47–53
Lebaudy A, Vavasseur A, Hisy E, Dreyer I, Leonhards N, Thibaud JB, Very AA, Simonneau T, Sentenac H (2008) Plant adaptation to fluctuating environment and biomass production are strongly dependent on guard cell potassium channels. Proc Natl Acad Sci USA 105:5271–5276
Lemström K (1904) Electricity in agriculture and horticulture. Electrician Publications, London
Liawruangrath S, Oungpipat W, Watanesk S, Liawruangrath B, Dongduen C, Purachat P (2001) Asparagus-based amperometric sensor for fluoride determination. Anal Chim Acta 448:37–46
Lindley J (1854) New plant *Vallota miniata*. The Gardener's Chronicle 8:119
Liubimova MN, Deminovskaya NS, Fedorovich IB (1964) The part played by ATP in the motor function of *Mimosa pudica* leaf. Biokhimia (Moscow) 29:774–779
Liubimova-Engel'gardt MN, Burnasheva SA, Fain FS, Mitina NA, Poprykina IM (1978) *Mimosa pudica* adenosine triphosphatase. Biokhimia (Moscow) 43:748–760
Maffei ME, Mithofer A, Boland W (2007) Insect feeding on plants: rapid signals and responses preseding the induction of phytochemical release. Phytochemistry 68:2946–2959
Markin VS, Sachs F (2004) Thermodynamics of mechanosensitivity. physical. Biol 1:110–124
Markin VS, Volkov AG, Jovanov E (2008) Active movements in plants: mechanism of trap closure by *Dionaea muscipula* Ellis. Plant Signal Behav 3:778–783
Martinac B, Kloda A (2003) Evolutionary origins of mechanosensitive ion channels. Progress Biophys Mol Biol 82:11–24
Martinac B (2004) Mechanosensitive ion channels: molecules of mechanotransduction. J Cell Sci 117:2449–2460
Mayer WE, Flash D, Raju MVS, Starrach N, Wiech E (1985) Mechanics of circadian pulvini movements in *Phaseolus coccineus* L. shape and arrangement of motor cells, micellation of motor cell walls, and bulk moduli of extensibility. Planta 163:381–390
Mei Y, Ran L, Ying X, Yuan Z, Xin S (2007) A sequential injection analysis/chemiluminescent plant tissue-based biosensor system for the determination of diamine. Biosens Bioelectron 22:871–876
Morita MT, Tasaka M (1996) Gravity sensing and signaling. Curr Opin Plant Biol 7:712–718
Mwesigwa J, Collins DJ, Volkov AG (2000) Electrochemical signaling in green plants: effects of 2,4-dinitrophenol on resting and action potentials in soybean. Bioelectrochem 51:201–205
Pal M, Roychaudhury A, Pal A, Biswas S (1990) A novel tubulin from *Mimosa pudica*. Eur J Biochem 192:329–335

Pfeffer W (1905) The physiology of plants. Clarendon Press, Oxford
Quail PH (1997) An emerging molecular map of the phytochromes. Plant, Cell Environ 20:657–665
Quin W, Zhang Z, Peng Y (2000) Plant tissue-base chemiluminescence flow biosensor for urea. Analytica Chim Acta 407:81–86
Racusen R, Satter RL (1975) Rhytmic and phytochrome-regulated changes in transmembrane potential in *Samanea pulvini*. Nature 255:408–410
Regel E (1864) *Clivia miniata* Lindl. amaryllideae. Gartenflora 14:131–134
Ricca U (1916) Soluzione d'un problema di fisiologia. La propagazione di stimulo nella "*Mimosa.*" Nuovo Giornale Botanico Italiano Nuovo Serie 23:51-170
Ritter JW (1811) Electrische Versuche an der *Mimosa pudica* L. In Parallel mit gleichen Versuchen an Fröschen. Denkschr Köningl Akad Wiss (München) 2:345-400
Sack FD (1991) Plant gravity sensing. Intl Rev Cytol 127:193–252
Samejima M, Sibaoka T (1982) Membrane potentials and resistance of excitable cells in the petiole and main pulvinus of *Mimosa pudica*. Plant Cell Physiol 23:459–465
Satter RL (1990) Leaf movements: an overview of the field. In: Satter RL, Gorton HL, Vogelmann TC (eds) The pulvinus: motor organ for leaf movement. The American Society of Plant Physiologists, Rockville, pp 1–9
Schildknecht H, Bender W (1983) Chemonastisch wirksame leaf factors aus *Mimosa pudica*. Chemische Zeitung 107:111–114
Schildknecht H, Meier-Ausgenstein W (1990) Role of turgorins in leaf movement. In: Satter RL, Gorton HL, Vogelman TC (eds) The pulvinus: motor organ for leaf movement. American Society of Plant Physiologists, Rockville, pp 101–129
Shimmen T (2006) Electrophysiology in mechanosensing and wounding response. In: Volkov AG (ed) Plant electrophysiology. Springer, Berlin, pp 319–339
Scott BIH, Gulline HF (1975) Membrane changes in a circadian system. Nature 254:69–70
Shigematsu H, Toko K, Matsuno T, Yamafuji K (1994) Early gravi-electrical responses in bean epicotyls. Plant Physiol 105:875–880
Short TW, Briggs WR (1994) The transduction of blue light signals in higher plants. Ann Rev Plant Physiol Plant Mol Biol 45:143–171
Shvetsova T, Mwesigwa J, Labady A, Kelly S, Thomas D'J, Lewis K, Volkov AG (2002) Soybean electrophysiology: effects of acid rain. Plant Sci 162:723–731
Shvetsova T, Mwesigwa J, Volkov AG (2001) Plant electrophysiology: FCCP induces fast electrical signaling in soybean. Plant Sci 161:901–909
Sidwell JS, Rechnitz GA (1986) Progress and challenges for biosensors using plant tissue materials. Biosensors 2:221–233
Sinukhin AM, Britikov EA (1967) Action potentials in the reproductive system of plant. Nature 215:1278–1280
Stankovic B (2006) Electrophysiology of plant gravitropism. In: Volkov AG (ed) Plant electrophysiology. Springer, Berlin
Stoeckel H, Takeda K (1993) Plasmalemmal, voltage-dependent ionic currents from excitable pulvinar motor cells of *Mimosa pudica*. J Membr Biol 131:179–192
Tamiya T, Miyasaki T, Ishikawa H, Iruguchi N, Maki T, Matsumoto JJ, Tsuchiya T (1988) Movement of water in conjunction with plant movement visualized by NMR imaging. J Biochem 104:5–8
Tanada T, Vinten-Johansen C (1980) Gravity induces fast electrical field change in soybean hypocotyls. Plant, Cell Environ 3:127–130
Tasaka M, Kato T, Fukaki H (1999) The endodermis and shoot gravitropism. Trends Plant Sci 4:103–107
Taya M (2003) Bio-inspired design of intelligent materials. Proc SPIE 5051:54–65
Temmei Y, Uchida S, Hoshino D, Kanzawa N, Kuwahara M, Sasaki S, Tsuchiya T (2005) Water channel activities of *Mimosa pudica* plasma membrane intrinsic proteins are regulated by direct interaction and phosphorylation. FEBS Lett 579:4417–4422
Thomas B, Vince-Prue D (1997) Photoperiodism in plants. Academic, San Diego

Toriyama H, Jaffe M (1972) Migration of calcium and its role in the regulation of seismonasty in the motor cell of *Mimosa pudica* L. Plant Physiol 49:72–81

Toriyama H (1955) Observational and experimental studies of sensitive plants. VI. the migration of potassium in the primery pulvinus. Cytologia 20:367–377

Varin L, Chamberland H, Lafontaine JG, Richard M (1997) The enzyme involved in sulfation of the turgorins, gallic acid 4-O-(β-D-glucopyranosyl-6'-sulfate) is pulvini-localized in *Mimosa pudica*. Plant J 12:831–837

Volkov AG (1989) Oxygen evolution in the course of photosynthesis. Bioelectrochem Bioenerg 21:3–24

Volkov AG (2000) Green plants: Electrochemical interfaces. J Electroanal Chem 483:150–156

Volkov AG (ed) (2006a) Plant electrophysiology. Springer, Berlin

Volkov AG (2006b) Electrophysiology and phototropism. In: Balushka F, Manusco S, Volkman D (eds) Communication in plants neuronal aspects of plant life. Springer, Berlin, pp 351–367

Volkov AG, Deamer DW, Tanelian DL, Markin VS (1998) Liquid interfaces in chemistry and biology. Wiley, New York

Volkov AG, Adesina T, Markin VS, Jovanov E (2007a) Closing of Venus flytrap by electrical stimulation of motor cells. Plant Signal Behav 2:139–144

Volkov AG, Adesina T, Markin VS, Jovanov E (2008a) Kinetics and mechanism of *Dionaea muscipula* trap closing. Plant Physiol 146:694–702

Volkov AG, Baker K, Foster JC, Clemmens J, Jovanov E, Markin VS (2011a) Circadian variations in biologically closed electrochemical circuits in *Aloe vera* and *Mimosa pudica*. Bioelectrochem 81:39–45

Volkov AG, Carrell H, Adesina T, Markin VS, Jovanov E (2008b) Plant electrical memory. Plant Signal Behav 3:490–492

Volkov AG, Carrell H, Baldwin A, Markin VS (2009a) Electrical memory in Venus flytrap. Bioelectrochem 75:142–147

Volkov AG, Carrell H, Markin VS (2009b) Biologically closed electrical circuits in Venus flytrap. Plant Physiol 149:1661–1667

Volkov AG, Collins DJ, Mwesigwa J (2000) Plant electrophysiology: pentachlorophenol induces fast action potentials in soybean. Plant Sci 153:185–190

Volkov AG, Coopwood KJ, Markin VS (2008c) Inhibition of the *Dionaea muscipula* Ellis trap closure by ion and water channels blockers and uncouplers. Plant Sci 175:642–649

Volkov AG, Deamer DW, Tanelian DI, Markin VS (1997) Liquid interfaces in chemistry and biology. J Wiley, New York

Volkov AG, Dunkley T, Labady A, Brown C (2005) Phototropism and electrified interfaces in green plants. Electrochim Acta 50:4241–4247

Volkov AG, Dunkley TC, Labady AJ, Ruff D, Morgan SA (2004a) Electrochemical signaling in green plants induced by photosensory systems: molecular recognition of the direction of light. In: Bruckner-Lea C, Hunter G, Miura K, Vanysek P, Egashira M, Mizutani F (eds) Chemical sensors vi: chemical and biological sensors and analytical methods. Pennington, The Electrochemical Society, pp 344–353

Volkov AG, Dunkley TC, Morgan SA, Ruff D, Boyce Y, Labady AJ (2004b) Bioelectrochemical signaling in green plants induced by photosensory systems. Bioelectrochem 63:91–94

Volkov AG, Foster JC, Ashby TA, Walker RK, Johnson JA, Markin VS (2010a) *Mimosa pudica*: electrical and mechanical stimulation of plant movements. Plant, Cell Environ 33:163–173

Volkov AG, Foster JC, Baker KD, Markin VS (2010b) Mechanical and electrical anisotropy in *Mimosa pudica*. Plant Signal Behav 5:1211–1221

Volkov AG, Foster JC, Jovanov E, Markin VS (2010c) Anisotropy and nonlinear properties of electrochemical circuits in leaves of *Aloe vera* L. Bioelectrochem 81:4–9

Volkov AG, Foster JC, Markin VS (2010d) Molecular electronics in pinnae of *Mimosa pudica*. Plant Signal Behav 5:826–831

Volkov AG, Foster JC, Markin VS (2010e) Signal transduction in *Mimosa pudica*: Biologically closed electrical circuits. Plant, Cell Environ 33:816–827

Volkov AG, Foster JC, Markin VS (2010f) Molecular electronics in pinnae of *Mimosa pudica*. Plant Signal Behav 5:1–6
Volkov AG, Foster JC, Markin VS (2011b) Anisotropy and nonlinear properties of electrochemical circuits in leaves of *Aloe vera* L. Bioelectrochem 81:4–9
Volkov AG, Haack RA (1995a) Bioelectrochemical signals in potato plants. Russ J Plant Physiol 42:17–23
Volkov AG, Haack RA (1995b) Insect induced bioelectrochemical signals in potato plants. Bioelectrochem Bioenerg 35:55–60
Volkov AG, Labady A, Thomas D'J, Shvetsova T (2001a) Green plants as environmental biosensors: electrochemical effects of carbonyl cyanide 3-chlorophenylhydrazone on soybean. Analytical Sci 17:i359–i362
Volkov AG, Lang RD, Volkova-Gugeshashvili MI (2007b) Electrical signal in *Aloe vera* induced by localized thermal stress. Bioelectrochem 71:192–197
Volkov AG, Mwesigwa J (2001a) Electrochemistry of soybean: effects of uncouplers, pollutants, and pesticides. J Electroanal Chem 496:153–157
Volkov AG, Mwesigwa J (2001b) Interfacial electrical phenomena in green plants: action potentials. In: Volkov AG (ed) Liquid interfaces in chemical, biological, and pharmaceutical applications. M Dekker, New York, Pp, pp 649–681
Volkov AG, Mwesigwa J, Shvetsova T (2001b) Soybean as an environmental biosensor: action potentials and excitation waves. In: Butler M, Vanysek P, Yamazoe N (eds) Chemical and biological sensors and analytical methods, vol II. The Electrochemical Society, Pennington, pp 229–238
Volkov AG, Pinnock MR, Lowe DC, Gay MS, Markin VS (2011c) Complete hunting cycle of *Dionaea muscipula*: consecutive steps and their electrical properties. J Plant Physiol 168:109–120
Volkov AG, Wooten JD, Waite AJ, Brown CR, Markin VS (2011d) Circadian rhythms in electrical circuits of *Clivia miniata*. J Plant Physiol 168:1753–1760
Wåhlin L (1986) Atmospheric electrostatics. Wiley, New York
Weintraub M (1951) Leaf movements in *Mimosa pudica* L. New Phytol 50:357–382
Wijesuriya DC, Rechnitz GA (1993) Biosensors based on plant and animal tissues. Biosens Bioelectron 8:155–160
Yamashiro S, Kameyama K, Kanzawa N, Tamiya T, Mabuchi I, Tsuchia T (2001) The gelsolin/fragmin family protein indentified in the higher plant *Mimosa pudica*. J Biochem 130:243–249
Yao H, Xu Q, Yuan M (2008) Actin dynamics mediates the changes of calcium level during the pulvinus movement of *Mimosa pudica*. Plant Signal Behav 3:954–960
Zhu L, Li Y, Zhu G (2004) A novel renewable plant tissue-based electrochemiluminescent biosensor for glycolic acid. Sens Actuators B 98:115–121

Chapter 8
Generation, Transmission, and Physiological Effects of Electrical Signals in Plants

Jörg Fromm and Silke Lautner

Abstract This review explores the relationship between electrical long-distance signaling and the potential consequences for physiological processes in plants. Electrical signals such as action potentials (APs) and variation potentials (VPs) can be generated by spontaneous changes in temperature, light, touch, soil water content, by electrical as well as chemical stimulation or by wounding. An AP is evoked when the stimulus is sufficiently great to depolarize the membrane to below a certain threshold, while VPs are mostly induced by wounding, which induces a hydraulic wave transmitted through the xylem, thereby causing a local electrical response in the neighboring symplastic cells. Once generated, the signal can be transmitted over short distances from cell-to-cell through plasmodesmata, and after having reached the phloem it can also be propagated over long distances along the sieve tube plasma membrane. Such electrical messages may have a large impact on distant cells, as numerous well-documented physiological effects of long-distance electrical signaling have been shown. Electrical signals, for instance, affect phloem transport as well as photosynthesis, respiration, nutrient uptake, and gene expression.

8.1 Introduction

Over the past years, focus on electrophysiology research has been shifted strongly toward long-distance signaling. Curiously, plants possess most of the chemistry of the neuromotoric system in animals, i.e., neurotransmitter, such as acetylcholine,

---

J. Fromm (✉) · S. Lautner
Institute for Wood Biology, Universität Hamburg, Leuschnerstrasse 91,
21031 Hamburg, Germany
e-mail: j.fromm@holz.uni-hamburg.de

cellular messengers like calmodulin, cellular motors e.g., actin and myosin, voltage-gated ion channels and sensors for touch, light, gravity, and temperature. Although this cellular equipment has not reached the same great complexity as is the case in nerves, a very simple neural network has been formed within the phloem allowing plants to communicate successfully over long distances. The reason why plants have developed pathways for electrical signal transmission most probably lies in the necessity to rapidly respond to environmental stress factors. Different environmental stimuli evoke specific responses in living cells which have the capacity to transmit a signal to the responding region. In contrast to chemical signals such as hormones, electrical signals are able to transmit information over long distances very quickly: most of the plant action potentials (AP) studied so far have a velocity in the range of 0.005–0.2 m s^{-1} (Fromm and Lautner 2007).

As regards to the origin of the neuronal system in plants it appears unlikely that it was adopted from animal system. In our search for the common evolutionary roots of AP in plants and animals we need to look at unicellular ancestors which have no need of transmitting signals over long distances. Hence, the function of electrical transmission has most probably evolved at a later evolutionary stage. The assumption is that in the course of evolution the development of plants and animals branched off into different directions. Since cellular excitability was found to exist in primitive organisms, it is obvious that both plants and animals inherited their basic neuronal capabilities from their bacterial ancestors (Simons 1992). Szmelcman and Adler (1976) observed changes in membrane potential during bacterial chemotaxis. Even the sensitivity to mechanical touch is known to be an early evolutionary achievement. Martinac et al. (1987) detected pressure-sensitive ion channels in *Escherichia coli*, suggesting that these channels have an osmotic function. For the early evolution of action potentials, an osmotic function can also be assumed in unicellular algae such as *Acetabularia* (Mummert and Gradmann 1976). A mechanosensitive ion channel was also found in the yeast plasma membrane (Gustin et al. 1988) providing convincing evidence that plants inherited mechanical sensitivity from bacterial ancestors in the course of millions of years of evolution. The characean algae, which include *Chara* and *Nitella* are also known to be ancestors of higher plants. In *Nitella*, AP were observed in the internodal cells in 1898 by Hörmann who used extracellular electrodes long before APs were observed in isolated nerve cells by Adrian and Bronk (1928). Characean internodal cells respond to electrical stimulation in a manner similar to the contraction response displayed by skeletal muscles following electrical stimulation by nerve cells. In characean cells, electrical stimulation causes the cessation of protoplasmic streaming which is incited by the same interactions between actin and myosin that cause contraction in muscles (Hörmann 1898). In the course of evolution, once plants had gained and settled on dry land their excitability and neuronal capability were used to develop numerous survival tactics. For instance, one important step was the development of stomatal guard cells rapidly responding to environmental changes, while another was the development of an electrical communication system which uses the phloem to transmit information over long distances within the plant body (Fromm and Lautner 2005, 2007).

8.2 Generation of Electrical Signals

Electrical signals can be generated at almost any site of the symplastic continuum by environmental stimuli like changes in temperature, touch, or wounding. Even acid rain can induce AP (Shvetsova et al. 2002), as well as irradiation at various wavelengths (Volkov et al. 2004). Upon perception, electrical signals can be propagated via plasmodesmata to other cells of the symplast (van Bel and Ehlers 2005). With regard to APs, the plasma membrane is depolarized first, a process known as formation of the receptor potential, e.g., by mechanical stimulation as observed in *Chara* (Kishimoto 1968). The receptor potential is defined as an electrical replica of the stimulus lasting for the period of time that the stimulus is present. An action potential is evoked when the stimulus is great enough to depolarize the membrane to below a certain threshold. Observations in response to cold shock have shown that both mechanosensitive Ca^{2+} channels can be modulated (Ding and Pickard 1993) and also depolarization-activated Ca^{2+} channels can open to induce Ca^{2+} influx (Furch et al. 2009; White 2004, 2009; White and Ridout 1999). During cold sensing microtubules have a dual function as primary sensors as well as amplifiers (Nick 2008). In response to cold, microtubules disassemble and release the constraints upon the activity of the ion channels such that calcium can enter the cell. Interaction with calmodulin will facilitate further disassembly of microtubules and thus trigger a positive feedback loop (Nick 2008). Subsequently, Cl^- efflux can be induced and an AP is generated (Fig. 8.1).

In contrast to action potentials, variation potentials (VP) depend on a spontaneous wound-induced relaxation of the hydrostatic pressure in the xylem. As a consequence, the neighboring living cells intake water and induce trains of local turgor changes (Stahlberg and Cosgrove 1997; Davies and Stankovic 2006). The variation potential is generated by the sudden increase in turgor which elicits gating of mechanosensitive Ca^{2+} channels and the knockout of plasma membrane proton pumps (Stahlberg et al. 2006). As wounding causes Ca^{2+} signals a rise of the cytosolic Ca^{2+} level can also affect the activity of further ion channels such as the slow vacuolar (SV) channel which is regulated by Ca^{2+} (Beyhl et al. 2009; Rienmüller et al. 2010). In addition to the hydraulic wave, wound-induced chemicals are able to cause a local electrical response in the surrounding cells via ligand-activated channels.

8.3 Transmission of Electrical Messages

We will explore here the characteristics of electrical signals, including their means of transmission and the electrical properties of the phloem. Those molecular, biophysical, and structural basics are known to have a large impact on long-distance signaling and its physiological responses.

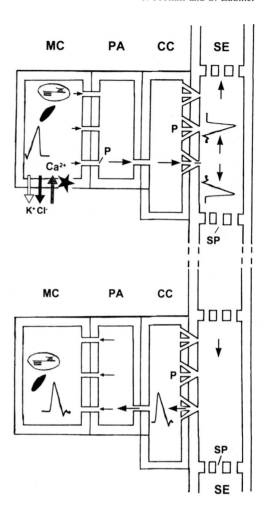

Fig. 8.1 Electrical signaling in higher plants. Stimulation by cold-shock or touch (*star*) induces calcium influx into a living cell, e.g., a mesophyll cell (*MC, above*). After the membrane potential is depolarized below a certain threshold level, an action potential is elicited by chloride and potassium efflux. The signal is propagated over short distances through plasmodesmal (*P*) networks and, after it passed the few plasmodesmata between sieve element/companion cells (*SE/CCs*) and phloem parenchyma cells (*PA*), will enter the SE/CC-complex to be transmitted over long distances. Sieve pores (*SP*) with their large diameters present low-resistance corridors for a rapid propagation of electrical signals along the SE plasma membrane. Such signals can leave the phloem at any site via plasmodesmata (*below*) to affect certain physiological processes in the neighboring tissue

8.3.1 Types of Signals

In plants, two main types of electrical signals exist. AP can be induced by sudden changes in temperature, light, touch, irrigation as well as by chemical and electrical stimulation. An action potential usually has an all-or-nothing character, and it travels with constant velocity and magnitude (Zawadzki et al. 1991). Its speed of transmission is in the range of 0.5–20 cm s^{-1} in most species and it is characterized by a large transient depolarization and spreads passively from the excited region of plasma membrane to the neighboring nonexcited region. The ion transport processes which create the conditions necessary for the generation of an action potential were investigated intensively in members of the green algal family Characeae (Tazawa et al. 1987). One of the ion transport mechanisms responsible for depolarization is based on chloride, the efflux of which increases upon membrane stimulation (Gaffey and Mullins 1958; Oda 1976). Another ion involved in plasma membrane excitation is calcium. Studies have shown that both the peak of the action potential (Hope 1961) and the inward current (Findlay 1961, 1962) are dependent on the apoplastic calcium concentration. Several publications prove that both chloride and calcium are involved in the formation of an individual action potential (e.g., Beilby and Coster 1979; Lunevsky et al. 1983). In addition to these ions, it was found that potassium efflux from the cell increases upon stimulation of the membrane (Spyropoulos et al. 1961; Oda 1976). These ion shifts during an action potential were confirmed in higher plants, such as trees by a method which uses inhibitors of ionic channels as well as energy-dispersive X-ray microanalysis (Fromm and Spanswick 1993). Results also indicated that calcium influx as well as potassium and chloride efflux are involved in the generation of action potentials. When AP were induced by electrical stimulation in willow it became clear that the required stimulus depends on both, its intensity as well as its duration (Fromm and Spanswick 1993). An increase in stimulus strength does not produce any change in the amplitude nor in the form of the action potential once it has been induced, showing that it conforms to the all-or-nothing law. Concerning refractory periods, they were found to be much longer in plants than in animal systems, with durations between 1 s in the Venus flytrap (Volkov et al. 2008c) and 50 s in maize (Fromm and Bauer 1994).

In contrast to action potentials, VPs which in the literature are also found to be called slow wave potentials, show longer and delayed repolarizations and a large range of variation. Typically they are slower propagating than APs, having a speed of 0.5–5 mm in most species. They can be generated by wounding, organ excision, or flaming, are non-self-perpetuating and vary with the intensity of the stimulus. VPs are induced in xylem parenchyma cells by a local change to either a hydraulic wave or a chemical stimulus.They are characterized by amplitudes and velocities that decrease with increasing distance from the injured site and by their dependence on xylem tension. When xylem tension becomes negligible at saturating humidity, VPs cannot be evoked. Their ionic basis differs from the one underlying APs because they are induced by a transient shutdown of a P-type H^+ ATPase (Stahlberg et al. 2006) and a turgor-dependent gating of mechanosensitive Ca^{2+} channels.

8.3.2 Means of Signal Transmission

In general, electrical signals allow the rapid transmission of information via plasmodesmata (Fig. 8.1). Electrical coupling via plasmodesmata was demonstrated in a variety of species such as *Nitella* (Spanswick and Costerton 1967); *Elodea* and *Avena* (Spanswick 1972) and *Lupinus* (van Bel and van Rijen 1994), indicating that plasmodesmata are relays in the signaling network between neighboring cells. However, long distances between different organs can be bridged more rapidly only via low resistance connections extending continuously throughout the whole plant. The sieve tube system seems to fulfill these conditions perfectly, for its unique structure of the sieve tube members appears to be well suitable for the transmission of electrical signals due to the relatively large, unoccluded sieve plate pores, the continuity of the plasma membrane and ER (Evert et al. 1973), as well as to the lack of vacuoles. Moreover, the low degree of electrical coupling in lateral direction caused by only few plasmodesmata at the interface between companion cells and phloem parenchyma cells (Kempers et al. 1998) facilitates long-distance signaling. However, those lateral plasmodesmata are making electrical signaling from neighboring cells to the sieve elements/companion cells possible (SE/CC, Fig. 8.1). Thus, signal transmission within the plant depends on the electrical conductance of plasmodesmata in lateral direction as well as on the high degree of electrical coupling via the sieve pores in longitudinal direction. Regarding VPs, the processes between the stimulation of mechanosensitive channels in xylem parenchyma cells and the transmission of the generated VP into and along sieve elements have not been studied in detail so far. Future work has also to show at what stage the VP extinguishes along the phloem pathway.

Strong evidence has accumulated that electrical transmission in sieve elements also occurs in species that do not perform rapid leaf movements as e.g., in Mimosa. In zucchini plants electrical signal transmission via sieve tubes between a growing fruit and the petiole of a mature leaf reached maximum velocities of 10 cm s^{-1} (Eschrich et al. 1988). This is in a similar velocity range as the movement of the action potential in sieve tubes of *Mimosa pudica* (Fromm and Eschrich 1988b). It is obvious that no chemical substance is capable of moving so fast in the assimilate flow. By contrast, hydraulic signals might transmit stimulations, but they would not be able to carry encoded plus- or minus-signals for hyperpolarization or depolarization, respectively as shown in poplar sieve tubes (Lautner et al. 2005). Recently, Mencuccini and Hölttä (2010) have brought to the forefront the potentially important role of phloem pressure-concentration waves in the coupling of photosynthesis and belowground respiration. However, their conclusion that pressure-concentration waves traveling at relatively high speed along the phloem can carry important physiological information that affects soil 'autotrophic' respiration was questioned (Kayler et al. 2010) because the mechanisms behind pressure-concentration waves are not clear.

In the wounded tomato plant the pathway for systemic electrical signal transmission is also associated with the phloem (Rhodes et al. 1996), indicating that it regulates the induction of proteinase inhibitor activity in parts of the shoot distant from the wound (Wildon et al. 1992).

8.3.3 The Aphid Technique as a Tool for Measuring Electrical Signals in the Phloem

Since the phloem is located inside the plant body several cell layers distant from the plant surface, experiments on electrical signaling via the phloem of intact plants are difficult to perform. Microelectrode measurements in combination with dye solutions injected into the cell to be measured after obtaining electrophysiological results is a time-consuming technique because the measured cell type can only be roughly estimated at the beginning and very often the microelectrode tip was not properly inserted in the phloem as revealed by microscopic checks after the experiment. Microelectrodes brought into contact with sieve tube exudates that appear at the cut end of an aphid stylet (Wright and Fisher 1981; Fromm and Eschrich 1988b), enabled us to monitor the membrane potential of sieve tubes and its changes after plant stimulation (Fig. 8.2a, b). The successful use of aphid stylets to measure electrical signals within the sieve tubes depends on their functioning as an effective salt bridge between the sieve tube cytoplasm and the microelectrode. Sieve tube exudates typically contain high K^+ concentrations; measurements on barley leaves gave values ranging from 50 to 110 mM (Fromm and Eschrich 1989). The stylet's food canal dimensions can be used to roughly calculate its electrical resistance. Using an average area of 6 μm^2 and assuming the canal to be filled with 100 mM KCl, its resistance would be about 2.6×10^9 Ω (Wright and Fisher 1981). Although this value is about three times greater than the typical resistance of a glass microelectrode, it is still within the measuring capacities of the electrometer used (input impedance $>10^{12}\Omega$). The stylets are embedded in hardened saliva, which insulates electrically.

For instance, *M. pudica*, a classic example for the conductance of rapid excitation in higher plants, can be stimulated by wounding, touch, chemically, or electrically (Volkov et al. 2010). When a microelectrode tip was brought into contact with the stylet stump at the petiolus with its cut end sealed into saline solution to which the Ag/AgCl reference electrode was connected, a resting potential of -160 mV was established, well in line with values found in other species and by other methods (Eschrich et al. 1988; van Bel and van Rijen 1994). Cooling the apical end of the petiolus evoked a rapidly moving action potential transmitted basipetally within the sieve tubes at the rate of up to 3–5 cm s^{-1} (AP, Fig. 8.2b). In contrast, wounding by flaming induced a variation potential with irregular form and of long duration (VP, Fig. 8.2b). In Mimosa, both action and VP become immediately evident as bending pulvini cause impressive movements of the paired leaflets. Microautoradiography of the petiolus showed the localization of the phloem by ^{14}C-labeled photoassimilates from the leaves exposed to $^{14}CO_2$ (Fig. 8.2c). The vascular bundles are surrounded by a sclerenchyma sheath in order to restrict electrical signaling to the phloem. When a phloem-transmitted action potential reaches the pulvinus which has no sclerenchyma (Fig. 8.2d, Fromm and Eschrich 1988a), it is transmitted laterally via plasmodesmata into the cells of the motor cortex. The latter possess voltage-gated ion channels which respond to the signal, causing ion efflux associated with water efflux which lead to leaf movements (Fromm and Eschrich 1988c).

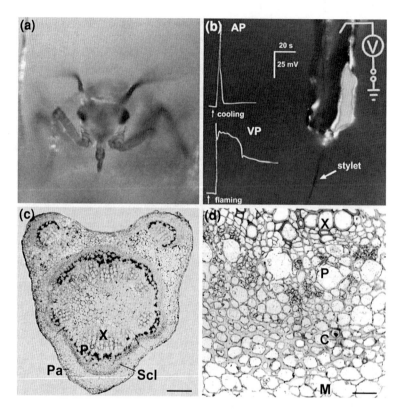

Fig. 8.2 Transmission of AP and VP in sieve tubes of *Mimosa*. **a** Front-view of *Rhopalosiphum padi* sucking at the base of a petiolus with its stylet inserted into a sieve element (x32). **b** After the aphid separated from its stylet by a laser pulse, the stylet stump exuded sieve tube sap to which the tip of a microelectrode was attached (x400). Cooling the apical end of the petiolus evoked an action potential (*AP*) while flaming triggered a variation potential (*VP*) transmitted basipetally within the sieve tubes. **c** Microautoradiography of the petiolus. ^{14}C-labeled photoassimilates from the leaves accumulated in the phloem (*P*) which is surrounded by a sclerenchyma sheath (*Scl*) in order to restrict electrical signaling to the phloem. Bar = 150 μm. **d** Microautoradiograph of a cross section of the primary pulvinus at the base of the petiolus. Labeled photoassimilates are restricted to the phloem strands (*P*). Since sclerenchyma tissue is absent, electrical signals can be transmitted laterally from the phloem via living collenchyma cells (*C*) to the motor cells (*M*) which cause the leaf movements by either losing or gaining turgor. X, xylem; Pa, parenchyma. Bar = 30 μm

8.3.4 Electrical Properties of the Phloem

The phloem presents a network for assimilate allocation as well as chemical and electrical communication within the plant. Concerning assimilate transport, osmolytes like sucrose generate the hydrostatic pressure which drives nutrient and water flow between the source and the sink phloem. Proton-coupled sucrose symporters, such as ZmSUT1, localized to the sieve tube and companion cell plasma membrane are capable of mediating both, the sucrose uptake into the phloem in mature leaves and the desorption of sugar from the phloem into sink tissues. The use of patch-clamp techniques revealed that the ZmSUT1-mediated sucrose-coupled proton current depended on the direction of the sucrose- and pH-gradient as well as on the membrane potential across the transporter (Carpaneto et al. 2005). Concerning the membrane potential it has been shown that a sink-source-regulated and sugar-inducible K$^+$ channel (VFK1) dominates the electrical properties of the sieve tube plasma membrane (Ache et al. 2001). The source site of phloem cells is characterized by K$^+$ concentrations of about 100 mM within the cytoplasm and 10 mM in the apoplast (Mühling and Sattelmacher 1997), resulting in an equilibrium potential for potassium ions (EK) of around -60 mV. Since the membrane potential of the SE/CC is between -130 and -200 mV (Ache et al. 2001; van Bel 1993) and thus more negative than EK, VFK1 loads K$^+$ into the phloem. Furthermore, K$^+$ channels of the AKT2/3 family have been identified as photosynthate-induced phloem channels. From studies of an AKT2/3 loss-of-function mutant it was shown that this mutant exhibited reduced K$^+$ dependence of the phloem potential and that AKT2/3 regulates sucrose/H$^+$ symporters via the membrane potential (Deeken et al. 2002). Furthermore, there is an electrogenic component of the sieve tube membrane potential, the magnitude of which is substantially greater than that predicted for EK (Wright and Fisher 1981). With regard to calcium, dihydropyridine-sensitive Ca^{2+} channels have been localized in the sieve element plasmalemma from *Nicotiana tabacum* and *Pistia stratiotes* (Volk and Franceschi 2000) as well as from *Vicia faba* (Furch et al. 2009), indicating that Ca^{2+} channels are also abundant in sieve elements. In general, non-specific cation channels permeable to Ca^{2+} can be divided into three main groups in plants: hyperpolarization-activated Ca^{2+} channels (Demidchik and Maathuis 2007; Hamilton et al. 2000; Kiegle et al. 2000), depolarization-activated Ca^{2+} channels (Thuleau et al. 1994; Thion et al. 1998; White 2009), and mechanosensitive channels (Cosgrove and Hedrich 1991; Ding and Pickard 1993; Dutta and Robinson 2004). The latter respond to several external stimuli that affect membranes by physical forces associated with turgor changes and mechanical perturbation and play an important role during VP. Also in sieve elements (Thorpe et al. 2010) mechanosensitive Ca^{2+} channels are believed as sensors of cold shocks. The fully developed mictoplasmic cytoskeleton (Thorsch and Esau 1981; Evert 1990) involved in mechanosensitive channel activation is presumably absent, therefore the activation of these channels in sieve elements is of particular biophysical interest. Apart from mechanosensitive channels, depolarization-activated Ca^{2+} channels (Carpaneto et al. 2007) are most likely involved in the induction of Ca^{2+} signatures in response to low temperature shocks (White 2009; Plieth et al. 1999) and could be important for triggering action potentials. The so-called maxi cation channel (White 2009; White and Ridout 1999; White 2004) is a depolarization-activated Ca^{2+} channel which induces Ca^{2+} influx after cold shock. Activation of such channels in the plasma membrane of sieve elements in response to cold shock is associated with forisome expansion and inhibition of phloem transport. Since these channels are probably involved in the generation of action potentials, they are a key subject of further research.

8.4 Physiological Effects of Electrical Signals

8.4.1 Regulation of Rapid Leaf Movements

M. pudica, the so-called sensitive plant, responds to various stimuli, such as touch, wounding, spontaneous temperature, or light changes and chemical as well as electrical stimuli (Volkov et al. 2010) with rapid leaf movements. Both AP and VP cause turgor changes in primary and tertiary pulvini that cause the leaves to move and fold together, making them look dead and unappealing to a would-be herbivore (Table 8.1, Fromm 1991; Fromm and Eschrich 1990). Apart from *Mimosa*, insectivorous plants such as *Dionaea* and *Drosera*, which live in nitrogen-depleted areas, use APs within specialized leaf traps to catch insects in order to secure their nitrogen supply (Table 8.1, Sibaoka 1969; Williams and Pickard 1972a, b). After mechanical pressure of a trigger hair of *Dionaea*, calcium is released into the cytosol of the sensor cells (Hodick and Sievers 1988) and an AP is generated. To preserve the plant against accidental stimulation, the trap closes only if any of the trigger hairs is bent no later than 40 s after the first stimulation. Kinetics and mechanism of *Dionaea* trap closing was studied in detail during the last years by Forterre et al. (2005) and Volkov et al. (2008a, b, 2009a, b), indicating that the most spectacular phase of the traps rapid closing and catching of insects is driven by the release of elastic energy stored in the leaves of the trap.

8.4.2 Electrical Signaling and its Impact on Phloem Transport

During passage of electrical signals through the phloem one may speculate that such signals could affect phloem transport and/or phloem loading as well as unloading. In maize leaves it turned out that both electrical stimulation as well as cold-stimulation induce AP with amplitudes higher than 50 mV that are propagated basipetally in sieve tubes at speeds of 3–5 cm s^{-1} (Fromm and Bauer 1994). Stimulation with ice water has been reported to induce AP in a number of plant species, including *Biophytum* (Sibaoka 1973) as well as pumpkin and tomato (Van Sambeek and Pickard 1976). The fact that Woodley et al. (1976) observed that localized chilling temporarily stops or reduces translocation of ^{14}C in sunflowers for 10–15 min and that this reduction in translocation corresponds closely to electrical changes measured along the stem gave rise to the idea of a possible relationship between AP and the cold-shock-induced inhibition of phloem transport. In addition, Minchin and Thorpe (1983) showed that rapid temperature drops of only 2.5°C caused a brief abeyance of phloem transport in *Ipomea purpurea*, *Phaseolus vulgaris*, and *Nymphoides geminata*, a phenomenon not observed when the temperature was reduced at a slower rate. In maize leaves, rapid cold-shock treatments cause sieve elements to trigger AP while phloem transport in distant leaf parts is strongly reduced, as shown by autoradiography at a

8 Generation, Transmission, and Physiological Effects

Table 8.1 Published physiological responses of electrical signals in plants

Stimulus	Signal	Plant	Physiological response	Reference(s)
Mechanical	AP	*Dionaea*	Trap closure release of digestive enzymes	Sibaoka 1969; Volkov et al. 2008a, 2008b, 2009a, b; Forterre et al. 2005
Mechanical	AP	*Drosera*	Tentacle movement to wrap around the insect	Williams and Pickard 1972a, b
Cold-shock, mechanical	AP	*Mimosa*	Regulation of leaf movement	Fromm and Eschrich 1988a, b, c; Sibaoka 1969
Electrical	AP	*Chara*	Cessation of cytoplasmic streaming	Hayama, Shimmen, and Tazawa 1979
Electrical	AP	*Conocephalum*	Increase in respiration	Dziubinska et al. 1989
Pollination	AP	*Incarvillea, Hibiscus*	Increase in respiration	Sinyukhin and Britikov 1967; Fromm, Hajirezaei and Wilke 1995
Re-irrigation	AP	*Zea*	Increase in gas exchange	Fromm and Fei 1998; Grams et al. 2007
Cold shock	AP	*Zea*	Reduction in phloem transport	Fromm and Bauer 1994
Increase in light intensity	ES	*Zea*	Effects on K$^+$ and H$^+$ transport at the roots	Shabala et al. 2009
Electrical, cooling	AP	*Luffa*	Decrease of elongation growth of the stem	Shiina and Tazawa 1986
Electrical	AP	*Lycopersicon*	Induction of *pin2* gene expression	Stankovic and Davies 1996
Heating	ES	*Lycopersicon*	Induction of *pin2* gene expression	Wildon et al. 1992
Heating	VP	*Lycopersicon*	Increase in inositol phosphate and Rubisco transcript	Davies 2004
Heating	VP	*Vicia*	Increase in respiration	Filek and Koscielniak 1997
Heating	ES	*Vicia*	Sieve plate occlusion by dispersion of forisomes and callose production	Furch et al. 2007
Heating	ES	*Cucurbita*	Coagulation of sieve element proteins and callose deposition, stop of mass flow	Furch et al. 2010
Heating	VP	*Solanum*	Induction of jasmonic acid biosynthesis and *pin2* gene expression	Fisahn et al. 2004
Wounding	VP	*Pisum*	Inhibition of protein synthesis, formation of polysomes	Davies, Ramaiah and Abe 1986; Davies and Stankovic 2006
Heating	VP	*Mimosa, Populus, Zea*	Transient reduction of photosynthesis	Koziolek et al. 2004; Lautner et al. 2005; Grams et al. 2009

AP action potential; *VP* variation potential; *ES* electrical signal distance of over 15 cm from the site of cold-stimulation (Table 8.1, Fromm and Bauer 1994). Evidence of a link between electrical signaling and the reduction of phloem transport was documented based on the decrease in symplastic K^+ and Cl^- concentration. In *Luffa cylindrica* AP affected elongation growth of the stem, most likely by K^+ and Cl^- efflux which reduced cell turgor and caused growth retardation (Table 8.1, Shiina and Tazawa 1986). Since the concentrations of either ion are also reduced in the sieve element cytoplasm after stimulation (Fromm and Bauer 1994), decreased cell turgor may have caused the reduction in phloem translocation, since the latter requires the intracellular movement of water as a transport medium. On the other hand, the reduction in phloem translocation may also have been caused by sieve plate occlusion after Ca^{2+} rise above a critical threshold during electrical signaling. Hence, both callose production as well as protein expansion can be involved in stoppage of phloem transport in response to electrical signaling (Table 8.1). Strong evidence exists that sieve-plate occlusion is a dual event, since sieve elements were found to occlude quicker by proteins than the onset of callose production (Furch et al. 2010, 2007). Dependent on the rise of Ca^{2+} concentration first specialized protein bodies, so-called forisomes (Knoblauch et al. 2001), disperse abruptly coincident with transmission of an electrical signal in *V. faba* (Furch et al. 2007). In contrast to *Vicia*, in intact *Cucurbita maxima* plants a different occlusion mechanism was found: sieve element-proteins coagulate several centimeters away from the stimulation site (Furch et al. 2010). Several minutes after stimulation callose deposition reached its maximum, followed by a slower degradation (Furch et al. 2010). Such a combined occlusion mechanism guarantees very quick and effective sieve tube sealing after damage and should prevent sieve tubes from loss of sap.

8.4.3 The Role of Electrical Signals in Root-to-Shoot Communication of Water-Stressed Plants

Electrical signaling between roots and shoots of plants growing in drying soil has evoked considerable interest in recent years. Since plants growing in drying soil showed stomatal closure and leaf growth inhibition before reductions in leaf turgor were measured, non-hydraulic signals from roots may serve as a sensitive link between soil water changes and shoot responses (Davies and Zhang 1991). Thus, stomata appear to be able to receive information on the soil water status independent of the leaf water potential. Evidence that the nature of this information is chemical was obtained by analyzing the xylem sap from unwatered plants, indicating the involvement of ion content, pH, amino acids, and hormones (Schurr and Gollan 1990). Since the velocity of a chemical substance in the phloem is relatively slow, typically shown to be 50–100 cm h^{-1} (Canny 1975), the question remained open, how the leaf is capable of responding so rapidly to the changing water status of the soil. To prove which types of signals are involved in rapid root-to-shoot communication electric potential as well as cell turgor was detected on maize plants, subjected to a drying cycle before re-irrigation. Evidence of electrical as well as hydraulic root-to-shoot signaling was obtained by both methods on 80–120 cm tall plants (Table 8.1, Fromm and Fei 1998; Grams et al. 2007). First, plants were exposed to drought conditions by decreasing the soil water content to 40–50% of field capacity. The decrease in water content resulted in a decline in net CO_2 uptake and transpiration rate while the electrical potential difference between two surface points showed a diurnal rhythm, which seemed to be correlated with the soil water status (Fromm and Fei 1998). Second, after soil drying re-irrigation of the plants initiated both a hydraulic signal and an action potential, followed by a two-phase response of the net CO_2 uptake rate and stomatal conductance of leaves (Grams et al. 2007). While the transitional first phase was characterized by a rapid decrease in both levels, in the second phase both parameters (CO_2 uptake rate and stomatal conductance) gradually increased to levels above those of drought-stressed plants. Elimination of either the hydraulic signal by compensatory pressure application to the roots, or of the action potential by cooling of the leaf blade showed that the two signals (1) propagated independently from each other and (2) triggered the two-phase response in gas exchange. Experiments with dye solution showed that the increase in gas exchange could not be triggered by water ascent (Fromm and Fei 1998). The results provided evidence that the hydraulic signal triggered a hydropassive decrease in stomatal aperture while the action potential initiated the increase in CO_2 uptake rate and stomatal conductance.

In these studies the use of aphid stylets as 'bioelectrodes' also revealed that sieve tubes served as a pathway for electrical signaling. The membrane potential of the sieve tubes responded rapidly upon watering the dried plants as well as after inducing spontaneous water stress to the roots by polyethylene glycol (Fromm and Fei 1998). Results therefore lead to the suggestion that electrical root-to-shoot communication plays an essential role in the regulation of photosynthesis of drought-stressed plants. On the other hand, studies on maize also presented evidence for a rapid electrical shoot-to-root communication which affects ion uptake at the roots (Table 8.1, Shabala et al. 2009). Exposure of the shoot to light induces short-term effects on net K^+ as well as H^+ transport at the root surface while the electrical potential between xylem vessels and an external electrode responded within seconds before a decrease in xylem pressure occurs (Shabala et al. 2009).

8.4.4 The Role of Electrical Signalling During Fertilization

Strong evidence also exists that electrical signals evoke specific responses of the ovary during the processes of pollination and fertilization. As regards pollination, two different kinds of electric potential changes were measured in the style of flowers. First, Sinyukhin and Britikov (1967) recorded an action potential in the style of *Lilium martagon* and *Incarvillea grandiflora* a few minutes after placing pollen on the stigma lobes. Furthermore, an action potential was detected after mechanical irritation of the *Incarvillea* lobe, causing closure of the stigma lobes without further transmission. In both species, the pollen-induced AP propagated towards the ovary to stimulate the oxygen consumption by 5–11%, 60–90 s after arrival of the action potential. At this moment, most likely postpollination effects are triggered to start, such as the induction of ovary enlargement and wilting of the corolla, which occur long before fertilization. Second, electrical potential changes were measured in the style of *Lilium longiflorum* flowers 5–6 h after pollination (Spanjers 1981). No signals can be detected when applying killed pollen or pollen of other species. In *Hibiscus rosa-sinensis*, different stimuli applied to the stigma of flowers evoke specific electrical signals each that propagate toward the ovary at speeds of 1.3–3.5 cm s^{-1} (Fromm et al. 1995). To investigate the first reactions of the ovarian metabolism, various metabolites were analyzed 10 min after stimulating the stigma by pollen, wounding or cold-shock. Self- as well as cross-pollination hyperpolarized the resting potential of style cells 50–100 s later, followed by a series of 10–15 action potentials. 3–5 min after pollination, the ovarian respiration rate increased transiently by 12%, with the levels of ATP, ADP, and starch rising significantly (Table 8.1, Fromm et al. 1995). By contrast, cold-shock of the stigma caused a single action potential, whereas wounding generated a strong depolarization of the membrane potential with an irregular form and a lower transmission rate. Either treatment caused a spontaneous decrease in the ovarian respiration rate, as well as reduced metabolite concentrations in the ovary. Since there was no evidence that a chemical substance had been transported within 10 min over a distance of 8–10 cm from the stigma to the ovary, the metabolism must have responded to the electrical signals (Fromm et al. 1995). In the light of these results the question came up how an electrical signal is able to cause the biochemical response. Most likely the latter may be achieved through subcellular changes of K^+, Cl^-, and Ca^{2+} ions which are responsible for the generation of action potentials. According to Davies (1987) local changes in ion concentration can lead to modified activities of enzymes in the cell wall, the plasmalemma, and the cytoplasm. This kind of mechanism may also be involved in the fluctuation of the starch level of the ovary after stigma stimulation. The biochemical regulation of starch synthesis is centered almost exclusively on ADP-Glc-pyrophosphorylase (Preiss et al. 1985). The characteristics of this enzyme in ovaries will therefore be analyzed in future to gain a better understanding of the biochemical role of electrical signaling during fertilization.

8.4.5 The Role of Electrical Signalling in the Regulation of Photosynthesis

Most of the work on functions for electrical signals in plants focused on responses evoked by heat stimulation and evidence exists of their role in transcription, translation, and respiration (Stankovic and Davies 1997; Davies 2004). Recently, evidence was found for a link between electrical signaling and photosynthetic response in *Mimosa* (Table 8.1, Koziolek et al. 2004), poplar (Lautner et al. 2005) and maize (Grams et al. 2009). Flaming of a Mimosa leaf pinna evoked a variation potential that travels at a speed of 4–8 mm s^{-1} into the neighboring pinna of the leaf to transiently reduce the net-CO_2 uptake rate. Shortly after that, the PSII quantum yield of electron transport is also reduced. Two-dimensional image analysis of the chlorophyll fluorescence signal revealed that the yield reduction spreads acropetally through the pinna and via the veins through the leaflets. The results provide evidence of the role of electrical signals in the regulation of photosynthesis because the high speed of the signals rules out the involvement of a slow-moving chemical signal. In addition to the photosynthetic response, it was shown that wounding causes lateral chloroplast movement within 10 min after wounding in *Elodea canadensis* (Gamalei et al. 1994). The time course of chloroplast movement coincides with rapid changes in the membrane potential with low amplitudes (humming, 4–7 mV), recorded by microelectrodes impaled into the midrib of the attached leaf.

To gain a deeper understanding of the role of electrical signaling in photosynthesis, poplar as well as maize plants were stimulated by flaming. In poplar, depolarizing signals travel over long distances across the stem from heat-wounded leaves to adjacent leaves where the net-CO_2 uptake rate is temporarily depressed towards compensation (Table 8.1, Lautner et al. 2005). Surprisingly, signals induced by cold-shock did not affect photosynthesis. In coincidence with the results on *Mimosa*, electrical signaling also significantly reduced the quantum yield of electron transport through PSII in poplar. Cold-blocking of the stem proved that the electrical signal transmission via the phloem becomes disrupted, causing the leaf gas exchange to remain unaffected. Furthermore, calcium-deficient trees showed a marked contrast inasmuch as the amplitude of the electrical signal was distinctly reduced, concomitant with the absence of a significant response in leaf gas exchange upon flame-wounding (Lautner et al. 2005). In conclusion, these experiments showed that the heat-induced VP depends on calcium, moves within the phloem and triggers the photosynthetic response in distant leaves.

Similar results were obtained in monocotyledonous plants. In maize plants, combined measurements of electric potential with those of chlorophyll fluorescence and leaf gas exchange indicated flaming of the leaf tip to evoke a VP that traveled through the leaf while reducing the net-CO_2 uptake rate and the photochemical quantum yield of both photosystems (Table 8.1, Grams et al. 2009). Chlorophyll fluorescence analysis of PS II showed that the yield reduction spread basipetally via the veins through the leaf at a velocity of 1.6 mm s^{-1} while the transmission speed in the intervein region was c. 50 times slower. Passage of the signal through the veins was confirmed because PS I, which is present in the bundle sheath cells around the leaf vessels, was affected first. Accordingly, passage of the signal along the veins represents a path with higher speed than within the intervein region of the leaf. During the variation potential, cytoplasmic pH decreased transiently from 7.0 to 6.4, while apoplastic pH increased

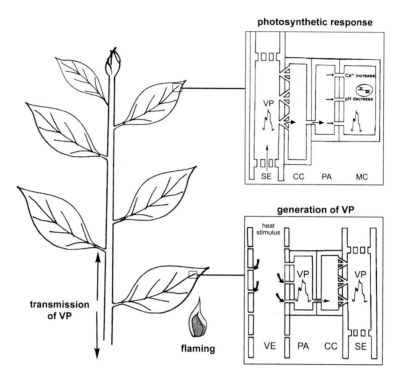

Fig. 8.3 Photosynthetic response of electrical signaling in plants. After heat stimulation of a leaf a variation potential is evoked and transmitted via the phloem to distant leaves where first the net CO_2 uptake rate decreased transiently and second the quantum yield of electron transport through PSII is temporarily reduced. It is assumed that the cytosolic increase in calcium and decrease in pH which generate the VP play a significant role in the photosynthetic response. companion cell (*CC*); mesophyll cell (*MC*); parenchyma cell (*PA*); sieve element (*SE*); vessel (*VE*)

transiently from 4.5 to 5.2 (Grams et al. 2009), indicating an inhibition of the proton pump within the plasma membrane. Furthermore, photochemical quantum yield of isolated chloroplasts was strongly affected by pH changes in the surrounding medium, indicating a putative direct impact of the variation potential via changes of cytosolic pH on photosynthesis (Fig. 8.3). Also, in *Chara corallina* evidence was presented that electrical signals evoked at the plasma membrane are transmitted to the thylakoid membranes and that fluorescence changes derive from an increase in pH gradient at the thylakoid membrane (Bulychev and Kamzolkina 2006). Further research has to be done on the responsiveness of the various types of molecules that are involved in electron transport as well as on enzymes involved in the uptake of CO_2 during electrical signaling. Obviously, the involved ion fluxes such as protons and calcium within a VP and/or the amplitude and duration of the VP play a key role in the generation of the photosynthetic response.

8.4.6 Effects of Electrical Signals on Gene Expression

Numerous studies have shown possible effects of electrical signals on gene expression. For example in *Arabidopsis*, several types of genes are up-regulated by touch, implicating Ca^{2+} signaling, cell wall modification and disease resistance as potential downstream responses (McCormack et al. 2006). In tomato, wounding induces the proteinase inhibitor gene (*pin2*) expression in distant parts of the plant. First evidence that the systemic wound signal is a propagated electrical signal was given by Wildon et al. (1992). Local flaming evoked rapid changes in membrane potential in distant tissue, much faster than the transport of chemical signals within the phloem could provide. Interestingly, transcript accumulation began after arrival of the electrical signal (Table 8.1, Wildon et al. 1992). In 1996, Stankovic and Davies showed that both, AP as well as VP, can induce *pin2* gene expression in tomato plants. Further changes in gene expression in tomatoes induced by flaming were reported from Davies (2004). Ultra-rapid changes in the level of a putative second messenger, inositol phosphate (IP3), as well as a Rubisco small subunit occur after flame wounding (Table 8.1). A systemic transient increase in IP3 levels appeared after 75–120 s, while the Rubisco SS transcript increased transiently after 90–150 s (Davies 2004). As described by Fisahn et al. (2004), the transient increase in cytosolic Ca^{2+} during the action potential is required for induction of PIN II gene expression and jasmonic acid biosynthesis.

8.5 Long-Distance Electrical Signaling in Woody Plants

Particularly in trees, electrical communication over long distances between different organs is important to coordinate physiological activities. Due to environmental changes, different electrical signals can be evoked in the symplast and transmitted to distant organs, with concomitant specific effects on various physiological processes. In several species lacking rapid leaf movements it was observed that the longitudinal reach of electrical signals is limited and may not extend over the whole plant. Therefore, it was suggested that, in contrast to animal APs, plant APs do not seem to be primarily engaged in propagating electrochemical signals but rather in gating ion channels (Pyatygin et al. 2008). However, in *Vitis* strong evidence is given that a flame-induced VP travels within the phloem over distances up to 1.5 m and in Bauhinia, APs induced by re-irrigation after drought travel over distances of 4 m in acropetal direction (Fromm, unpublished results). A fast communication process between the root system and the canopy is of great advantage in order to adapt physiological processes to changing environmental conditions. For example, hormone-induced APs in the roots were shown to propagate throughout willow plants at velocities of 2–5 cm s^{-1} in order to affect the gas exchange of the leaves (Fromm and Eschrich 1993). To gain a deeper understanding of the role of electrical signaling in the photosynthesis of trees, poplar shoots were stimulated by flaming and triggered VPs travel across the shoot to adjacent leaves were the net CO_2 uptake rate as well as the quantum yield of electron transport through PS II are temporarily reduced (Lautner et al. 2005). In addition, continuous and automatic measurements of electric potential differences between tree tissues can be effectively used to study long-term electrical events. In several fruit-bearing trees, systemic patterns of electric potential differences, monitored continuously in the sapwood at various positions in the trunk, were obtained during day–night cycles and at different conditions of soil water availability (Gurovich and Hermosilla 2009; Oyarce and Gurovich 2010). Results are expected to quantitatively define early detection of water stress and the adaptive response of trees to transient soil water availability.

8.5.1 Membrane Potential, Electrical Signals and Growth of Willow Roots

Since willow roots were shown to respond to externally applied hormones with propagating APs (Fromm and Eschrich 1993), it was an important challenge to measure the magnitude of the current that flows during action potentials. With the use of the vibrating probe technique it was possible to quantify the current, the sensitivity of the probe being in the range of $\mu A\ cm^{-2}$, i.e., sufficiently sensitive to measure ion fluxes of $pmol\ cm^{-2}\ s^{-1}$ (Fromm et al. 1997). Therefore, microelectrode recordings and vibrating probe measurements were used in tandem to correlate changes in membrane potential with changes in endogenous current. Transient depolarizations were elicited in root cortex cells by spermine, while abscisic acid caused a transient hyperpolarization. For the latter it is assumed that K^+ leaves the cortex cells, similar to the K^+ efflux measured in guard cells (Mansfield et al. 1990). All changes in membrane potential were accompanied by transient responses of the endogenous current. These responses suggested that first anions, then cations leave the root during spermine-induced depolarizations. From the changes in the endogenous current an apparent efflux of anions (presumably Cl^-) and cations (presumably K^+) of 200–700 $pmol\ cm^{-2}$ per signal was calculated (Fromm et al. 1997). Furthermore, it was possible to demonstrate the effect of the growth regulators spermine and abscisic acid on root growth. The mean growth rate of roots increased by up to 30% after application of spermine, while it almost came to a standstill after treatment with abscisic acid.

8.5.2 Electrical Properties of Wood-Producing Cells

In the course of the evolutionary process plants found it necessary to develop wood in order to increase their mechanical strength so as to be able to reach tree heights of 100 m and more. Extensive literature exists addressing wood anatomy, chemistry, and physical properties. However, we have only just begun to form an understanding of the molecular and electrophysiological mechanisms of cambial activity and wood formation, a field now considered a main research area in tree physiology. One of the main model tree species for basic wood research is poplar. Because of its suitability for genetic transformation and its ease of vegetative propagation poplar has become the commonly used model tree species in Europe and the United States. To give a description of the electrophysiological processes in wood formation biophysical and molecular techniques have been used to analyze K^+ transporters of poplar. K^+ transporters homologous to those of known function in *Arabidopsis* phloem- and xylem-physiology were isolated from a poplar wood EST library and the expression profile of three distinct K^+ channel types was analysed by quantitative RT-PCR (Langer et al. 2002). Thus, it was found that the *P. tremula* outward rectifying K^+ channel (PTORK) and the *P. tremula* K^+ channel 2 (PTK2) correlated with the seasonal wood production. Both K^+ channel genes are expressed in young poplar twigs, and while PTK2 was predominantly found in the phloem fraction, PTORK was detected in both phloem and xylem fractions. Following the heterologous expression in *Xenopus* oocytes the biophysical properties of the different channels were determined. PTORK, upon membrane depolarization mediates potassium release, while PTK2 is almost voltage-independent, carrying inward K^+ flux at hyperpolarized potential and K^+ release upon depolarization (Langer et al. 2002). In addition, in vivo patch-clamp studies were performed on isolated protoplasts from PTORK and PTK2 expressing suspension cultures. Poplar branches were therefore induced to build callus and the resulting meristematic tissues were used to generate suspension cultures. Protoplasts were isolated and the plasma membrane potassium conductances were compared with the electrical properties of Xenopus oocytes expressing PTORK and PTK2 individually. Concerning PTORK it was shown that the properties of this channel are similar in both experimental systems and also to other plant depolarization-activated K^+ release channels (Ache et al. 2000; Gaymard et al. 1998; Langer et al. 2002). In addition, PTORK could be localized immunologically in young fibers and ray cells of poplar wood (Arend et al. 2005). In coincidence with the activity of the K^+ channels a plasma membrane H^+-ATPase, generating the necessary H^+ gradient (proton-motive force) for the uptake of K^+ into xylem cells, was localized in the poplar stem using specific antibodies (Arend et al. 2002, 2004). Since potassium is the most abundant cation in plants, playing a central role in many aspects of plant physiology, we conclude that K^+ channels are involved in the regulation of K^+-dependent wood formation. Since seasonal changes in cambial potassium content correlate strongly with the osmotic potential of the cambial zone (Wind et al. 2004), potassium is likely to play a key role in the regulation of wood formation due to its strong impact on osmoregulation in expanding cambial cells (Ache et al. 2010; Fromm 2010). On the other hand, since PTORK appears in the plasma membrane of sieve elements of the phloem as well as in xylem rays, this channel may play a role in the generation of electrical signals within the poplar phloem and xylem. However, the potential role of xylem ray cells in radial transmission of electrical signals within the tree stem will still have to be proved in the future.

8.6 Conclusion

A large amount of results on electrical behavior in plants has accumulated in the last 10 years, indicating that a fundamental property of plants is the generation and transmission of electrical messages caused by environmental changes through their different cells and organs. Why is there a need for electrical signals in plants? Obviously, electrical signals have some properties that chemical signals do not have. Speed is not only very important for the capture of insects but also for the effective response after an insect or fungal attack, for the plant needs to put up systemic defenses as quickly as possible to prevent spread e.g., of pathogens (Davies 2004). In addition, electrical signals in plants are designed to inform as much of the planty body as quickly as possible so that all the intervening tissue is informed and can respond immediately (Davies 1993). The passage of electrical signals can occur over long as well as over short distances to control the physiological behavior of the plant. Obviously, plants have developed a rudimentary neural network which is able to respond to a variety of environmental stimuli that can be both, abiotic as well as biotic. Due to the impulses generated by environmental changes both, action and VPs serve as information carriers. The primary step in signal perception is an opening of plasmalemmal ion channels, leading to ion fluxes which generate action or VPs. In animals as well as plants similar molecules are used to drive physiological responses and astounding similarities exist between APs in plants and animals. The generation of APs in plants follows the all-or-nothing law, too, (Shiina and Tazawa 1986, Fromm and Spanswick 1993) and plant APs also show refractory periods. In higher plants, electrical signals are transmitted from cell to cell via plasmodesmata over short distances, while propagation over long-distances along the plasma membrane of sieve tubes occurs through the successive opening and closure of ion channels (Fig. 8.1). Calcium, chloride, and potassium channels are involved in the generation of APs and several ion channels postulated to be involved in electrical transmission were identified in the phloem and in the xylem (Ache et al. 2001; Langer et al. 2002).

Concerning the physiological functions of electrical signals numerous competences have already been investigated. Apart from the role of action and VPs in carnivorous plants and *Mimosa*, a concrete relationship between electrical stimulation and the increased production of proteinase inhibitors was proven to exist in tomato (Stankovic and Davies 1997). Other work showed that APs regulate respiration (Dziubinska et al. 1989), phloem transport (Fromm and Bauer 1994), fertilization (Sinyukhin and Britikov 1967; Fromm et al. 1995), and photosynthesis (Koziolek et al. 2004; Lautner et al. 2005). So far, we have seen but glimpses of a complex electrical long-distance signaling system in plants and obtained evidence of the role played by electrical signals in the daily processes of plant life. It is to be expected that future improvements in investigation methods will reveal more aspects of the signaling complexity and its physiological responses. Future research in this field will need to focus especially on the molecular controlling points of electrical signaling, which are essential for a better understanding of whole plant physiology.

References

Ache P, Becker D, Ivashikina N, Dietrich P, Roelfsema MRG, Hedrich R (2000) GORK, a delayed outward rectifier expressed in guard cells of *Arabidopsis thaliana*, is a K$^+$ selective, K$^+$ sensing ion channel. FEBS Lett 486:93–98

Ache P, Becker D, Deeken R, Dreyer I, Weber H, Fromm J, Hedrich R (2001) VFK1, a *Vicia faba* K$^+$ channel involved in phloem unloading. Plant J 27:571–580

Ache P, Fromm J, Hedrich R (2010) Potassium-dependent wood formation in poplar: seasonal aspects and environmental limitations. Plant Biol 12:259–267

Adrian ED, Bronk DW (1928) The discharge of impulses in motor nerve fibres. I. Impulses in single fibres of the phrenic nerve. J Physiol 66:81–101

Arend M, Weisenseel MH, Brummer M, Osswald W, Fromm J (2002) Seasonal changes of plasma membrane H$^+$-ATPase and endogenous ion current during growth in poplar plants. Plant Physiol 129:1651–1663

Arend M, Monshausen G, Wind C, Weisenseel MH, Fromm J (2004) Effect of potassium deficiency on the plasma membrane H$^+$-ATPase of the wood ray parenchyma in poplar. Plant Cell Environ 27:1288–1296

Arend M, Stinzing A, Wind C, Langer K, Latz A, Ache P, Fromm J, Hedrich R (2005) Polar-localised poplar K$^+$ channel capable of controlling electrical properties of wood-forming cells. Planta 223:140–148

Beilby MJ, Coster HGL (1979) The action potential in *Chara corallina*. II. Two activation-inactivation transients in voltage clamps of plasmalemma. Austr J Plant Phys 6:329–335

Beyhl D, Hörtensteiner S, Martinoia E, Farmer EE, Fromm J, Marten I, Hedrich R (2009) The fou2 mutation in the major vacuolar cation channel TPC1 confers tolerance to inhibitory luminal calcium. Plant J 58:715–723

Bulychev AA, Kamzolkina NA (2006) Effect of action potential on photosynthesis and spatially distributed H$^+$ fluxes in cells and chloroplasts of *Chara corallina*. Russ J Plant Physiol 53:5–14

Canny MJP (1975) Mass transfer. In: Zimmermann HM, Milburn JA (eds) Encyclopedia of plant physiology. Springer, Berlin, pp 139–153

Carpaneto A, Geiger D, Bamberg E, Sauer N, Fromm J, Hedrich R (2005) Phloem-localized, proton-coupled sucrose carrier ZmSUT1 mediates sucrose efflux under control of sucrose gradient and pmf. J Biol Chem 280:21437–21443

Carpaneto A, Ivashikina N, Levchenko V, Krol E, Zhu J-K, Hedrich R (2007) Cold transiently activates calcium-permeable channels in *Arabidopsis* mesophyll cells. Plant Physiol 143:487–494

Cosgrove DJ, Hedrich R (1991) Stretch-activated chloride, potassium, and calcium channels coexisting in plasma membranes of guard cells of *Vicia faba* L. Planta 186:143–153

Davies E (1987) Action potentials as multifunctional signals in plants: a unifying hypothesis to explain apparently disparate wound responses. Plant Cell Environ 10:623–631

Davies E (1993) Intercellular and intracellular signals in plants and their transduction via the membrane—cytoskeleton interface. Semin Cell Biol 4:139–147

Davies E (2004) New functions for electrical signals in plants. New Phytol 161:607–610

Davies E, Ramaiah KVA, Abe S (1986) Wounding inhibits protein synthesis yet stimulates polysome formation in aged, excised pea epicotyls. Plant Cell Physiol 27:1377–1386

Davies E, Stankovic B (2006) Electrical signals, the cytoskeleton, and gene expression: a hypothesis on the coherence of cellular processes to environmental insult. In: Baluska F, Mancuso S, Volkmann D (eds) Communication in plants—neuronal aspects of plant life. Springer, Berlin, pp 309–320

Davies WJ, Zhang J (1991) Root signals and the regulation of growth and development of plants in drying soil. Annu Rev Plant Physiol Plant Mol Biol 42:55–76

Deeken R, Geiger D, Fromm J, Koroleva O, Ache P, Langenfeld-Heyser R, Sauer N, May ST, Hedrich R (2002) Loss of the AKT2/3 potassium channel affects sugar loading into the phloem of *Arabidopsis*. Planta 216:334–344

Demidchik V, Maathuis FJM (2007) Physiological roles of non-selective cation channels in plants: from salt stress to signalling and development. New Phytol 175:387–404

Ding JP, Pickard BG (1993) Modulation of mechanosensitive calcium-selective cation channels by temperature. Plant J 3:713–720

Dutta R, Robinson KR (2004) Identification and characterization of stretch-activated ion channels in pollen protoplasts. Plant Physiol 135:1398–1406

Dziubinska H, Trebacz K, Zawadzki T (1989) The effect of excitation on the rate of respiration in the liverwort *Conocephalum conicum*. Physiol Plant 75:417–423

Eschrich W, Fromm J, Evert RF (1988) Transmission of electric signals in sieve tubes of zucchini plants. Bot Acta 101:327–331

Evert RF, Eschrich W, Eichhorn SE (1973) P-protein distribution in mature sieve elements of *Cucurbita maxima*. Planta 109:193–210

Evert R (1990) Dicotyledons. In: Behnke H-D, Sjolund RD (eds) Sieve elements—comparative structure, induction and development. Springer, Berlin, pp 103–137

Filek M, Koscielniak J (1997) The effect of wounding the roots by high temperature on the respiration rate of the shoot and propagation of electric signal in horse bean seedlings (Vicia faba L. Minor). Plant Science 123:39–46

Findlay GP (1961) Voltage-clamp experiments with *Nitella*. Nature 191:812–814

Findlay GP (1962) Calcium ions and the action potential in *Nitella*. Aust J Biol Sci 15:69–82

Fisahn J, Herde O, Willmitzer L, Pena-Cortes H (2004) Analysis of the transient increase in cytosolic Ca^{2+} during the action potential of higher plants with high temporal resolution: requirement of Ca^{2+} transients for induction of jasmonic acid biosynthesis and PINII gene expression. Plant Cell Physiol 45:456–459

Forterre Y, Skothelm JM, Dumals J, Mahadevan L (2005) How the venus flytrap snaps. Nature 433:421–425

Fromm J (1991) Control of phloem unloading by action potentials in *Mimosa*. Physiol Plant 83:529–533

Fromm J (2010) Wood formation of trees in relation to potassium and calcium nutrition. Tree Physiol 30:1140–1147

Fromm J, Bauer T (1994) Action potentials in maize sieve tubes change phloem translocation. J Exp Bot 45:463–469

Fromm J, Eschrich W (1988a) Transport processes in stimulated and non-stimulated leaves of *Mimosa pudica*. I. The movement of ^{14}C-labelled photoassimilates. Trees 2:7–17

Fromm J, Eschrich W (1988b) Transport processes in stimulated and non-stimulated leaves of *Mimosa pudica*. II. Energesis and transmission of seismic stimulations. Trees 2:18–24

Fromm J, Eschrich W (1988c) Transport processes in stimulated and non-stimulated leaves of *Mimosa pudica*. III. Displacement of ions during seismonastic leaf movements. Trees 2:65–72

Fromm J, Eschrich W (1989) Correlation of ionic movements with phloem unloading and loading in barley leaves. Plant Physiol Biochem 27:577–585

Fromm J, Eschrich W (1990) Seismonastic movements in Mimosa. In: Satter RL, Gorton HL, Vogelmann TC (eds) The pulvinus: motor organ for leaf movement. Americ Soc Plant Physiol, Rockville, pp 25–43

Fromm J, Eschrich W (1993) Electric signals released from roots of willow (*Salix viminalis* L.) change transpiration and photosynthesis. J Plant Physiol 141:673–680

Fromm J, Fei H (1998) Electrical signaling and gas exchange in maize plants of drying soil. Plant Sci 132:203–213

Fromm J, Lautner S (2005) Characteristics and functions of phloem-transmitted electrical signals in higher plants. In: Baluska F, Mancuso S, Volkmann D (eds) Communication in plants–neuronal aspects of plant life. Springer, Heidelberg, pp 321–332

Fromm J, Lautner S (2007) Electrical signals and their physiological significance in plants. Plant Cell Environm 30:249–257

Fromm J, Spanswick R (1993) Characteristics of action potentials in willow (*Salix viminalis* L.). J Exp Bot 44:1119–1125

Fromm J, Hajirezaei M, Wilke I (1995) The biochemical response of electrical signaling in the reproductive system of *Hibiscus* plants. Plant Physiol 109:375–384

Fromm J, Meyer AJ, Weisenseel MH (1997) Growth, membrane potential and endogenous ion currents of willow (*Salix viminalis*) roots are all affected by abscisic acid and spermine. Physiol Plant 99:529–537

Furch ACU, Hafke JB, Schulz A, van Bel AJE (2007) Ca^{2+}-mediated remote control of reversible sieve tube occlusion in *Vicia faba*. J Exp Bot 58:2827–2838

Furch ACU, van Bel AJE, Fricker MD, Felle HH, Fuchs M, Hafke JB (2009) Sieve element Ca^{2+} channels as relay stations between remote stimulus and sieve tube occlusion. Plant Cell 21:2118–2131

Furch ACU, Zimmermann MR, Will T, Hafke JB, van Bel AJE (2010) Remote-controlled stop of mass flow by biphasic occlusion in *Cucurbita maxima*. J Exp Bot 61:3697–3708

Gaffey CT, Mullins LJ (1958) Ion fluxes during the action potential in *Chara*. J Physiol 144:505–524

Gamalei YV, Fromm J, Krabel D, Eschrich W (1994) Chloroplast movement as response to wounding in *Elodea canadensis*. J Plant Physiol 144:518–524

Gaymard F, Pilot G, Lacombe B, Bouchez D, Bruneau D, Boucherez J, Michaux-Ferriere N, Thibaud JB, Sentenac H (1998) Identification and disruption of a plant shaker-like outward channel involved in K^+ release into the xylem sap. Cell 94:647–655

Grams TEE, Koziolek C, Lautner S, Matyssek R, Fromm J (2007) Distinct roles of electric and hydraulic signals on the reaction of leaf gas exchange upon re-irrigation in *Zea mays* L. Plant Cell Environ 30:79–84

Grams TEE, Lautner S, Felle HH, Matyssek R, Fromm J (2009) Heat-induced electrical signals affect cytoplasmic and apoplastic pH as well as photosynthesis during propagation through the maize leaf. Plant Cell Environ 32:319–326

Gurovich LA, Hermosilla P (2009) Electric signalling in fruit trees in response to water applications and light-darkness conditions. J Plant Physiol 166:290–300

Gustin MC, Zhou XL, Martinac B, Kung C (1988) A mechanosensitive ion channel in the yeast plasma membrane. Science 242:762–765

Hamilton DWA, Hills A, Kohler B, Blatt MR (2000) Ca^{2+} channels at the plasma membrane of stomatal guard cells are activated by hyperpolarization and abscisic acid. Proc Natl Acad Sci USA 97:4867–4972

Hayama T, Shimmen T, Tazawa M (1979) Participation of Ca^{2+} in cessation of cytoplasmic streaming induced by membrane excitation in Characeae internodal cells. Protoplasma 99:305–321

Hörmann G (1898) Studien über die Protoplasmaströmung bei den Characaean. Gustav Fischer Verlag, Jena

Hodick D, Sievers A (1988) The action potential of *Dionaea muscipula* Ellis. Planta 174:8–18

Hope AB (1961) Ionic relations of cells of *Chara corallina*. V. The action potential. Aust J Biol Sci 14:312–322

Kayler Z, Gessler A, Buchmann N (2010) What is the speed of link between aboveground and belowground processes? New Phytol 187:885–888

Kempers R, Ammerlaan A, van Bel AJE (1998) Symplasmic constriction and ultrastructural features of the sieve element/companion cell complex in the transport phloem of apoplasmically and symplasmically phloem-loading species. Plant Physiol 116:271–278

Kiegle E, Gilliham M, Haseloff J, Tester M (2000) Hyperpolarisation activated calcium currents found only in cells from the elongation zone of *Arabidopsis thaliana* roots. Plant J 21:225–229

Kishimoto U (1968) Response of *Chara* internodes to mechanical stimulation. Ann Rep Biol Works Fac Sci Osaka Univ 16:61–66

Knoblauch M, Peters WS, Ehlers K, van Bel AJE (2001) Reversible calcium-regulated stopcocks in legume sieve tubes. Plant Cell 13:1221–1230

Koziolek C, Grams TEE, Schreiber U, Matyssek R, Fromm J (2004) Transient knockout of photosynthesis mediated by electrical signals. New Phytol 161:715–722

Langer K, Ache P, Geiger D, Stinzing A, Arend M, Wind C, Regan S, Fromm J, Hedrich R (2002) Poplar potassium transporters capable of controlling K$^+$ homeostasis and K$^+$ dependent xylogenesis. Plant J 32:997–1009

Lautner S, Grams TEE, Matyssek R, Fromm J (2005) Characteristics of electrical signals in poplar and responses in photosynthesis. Plant Physiol 138: 2200–2209

Lunevsky VZ, Zherelova OM, Vostrikov IY, Berestovsky GN (1983) Excitation of Characeae cell membranes as a result of activation of calcium and chloride channels. J Membr Biol 72:43–58

Mansfield TA, Hetherington AM, Atkinson CJ (1990) Some current aspects of stomatal physiology. Annu Rev Plant Physiol Plant Mol Biol 41:55–75

Martinac B, Buechner M, Delcour AH, Adler J, Kung C (1987) Pressure-sensitive ion channel in *Escherichia coli*. Proc Natl Acad Sci USA 84:2297–2301

McCormack E, Velasquez L, Delk NA, Braam J (2006) Touch-responsive behaviours and gene expression in plants. In: Baluska F, Mancuso S, Volkmann D (eds) Communication in plants—neuronal aspects of plant life. Springer, Berlin, pp 249–260

Mencuccini M, Hölttä T (2010) The significance of phloem transport for the speed with which canopy photosynthesis and belowground respiration are linked. New Phytol 185:189–203

Minchin PEH, Thorpe MR (1983) A rate of cooling response in phloem translocation. J Exp Bot 34:529–536

Mühling KH, Sattelmacher B (1997) Determination of apoplastic K$^+$ in intact leaves by ratio imaging of PBFI fluorescence. J Exp Bot 48:1609–1614

Mummert E, Gradmann D (1976) Voltage dependent potassium fluxes and the significance of action potentials in *Acetabularia*. Biochim Biophys Acta 443:443–450

Nick P (2008) Microtubules as sensors for abiotic stimuli. In: Nick P (ed) Plant microtubules, 2nd edn. Springer, Berlin, pp 175–203

Oda K (1976) Simultaneous recording of potassium and chloride effluxes during an action potential in *Chara corallina*. Plant Cell Physiol 17:1085–1088

Oyarce P, Gurovich L (2010) Electrical signals in avocado trees. Plant Signaling Behav 5(1):34–41

Plieth C, Hansen U-P, Knight H, Knight MR (1999) Temperature sensing by plants: the primary characteristics of signal perception and calcium response. Plant J 18:491–497

Preiss J, Robinson N, Spilatro S, McNamara K (1985) Starch synthesis and its regulation. In: Heath R, Preiss J (eds) Regulation of carbon partitioning in photosynthetic tissue. Amer Soc Plant Physiol, Rockville, pp 1–26

Pyatygin SS, Opritov VA, Vodeneev VA (2008) Signaling role of action potential in higher plants. Russ J Plant Physiol 55:285–291

Rhodes J, Thain JF, Wildon DC (1996) The pathway for systemic electrical signal transduction in the wounded tomato plant. Planta 200:50–57

Rienmüller F, Beyhl D, Lautner S, Fromm J, Al-Rasheid KAS, Ache P, Farmer EE, Marten I, Hedrich R (2010) Guard cell-specific calcium sensitivity of high density and activity SV/TPC1 channels. Plant Cell Physiol 51(9):1548–1554

Schurr U, Gollan T (1990) Composition of xylem sap of plants experiencing root water stress: a descriptive study. In: Davies WJ, Jeffcoat B (eds) Importance of root to shoot communication in the response to environmental stress. Br Soc Plant Growth Regul, Bristol, pp 201–214

Shabala S, Pang J, Zhou M, Shabala L, Cuin TA, Nick P, Wegner LH (2009) Electrical signalling and cytokinins mediate effects of light and root cutting on ion uptake in intact plants. Plant Cell Environ 32:194–207

Shiina T, Tazawa M (1986) Action potential in *Luffa cylindrica* and its effects on elongation growth. Plant Cell Physiol 27:1081–1089

Shvetsova T, Mwesigwa J, Labady A, Kelly S, Thomas D, Lewis K, Volkov AG (2002) Soybean electrophysiology: effects of acid rain. Plant Sci 162:723–731

Sibaoka T (1969) Physiology of rapid movements in higher plants. Ann Rev Plant Physiol 20:165–184

Sibaoka T (1973) Transmission of action potentials in *Biophytum*. Bot Mag 86:51–61

Simons P (1992) The action plant. Movement and nervous behaviour in plants. Blackwell Publishing, Oxford

Sinyukhin AM, Britikov EA (1967) Action potentials in the reproductive system of plants. Nature 215:1278–1280

Spanjers AW (1981) Bioelectric potential changes in the style of *Lilium longiflorum* Thunb. after self- and cross-pollination of the stigma. Planta 153:1–5

Spanswick RM (1972) Electrical coupling between cells of higher plants: a direct demonstration of intercellular communication. Planta 102:215–227

Spanswick RM, Costerton JWF (1967) Plasmodesmata in *Nitella translucens*: structure and electrical resistance. J Cell Sci 2:451–464

Spyropoulos CS, Tasaki I, Hayward G (1961) Fractination of tracer effluxes during action potential. Science 133:2064–2065

Stahlberg E, Cosgrove DJ (1997) The propagation of slow wave potentials in pea epicotyls. Plant Physiol 113:33–41

Stahlberg R, Cleland RE, van Volkenburgh E (2006) Slow wave potentials—a propagating electrical signal unique to higher plants. In: Baluska F, Mancuso S, Volkmann D (eds) Communication in plants—neuronal aspects of plant life. Springer, Berlin, pp 291–308

Stankovic B, Davies E (1996) Both action potentials and variation potentials induce proteinase inhibitor gene expression in tomato. FEBS Lett 390:275–279

Stankovic B, Davies E (1997) Intercellular communication in plants: electrical stimulation of proteinase inhibitor gene expression in tomato. Planta 202:402–406

Szmelcman S, Adler J (1976) Change in membrane potential during bacterial chemotaxis. Proc Natl Acad Sci USA 73:4387–4391

Tazawa M, Shimmen T, Mimura T (1987) Membrane control in the Characeae. Annu Rev Plant Physiol 38:95–117

Thion L, Mazars C, Nacry P, Bouchez D, Moreau M, Ranjeva R, Thuleau P (1998) Plasma membrane depolarization-activated calcium channels stimulated by microtubule-depolymerizing drugs in wild-type *Arabidopsis thaliana* protoplasts, display constitutively large activities and a longer half-life in ton 2 mutant cells affected in the organization of the cortical microtubules. Plant J 13: 603–610

Thorpe MREH, Foeller J, van Bel AJE, Hafke JB (2010) Rapid cooling triggers forisome dispersion just before phloem transport stops. Plant Cell Environ 33:259–271

Thorsch J, Esau K (1981) Nuclear generation and the association of endoplasmic reticulum with the nuclear envelope and microtubules in maturing sieve elements of *Gossypium hirsutum*. J Ultrastr Res 74:195–204

Thuleau P, Moreau M, Schroeder JI, Ranjeva R (1994) Recruitment of plasma membrane voltage-dependent calcium-permeable channels in carrot cells. EMBO J 13:5843–5847

Van Bel AJE (1993) The transport phloem. Specifics of its functioning. Prog Bot 54:134–150

Van Bel AJE, Ehlers K (2005) Electrical signalling via plasmodesmata. In: Oparka KJ (ed) Plasmodesmata, Blackwell Publishing, Oxford, pp 263–278

Van Bel AJE, Van Rijen HVM (1994) Microelectrode-recorded development of symplasmic autonomy of the sieve element/companion cell complex in the stem phloem of *Lupinus luteus*. Planta 192:165–175

Van Sambeek JW, Pickard BG (1976) Mediation of rapid electrical, metabolic, transpirational and photosynthetic changes by factors released from wounds. I. Variation potentials and putative action potentials in intact plants. Can J Bot 54:2642–2650

Volk G, Franceschi VR (2000) Localization of a calcium-channel-like protein in the sieve element plasma membrane. Aust J Plant Physiol 27:779–786

Volkov AG, Dunkley TC, Morgan SA, Ruff D II, Boyce YL, Labady AJ (2004) Bioelectrochemical signaling in green plants induced by photosensory systems. Bioelectrochem 63:91–94

Volkov AG, Adesina T, Markin VS, Jovanov E (2008a) Kinetics and mechanism of *Dionaea muscipula* trap closing. Plant Physiol 146:694–702

Volkov AG, Coopwood KJ, Markin VS (2008b) Inhibition of the *Dionaea muscipula* Ellis trap closure by ion and water channel blockers and uncouplers. Plant Sci 175:642–649

Volkov AG, Adesina T, Jovanov E (2008c) Charge induced closing of *Dionaea muscipula* Ellis trap. Bioelectrochemistry 74:16–21

Volkov AG, Carrell H, Baldwin A, Markin VS (2009a) Electrical memory in Venus flytrap. Bioelectrochemistry 75:142–147

Volkov AG, Carrell H, Markin VS (2009b) Biologically closed electrical circuits in Venus flytrap. Plant Physiol 149:1661–1667

Volkov AG, Foster JC, Ashby TA, Walker RK, Johnson JA, Markin VS (2010) *Mimosa pudica*: Electrical and mechanical stimulation of plant movements. Plant Cell Environ 33:163–173

White PJ, Ridout MS (1999) An energy-barrier model for the permeation of monovalent and divalent cations through the maxi cation channel in the plasma membrane of rye roots. J Membr Biol 168:63–75

White PJ (2004) Calcium signals in root cells: the roles of plasma membrane calcium channels. Biol Plant 59:77–83

White PJ (2009) Depolarization-activated calcium channels shape the calcium signatures induced by low-temperature stress. New Phytol 183:6–8

Wildon DC, Thain JF, Minchin PEH, Gubb IR, Reilly AJ, Skipper YD, Doherty HM, Odonnell PJ, Bowles DJ (1992) Electrical signaling and systemic proteinase-inhibitor induction in the wounded plant. Nature 360:62–65

Williams SE, Pickard BG (1972a) Properties of action potentials in *Drosera* tentacles. Planta 103:193–221

Williams SE, Pickard BG (1972b) Receptor potentials and action potentials in *Drosera* tentacles. Planta 103:222–240

Wind C, Arend M, Fromm J (2004) Potassium-dependent cambial growth in poplar. Plant Biol 6:30–37

Woodley SJ, Fensom DS, Thompson RG (1976) Biopotentials along the stem of *Helianthus* in association with short-term translocation of ^{14}C and chilling. Can J Bot 54:1246–1256

Wright JP, Fisher DB (1981) Measurement of the sieve tube membrane potential. Plant Physiol 67:845–848

Zawadzki T, Davies E, Dziubinska H, Trebacz K (1991) Characteristics of action potentials in *Helianthus annuus*. Physiol Plant 83:601–604

Chapter 9
The Role of Plasmodesmata in the Electrotonic Transmission of Action Potentials

Roger M. Spanswick

Abstract The mounting evidence for the transmission of action potentials from cell to cell in a range of plants has exposed our lack of knowledge concerning the mechanism of transmission. While variation potentials (also known as slow wave potentials) involve chemicals released from damaged tissues and/or associated hydrodynamic changes, there is little or no evidence for the involvement of chemicals in the intercellular transmission of action potentials in plants. Plasmodesmata provide electrical connections between plant cells, as demonstrated by experiments in which current injected into one cell can produce a change in potential in a neighboring cell (electrical coupling). The evidence available to date supports a mechanism for electrotonic coupling of cells in transmission of action potentials rather than a direct transmission of excitation along the plasma membranes in the plasmodesmatal pores.

9.1 Introduction

It has become evident that action potentials in plants are not just a curiosity associated with the internodal cells of the Characean algae, the sensitive plant *Mimosa pudica*, or carnivorous plants such as *Drosera* or the Venus' flytrap. Action potentials have now been observed in a wide range of plants (Fromm and Lautner 2007; Pickard 1973). One question that appears to have received limited attention involves the details of the mechanism by which action potentials are transmitted from one cell to another.

R. M. Spanswick (✉)
Department of Biological and Environmental Engineering, Cornell University,
316 Riley-Robb Hall, Ithaca, NY 14853-5701, USA
e-mail: rms6@cornell.edu

In animals, two mechanisms of action potential transmission are well established: one is the *chemical transmission* that takes place at synapses, and the other is known as *electrotonic transmission*, which occurs when current can flow directly from cell-to-cell via gap junctions (Hille 2001). Although chemicals involved in chemical transmission at synapses, neurotransmitters such as γ-aminobutyrate (GABA), occur in plants (Cao et al. 2006; Murch 2006), there is no evidence that they are involved in transmission of action potentials; indeed, the term "plant synapse" appears to be confined to the context of intercellular chemical signaling (Baluška et al. 2005) without the involvement of action potentials.

The presence of plasmodesmata provides a route for the passage of electric current between cells (Spanswick and Costerton 1967; Spanswick 1972). It is generally assumed that plasmodesmata are important for the passage of action potentials, but there has been little or no recent consideration of the mechanism(s) involved. My purpose here is to raise the question of what exactly is necessary to achieve intercellular transmission of action potentials via plasmodesmata. First, because electric current is carried predominantly by ions in aqueous solutions, I shall briefly review the evidence which demonstrates that plasmodesmata provide a pathway for diffusion between plant cells.

9.2 The Structure of Plasmodesmata

The discovery of plasmodesmata is attributed to Tangl (1879) even though the diameters of plasmodesmata are an order of magnitude less than the resolution of the optical microscopes available at the time (Bell and Oparka 2011). However, it is possible that modification of the optical properties of the cell wall surrounding plasmodesmata (Taiz and Jones 1973; Badelt et al. 1994; Roy and Watada 1997) aided in detection of the structures.

The advent of electron microscopy has amply confirmed the presence of plasmodesmata and has led to work on the variety of structures (e.g., straight versus branched; Blackman and Overall 2001; Burch-Smith et al. 2011). It has also made it possible to develop hypotheses concerning the detailed structure of the plasmodesmatal pore (Beebe and Turgeon 1991; Bell and Oparka 2011; Ding et al. 1992; Overall and Blackman 1996), which is now recognized to be complex in a way that may account for size discrimination among molecules with regard to diffusion but with the presence of special mechanisms for the transport of macromolecules, including proteins and nucleic acids (Ding et al. 1999; Lucas et al. 2009; Zambryski and Crawford 2000).

9.3 The Symplasm as a Transport Pathway

The system of protoplasts connected via plasmodesmata (the symplasm) has been investigated extensively as a pathway for transport of solutes through tissues. Initial work on higher plants (see Spanswick 1974, 1976 for a summary) was hampered by a lack of resolution when using tracer techniques, although transport via this pathway could reasonably be inferred. The introduction of dye microinjection, electrophysiology, and the expression of fluorescent transgenic proteins has made it possible to demonstrate the symplasmic transport of nonelectrolytes, ions, and macromolecules.

9.3.1 Evidence for Intercellular Transport: Tracers and Fluorescent Dyes

Early work using dyes to demonstrate the effect of cytoplasmic streaming on transport (Bierburg 1909) also demonstrated intercellular transport in Characean cells, and this was later confirmed using radioactively labeled phosphate (Littlefield and Forsberg 1965) and chloride (Bostrom and Walker 1975; Tyree et al. 1974). Extensive work by Arisz and others did provide good evidence for symplasmic transport. In particular, Arisz (1960) used the aquatic plant *Vallisneria spiralis* to show that when short segments of the vascular bundles were removed from a section of a leaf, the remaining "parenchyma bridges" could support transport through that section of the leaf.

The introduction of techniques for the microinjection of fluorescent dyes into the cytoplasm of individual cells (Tucker 1982; Goodwin 1983) made it possible to demonstrate that small molecules could diffuse directly from cell to cell with a molecular mass cutoff of about 800. Subsequent work has demonstrated that larger molecules can move by diffusion through plasmodesmata under certain circumstances (McLean et al. 1997; Schönknecht et al. 2008). A large body of work has resulted from the study of the transport of viral nucleic acids (Ueki and Citovsky 2011), for which a specific mechanism involving "movement proteins" was demonstrated (Deom et al. 1990). This led to extensive work on the movement of macromolecules that is beyond the scope of this chapter. However, the expression of tobacco mosaic virus movement protein has been shown to increase the size exclusion limit of plasmodesmata to permit the diffusion of molecules with a molecular mass as high as 9,400 (Wolf et al. 1989) or green fluorescent protein (GFP) with a molecular mass of about 29,000 (Liarzi and Epel 2005).

9.3.2 Evidence for Intercellular Transport: Electrophysiology

The application of electrophysiological techniques to the study of plasmodesmata may be traced back to the pioneering work of Lou (1955) conducted in the laboratory of the muscle physiologist A.V. Hill in London in 1947–1948. Lou used pairs of giant internodal cells of *Nitella* cells that were isolated from the plant but remained connected to each other via the smaller cells in the node. There is a single layer of nodal cells across most of the area between the cells, and more cells are located around the periphery of the junction (Spanswick and Costerton 1967; Bostrom and Walker 1975; Fischer et al. 1976). The cells were immersed in oil and current was passed from one end of the system to the other using fine cotton threads as salt bridges to make contact with the cell wall surfaces. The change in electrical potential along the surface (Fig. 9.1) was measured as a function of distance using a separate pair of contacts. To interpret the experiment, it is necessary to appreciate that the internodal cells behave in a manner analogous to a leaky coaxial cable, with the plasma membrane acting as a "leaky" insulator that separates the outer conductor (the cell wall) from the inner conductor (the cytoplasm/vacuole). The relatively large gradient of electrical potential in the region where current is applied at the end of the cells reflects the relatively low conductance of the cell wall. As the distance from the site of current application increases, a fraction of the current enters the cell and flows along the interior of the cell, reducing the potential gradient due to the higher overall conductance of the combined parallel pathways. In the absence of open conducting plasmodesmatal connections in the node, all the current flowing through the vacuole and cytoplasm of the first cell would have to exit across the membrane of the first cell and would enter the second cell in a similar way to the entry into the first cell. This would increase the gradient of potential in the region of the node so that the additional change in potential ($\Delta\Psi_N$) would be about twice that of the initial change in potential at the site of current application ($\Delta\Psi_A$). The observation (Fig. 9.1) that $\Delta\Psi_N$ is less than $\Delta\Psi_A$ indicates that only a fraction of the intracellular current has exited the first cell and reentered the second cell. It appears that a significant fraction of the current flows directly from cell to cell without crossing the membranes. The resistance of the node is significant but requires a different method for accurate estimation.

9.3.3 Plasmodesmata as a Route for Intercellular Conduction of Electric Current

The use of microelectrodes to demonstrate the direct flow of current between cells in animal tissues provided a method to explore the connections between cells in an organ. For example, Kanno and Loewenstein (1964) demonstrated that current injected into a cell at one end of a *Drosophila* salivary gland produced a change in potential that could be detected with little attenuation in a neighboring cell, and was only reduced to 50% of the value in the cell into which the current was injected at a distance of about 9 cells from the point of current injection. This phenomenon is known as *electrical coupling*.

The application of this method to a pair of internodal cells of *Nitella translucens*, using four microelectrodes to make it possible to measure the resistance of the internodal cells, while taking into account the cable properties of the cells and

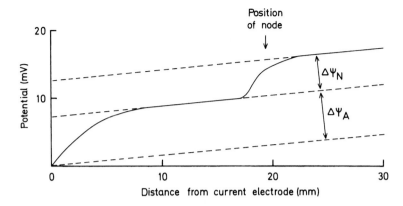

Fig. 9.1 Approximate representation of the potential gradient along a system consisting of a node and two adjacent cells of *Nitella* when a current is passed between external electrodes placed at either end of the preparation. External electrodes were also used to record the potential difference. (After Lou 1955; Spanswick 1976)

the changes in potential across the tonoplast, made it possible to calculate the specific resistance of the nodal region as well as the specific resistance of the cells (Spanswick and Costerton 1967). The resistance of the plasma membrane of the internodal cells was about 9 Ωm^2 while that of the node was only 0.17 Ωm^2, or more than two orders of magnitude less than the resistance of the four layers of plasma membrane that separate the two internodal cells. Nevertheless, an estimate of the resistance of the plasmodesmata based on the measured cross-sectional area of the node occupied by the lumens of the plasmodesmata and assuming an electrical conductivity based approximately on the ionic composition of the cytoplasm, yielded a value of another factor of 330 times smaller than the measured value. Comparable results were found for *Chara corallina* by Bostrom and Walker (1975). Thus, although the plasmodesmata clearly provide a low resistance pathway between the cells, and therefore a direct pathway for diffusion of ions or the passage of electric current, these data are consistent with the ultrastructural evidence that the interior of the plasmodesma is occupied by structures such as the desmotubule (Hepler 1982; Overall et al. 1982; also referred to as appressed endoplasmic reticulum), restricted diffusion of fluorescent dyes above a molecular mass of about 800 (Tucker 1982; Goodwin 1983) and the intercellular diffusion of chloride ions (Bostrom and Walker 1975; Tyree et al. 1974). Electrical coupling between internodal cells of the Characeae has also been demonstrated by Sibaoka (1966) and Skierczyńska (1968). The low value for the specific resistance of the node has also been confirmed (Bostrom and Walker 1975; Blake 1979; Coté et al. 1987; Ding and Tazawa 1989; Sibaoka and Tabata 1981; Reid and Overall 1992).

Application of similar electrophysiological methods to higher plant tissues is technically challenging, first due to the difficulty of inserting multiple microelectrodes in the tissue without dislodging microelectrodes already in place and, second, due to the complications of having contact between multiple cells. When current was injected into cell A of three cells, A, B and C, in a row along the edge of an *Elodea* leaf, the coupling ratio ($\Delta\Psi_B/\Delta\Psi_A$) between cell A and cell B (0.29) was lower than the coupling ratio ($\Delta\Psi_C/\Delta\Psi_B$) between cell B and cell C (0.72). This puzzling result was resolved by postulating that the tips of the microelectrodes used to inject current and measure the potential change in cell A were located in the vacuole. Thus all the current in cell A passed across the tonoplast and gave rise to a relatively large change in voltage compared to the situation in the adjoining cells. There, the current could flow around the vacuoles en route to exiting the cell across the plasma membrane or via the plasmodesmata to other cells in the leaf (Spanswick 1972). On the basis of this assumption, it was possible to simulate the pattern of coupling in an array of cells and calculate an approximate value for the specific resistance of the junction between the cells (0.0051 Ωm^2) that was much lower than the resistance of the plasma membrane (0.32 Ωm^2). By assuming that the plasmodesmata occupy 1% of the area of the cell walls and that the electrical conductivity of fluid within open plasmodesmata is equivalent to that of 100 mM KCl, it could be shown that the resistance of the plasmodesmata was probably in the range of 50–60 times that of open plasmodesmata. As with the results from Characean cells, this is consistent with the evidence that indicates that structures with the plasmodesmata would restrict the area available for the movement of ions.

Electrical coupling between higher plant cells was also demonstrated in oat coleoptiles (Spanswick 1972; Racusen 1976; Drake et al. 1978; Drake 1979), maize roots (Spanswick 1972) *Azolla* roots (Overall and Gunning 1982) and Arabidopsis root hair cells (Lew 1994). Due to the multicellular nature of the tissues, it is difficult to make quantitative estimates of the intercellular resistance (van Rijen et al. 1999). Holdaway-Clarke et al. (1996) avoided this problem by using pairs of cultured corn cells. A double-barreled microelectrode was used to inject current, measure the potential in the cell into which current was being injected, and to locate the site of injection, cytoplasm or vacuole, within the cell. The specific resistance of the junction between the cells was 0.038 Ωm^2. By measuring the number of plasmodesmata per unit area of cell wall, they were also able to calculate a value of 51 GΩ for the resistance of a single plasmodesma. This corresponds to a conductance of about 20 pS. A comparable value (4.7 pS) can be obtained from the data for *Nitella* from Spanswick and Costerton (1967). Thus the conductance of a plasmodesma appears to be within the range of values for ion channels (Hille 2001).

Gap junctions are the structures in animal cells equivalent to plasmodesmata in plants in that they provide a route for the passage of small molecules between cells and also permit electrical coupling between adjacent cells (Furshpan and Potter 1968; Loewenstein 1978). The conductance of single gap junctions also falls within the range of ion channels (Veenstra 2000). The fact that the conductance of gap junctions, acting as ion channels with a pore diameter narrowing to about 1.2–1.5 nm, is comparable to that of plasmodesmata (pore diameter about 40 nm), is consistent with the greater length of plasmodesmata and the presence of structures within the plasmodesma that decrease the conductance.

9.4 The Transmission of Action Potentials in Plants

In discussions of transmission of action potentials, it usually assumed that the presence of plasmodesmata can account for transmission without detailed discussion of the possible mechanisms involved. For a more thorough discussion, I shall consider the following possibilities: (1) The plasma membrane lining the plasmodesma contains channels that permit the propagation of an action potential by the same mechanism as a cylindrical cell such as a neuron or a Characean internodal cell. That is, the depolarization during an action potential creates a local current loop that depolarizes the neighboring region to a value below the threshold for triggering an action potential in that region, thus propagating the potential change along the cell (or plasmodesma in this case). (2) The local current generated externally by one cell might be sufficient to depolarize the membrane at the junction of two cells even in the absence of a direct connection via plasmodesmata. (3) The action potential might release sufficient ions, or other molecules, into the apoplast to depolarize neighboring cells below the threshold for the generation of a separate action potential. (4) Without continuous transmission of an action potential along the membranes of plasmodesmata, the current flowing from cell to cell due to electrical coupling of two cells might be sufficient to excite a neighboring cell electrotonically when an action potential is triggered in the first cell.

9.4.1 Can Transmission of the Action Potential Occur via Excitation of the Plasmodesmal Plasma Membrane?

It is conceivable that an action potential could propagate along a plasmodesmal pore via the same mechanism as along a cell. Since plasmodesmata have very small dimensions and are buried within cell walls, it is not feasible to test this hypothesis directly. However, it is possible to speculate as to whether the conditions exist for such a scenario to be realistic. One might ask, for example, whether there are ion channels located in the plasmodesmal plasma membrane and whether the cable properties of the plasmodesma are consistent with propagation of an action potential.

The protein composition of plasmodesmata is receiving increasing attention (Faulkner and Maule 2011; Fernandez-Calvino et al. 2011). One approach is to isolate cell walls and treat them as a source of trapped plasmodesmata. While there is an enrichment of proteins known to be associated with plasmodesmata, contamination with proteins from other locations in the cell would make it unlikely that low-abundance proteins, such as ion channels, could be definitively localized to the plasma membrane by this technique. There does not appear to be any information available regarding ion channel density in plant cells. However, if a rough estimate is made of the number of channels in a plasmodesma by assuming a channel density equal to the density of sodium channels in a squid giant axon

(330 channels μm^{-2}; Conti et al. 1975) and a plasmodesma of the dimensions of those in *Setcreasea* staminal hairs (288.8 nm long and 37.6 nm in diameter; Tucker 1982), which would give a surface area of 0.034 μm^2, the number of channels per plasmodesma would be about 11. A small number, such as this, would be subject to statistical fluctuation which might render individual plasmodesmata unreliable as routes for transport. However, this would be offset by the presence of multiple plasmodesmata in the cell wall. A potentially more serious problem, for which evidence is not yet available, is that channels might be excluded from the plasma membrane lining the plasmodesmal pore. Another transport protein, the H^+-ATPase, has been shown, using an immunochemical technique, to be absent or at a very low level in the membrane lining the plasmodesmal pore in the primary pulvinus cells of *M. pudica* L. (Fleurat-Lessard et al. 1995).

A further consideration is whether the cable properties of the plasmodesmal pore are such that an action potential will be propagated. A calculation of the length constant for a plasmodesma at the internodal/nodal cell junction in *N. translucens* (Spanswick and Costerton 1967) may be based on a plasmodesma with a length of 7.3 μm, a diameter of 50 nm, and a conductance of 4.7 pS. Assuming a specific membrane resistance of 2 Ωm^2 for the plasma membrane lining the plasmodesma, equal to the specific resistance of the plasma membrane for a cell in the light, yields a length constant in excess of 200 μm, suggesting that attenuation would be negligible along the length of the pore. The fact that action potentials are not transmitted between *Nitella* cells consistently may indicate that the plasmodesmal membrane is deficient in channels and has a high specific resistance.

9.4.2 Can the Local External Current Generated by an Action Potential in One Cell Produce a Depolarization in the Neighboring Cell Sufficient to Trigger a Separate Action Potential?

A few papers report the occurrence of action potentials in *Chara* cells that have been isolated but placed parallel to other cells in which action potentials have been triggered by electrical stimulation (Ping et al. 1990; Tabata 1990). The transmission is reported to be effective, although at reduced frequency, even at a distance of 1 cm (Ping et al. 1990). Although Ping et al. (1990) cite the use of freshly isolated cells, in which the membrane potential might be closer to the threshold than for cells given a chance to recover from being severed from their neighbors, it is puzzling that an action potential in one cell can be effective in depolarizing a separate cell at such a distance. Further investigation may be warranted. Also, further consideration of whether extracellular currents in the apoplast during an action potential in one cell could have an effect on a neighboring cell in intact tissues that would give rise to production of a separate action potential, a phenomenon known as ephaptic transmission (Jefferys 1995), would appear to be necessary.

9.4.3 Are Chemicals Involved in Intercellular Transmission of Action Potentials?

Although there have recently been attempts to extend the terminology of animal neurobiology to plants, including the concept of "the synapse" (Baluška et al. 2005), there is no evidence that synaptical transmission, involving the release of chemicals from synaptic vesicles at intercellular interfaces, is involved in the propagation of an action potential from cell-to-cell in plants. There remains the possibility that the increase in channel-mediated fluxes of ions during an action potential could change the concentrations in the apoplast sufficiently to depolarize a neighboring cell below the threshold for an action potential.

In considering the role of chemicals, it is important to distinguish between action potentials and variation potentials (also called *slow waves*). Both involve changes in electrical potential that can be recorded by surface electrodes attached to plant stems. However, action potentials must have the classical properties (all-or-none response, etc.) and can usually be produced by relatively mild stimulations, such as electrical stimulation (Fromm and Spanswick 1993; Pickard 1973), rapid chilling (Minorsky and Spanswick 1989), or relatively small mechanical forces (Shimmen 2003). Variation potentials, on the other hand, are produced by more severe treatments such as leaf scorching by a flame or mechanical injury to tissues (Stanković et al. 1998). The destruction of cells leads to release of chemicals, collectively referred to as Ricca's factor (Van Sambeek and Pickard 1976), to the apoplast. As these chemicals are carried along in the transpiration stream they will depolarize the membrane potential of cells they contact, which is likely one factor in the production of the variation potential (Stanković et al. 1998). As argued in detail by Malone (1996) changes in water movement will accompany such severe treatments and could give rise to propagated changes in hydraulic pressure. Pickard (2007) has reviewed the evidence for the presence of mechanosensitive Ca^{2+}-selective ion channels in plants. It is conceivable that pressure changes could give rise to changes in membrane potential and/or the generation of action potentials. Thus Malone (1996) raises the question of whether the electrical or the mechanical signals should be considered primary. There may well be room to establish cause and effect with regard to the role of electrical signals in specific instances. However, evidence for the occurrence and propagation of action potentials in a range of plants, not solely sensitive plants, continues to accumulate (Fromm and Lautner 2007). Although external supply of glutamate, a neurotransmitter in animals, has been demonstrated to trigger action potentials (Felle and Zimmermann 2007; Krol et al. 2007; Stolarz et al. 2010), it has not yet been demonstrated to be involved in the transmission of action potentials in vivo.

Some ambiguity exists in interpreting the results of amino acid applications because of the existence of proton cotransport systems for amino acids, which themselves produce a transient depolarization of the membrane potential when an amino acid is applied externally (Etherton and Rubinstein 1978). This, in itself could trigger an action potential by activating voltage-gated channels. The extent to which the large depolarization in oat coleoptile cells produced by glutamate (Kinraide and Etherton 1980) reflects the action of a mechanism in addition to the cotransport system remains to be determined, although mutants for glutamate receptors do have a reduced response of the membrane potential to amino acids (Stephens et al. 2008).

At this stage, systems equivalent to the chemical synapses responsible for the unidirectional transmission of action potentials in animals do not appear to be present in plants. However, it is likely that chemicals such as glutamate and GABA are released from damaged tissues into the xylem and contribute to the electrical changes responsible for variation potentials.

9.4.4 Is Propagation from Cell-to-Cell Electrotonic due to Flow of Current Between Cells via Plasmodesmata in the Absence of Excitation of the Plasma Membrane Within the Plasmodesmal Pore?

The observation that an action potential generated in an internodal cell of *Nitella* or *Chara* can trigger an action potential in a neighboring internodal cell (Sibaoka 1966; Spanswick and Costerton 1967; Sibaoka and Tabata 1981) provides a useful system for testing the mechanism by which transmission takes place. The fact that transmission does not occur in every instance supports the idea that the membranes within the plasmodesmata are not excitable, or that the membranes of the intervening nodal cells are not excitable. If an electrical stimulus is used that is just large enough to generate an action potential in one cell but not large enough to stimulate the neighboring cell, due to attenuation of the pulse as it passes through the plasmodesmata at the node, it appears that one of two things can happen (Fig. 9.2). The additional stimulus provided to the adjoining cell by the attenuated depolarization of the action potential may not be sufficient in all cases to trigger a separate action potential (Fig. 9.2b, lower trace). In other cases, the additional depolarization and its length may be sufficient to trigger a separate action potential in the neighboring cell (Fig. 9.2a, lower trace). On the basis of this observation (Spanswick and Costerton 1967; Spanswick 1974), and the later observations by Sibaoka and Tabata (1981) on *Chara braunii*, it was suggested that the transmission between the internodal cells is electrotonic. Sibaoka and Tabata (1981) point out that transmission of the action potential can occur in either direction, a symmetry that would not be expected at a classical synapse.

Fig. 9.2 Recordings of action potentials from two adjacent cells of *N. translucens*. **a** *Upper* trace: recording from the cell into which current was injected to produce the stimulus, *S*, which was followed by an action potential. *Lower* trace: recording from the adjacent internodal. Initially, the attenuated stimulus pulse, S_a, is followed by the attenuated signal of the action potential in the neighboring cell. In this case, a separate action potential was triggered. **b** *Upper* trace: recording from the cell to which the stimulus was applied. *Lower* trace: in this case, the attenuated stimulus and action potential was insufficient to trigger a separate action potential (Spanswick and Costerton 1967)

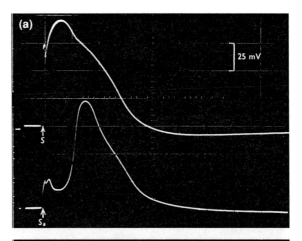
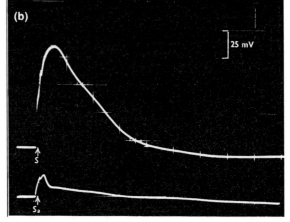

Also evident from the recordings of action potentials in *Nitella* (compare upper traces in Fig. 9.2a, b) and *Chara* is the feedback of the action potential in the second cell, when triggered, on the shape of the action potential in the first cell. Electrical coupling between cells in the stalks of *Drosera* tentacles are also postulated to account for the complex time courses of action potentials observed in some intracellular recordings from stalk cells (Williams and Spanswick 1976).

9.5 Conclusions

The evidence available to date indicates that the transmission of action potentials between plant cells is electrotonic. Since there are few studies that deal with the mechanisms of transmission, there is always the possibility that this conclusion will need to be modified or qualifications may need to be applied if it is shown that chemicals released by one cell modulate the electrotonic influence of a cell on its neighbor.

The long-distance transmission of action potentials in higher plants most likely occurs in the phloem (Fromm and Lautner 2007). The diameter of sieve plate pores is larger than that of plasmodesmata, so the mechanism of transmission in the phloem may be more difficult to determine; it may be difficult to distinguish between electrotonic transmission and direct propagation along the plasma membrane as it traverses the pores. While the propagation of action potentials in the phloem can be demonstrated using aphid stylets to access the interior of sieve tubes, detailed investigation of the mechanism of action potential transmission between sieve elements presents formidable technical challenges (van Bel et al. 2011).

The current chapter has emphasized the electrical properties of plasmodesmata as fixed quantities. However, there is a considerable body of work relating to the dynamic nature of the factors controlling plasmodesmatal conductance (Holdaway-Clarke 2005). These include the effects of Ca^{2+} (Holdaway-Clarke et al. 2000), cytoplasmic pH (Lew 1994), metabolites as judged by the effects of metabolic inhibitors (Cleland et al. 1994), and cytoskeletal interactions (Blackman and Overall 1998; White and Barton 2011), although these factors are not necessarily independent (Holdaway-Clarke 2005). This indicates that plasmodesmata may play a role in modulating the signaling pathways involving action potentials.

References

Arisz WH (1960) Symplasmitischer Salztransport in Vallineria-Blattern. Protoplasma 52:309–343
Badelt K, White RG, Overall RL, Vesk M (1994) Ultrastructural specialization of the cell wall sleeve around plasmodesmata. Am J Bot 81:1422–1427
Baluška F, Volkmann D, Menzel D (2005) Plant synapses: actin-based domains for cell-to-cell communication. Trends Plant Sci 10:106–111
Beebe DU, Turgeon R (1991) Current perspectives on plasmodesmata: structure and function. Physiol Plant 83:194–199
Bell K, Oparka K (2011) Imaging plasmodesmata. Protoplasma 248:9–25
Bierberg W (1909) Die Beduetung der Protoplasmarotation für der Stofftransport in den Pflanzen. Flora 99:52–80
Blackman LM, Overall RL (1998) Immunolocalization of the cytoskeleton to plasmodesmata of *Chara corallina*. Plant J 14:733–741
Blackman LM, Overall RL (2001) Structure and function of plasmodesmata. Austr J Plant Physiol 28:709–727
Blake IO (1979) The effect of cell excision and microelectrode perforation on membrane resistance measurements of *Nitella translucens*. Biochim Biophys Acta 554:62–67
Bostrom TE, Walker NA (1975) Intercellular transport in plants. I. The rate of transport of chloride and the electrical resistance. J Exp Bot 26:767–782
Burch-Smith TM, Stonebloom S, Xu M, Zambryski PC (2011) Plasmodesmata during development: re-examination of the importance of primary, secondary, and branched plasmodesmata structure versus function. Protoplasma 248:61–74
Cao J, Cole IB, Murch SJ (2006) Neurotransmitters, neuroregulators and neurotoxins in the life of plants. Can J Plant Sci 86:1183–1188
Cleland RE, Fujiwara T, Lucas WJ (1994) Plasmodesmal-mediated cell-to-cell transport in wheat roots is modulated by anaerobic stress. Protoplasma 178:81–85

Conti F, De Felice LJ, Wanke E (1975) Potassium and sodium ion current noise in the membrane of the squid giant axon. J Physiol 248:45–82

Coté R, Thain J, Fensom DS (1987) Increase in electrical resistance of plasmodesmata of *Chara* induced by an applied pressure gradient across nodes. Can J Bot 65:509–511

Deom CM, Schubert KR, Wolf S, Holt CA, Lucas WJ, Beachy RN (1990) Molecular characterization and biological function of the movement protein of tobacco virus in transgenic plants. Proc Nat Acad Sci USA 87:3284–3288

Ding B, Itaya A, Woo Y-M (1999) Plasmodesmata and cell-to-cell communication in plants. Int Rev Cytol 190:251–316

Ding B, Turgeon R, Parthasarathy MV (1992) Substructure of freeze-substituted plasmodesmata. Protoplasma 169:28–41

Ding D-Q, Tazawa M (1989) Influence of cytoplasmic streaming and turgor pressure gradient on the transnodal transport of rubidium and electrical conductance in *Chara corallina*. Plant Cell Physiol 30:739–748

Drake GA (1979) Electrical coupling, potentials, and resistances in oat coleoptiles: effects of azide and cyanide. J Exp Bot 30:719–725

Drake GA, Carr DJ, Anderson WP (1978) Plasmolysis, plasmodesmata, and the electrical coupling of oat coleoptile cells. J Exp Bot 29:1205–1214

Etherton B, Rubinstein B (1978) Evidence for amino acid-H^+ co-transport in oat coleoptiles. Plant Physiol 61:933–937

Faulkner C, Maule A (2011) Opportunities and successes in the search for plasmodesmal proteins. Protoplasma 248:27–38

Felle HH, Zimmermann MR (2007) Systemic signalling in barley through action potentials. Planta 226:203–214

Fernandez-Calvino L, Faulkner C, Walshaw J, Saalbach G, Bayer E, Benitez-Alfonso Y, Maule M (2011)Arabidopsis plasmodesmal proteome. PLoS ONE 6:e18880

Fischer RA, Dainty J, Tyree MT (1976) A quantitative investigation of symplasmic transport in *Chara corallina*. I. Ultrastructure of the nodal complex cell walls. Can J Bot 52:1209–1214

Fleurat-Lessard P, Bouché-Pillon S, Leloup C, Lucas WJ, Serrano R, Bonnemain J-L (1995) Absence of plasma membrane H^+-ATPase in plasmodesmata located in pit-fields of the young reactive pulvinus of *Mimosa pudica* L. Protoplasma 188:180–185

Fromm J, Lautner S (2007) Electrical signals and their physiological significance in plants. Plant Cell Environ 30:249–257

Fromm J, Spanswick RM (1993) Characteristics of action potentials in willow (*Salix viminalis* L.). J Exp Bot 44:1119–1125

Furshpan EJ, Potter DD (1968) Low resistance junctions between cells in embryos and tissue culture. Curr Topics Dev Biol 3:95–127

Goodwin PB (1983) Molecular size limit for movement in the symplast of the *Elodea* leaf. Planta 157:124–130

Hepler PK (1982) Endoplasmic reticulum in the formation of the cell plate and plasmodesmata. Protoplasma 111:121–133

Hille B (2001) Ionic channels of excitable membranes, 3rd edn. Sinauer Associates, Sunderland

Holdaway-Clarke TL (2005) Regulation of plasmodesmal conductance. In: Oparka K (ed) Plasmodesmata. Blackwell Publishing Ltd., Oxford, pp 279–297

Holdaway-Clarke TL, Walker NA, Hepler PK, Overall RL (2000) Physiological elevations in cytoplasmic free calcium by cold or ion injection result in transient closure of higher plant plasmodesmata. Planta 210:329–335

Holdaway-Clarke TL, Walker NA, Overall RL (1996) Measurement of the electrical resistance of plasmodesmata and membranes of corn suspension-culture cells. Planta 199:537–544

Jefferys JGR (1995) Nonsynaptic modulation of neuronal ctivity in the brain—Electric currents and extracellular ions. Physiol Rev 75:689–723

Kanno Y, Lowenstein WR (1964) Low-resistance coupling between gland cells. Some observations on intercellular contact membranes and intercellular space. Nature 201:194–195

Kinraide TB, Etherton B (1980) Electrical evidence for different mechanisms of uptake for basic, neutral, and acidic amino acids in oat coleoptiles. Plant Physiol 65:1085–1089

Krol E, Dziubinska H, Trebacz K, Koselski M, Stolarz M (2007) The influence of glutamic acid and aminoacetic acids on the excitability of the liverwort *Conocephalum conicum*. J Plant Physiol 164:773–784

Lew R (1994) Regulation of electrical coupling between *Arabidopsis* root hairs. Planta 193:67–73

Liarzi O, Epel BL (2005) Development of a quantitative tool for measuring changes in the coefficient of conductivity of plasmodesmata induced by developmental, biotic, and abiotic signals. Protoplasma 225:67–76

Littlefield L, Forsberg C (1965) Absorption and translocation of phosphorus-32 by *Chara globularis* Thuill. Physiol Plant 18:291–296

Loewenstein WR (1978) Cell-to-cell communication. Permeability, formation, genetics, and functions of the cell–cell membrane channel. In: Andreoli TE, Hoffman JF, Fanestil DD (eds) Membrane physiology. Plenum Press, New York, pp 335–356

Lou CH (1955) Protoplasmic continuity in plants. Acta Bot Sinica 4:183–222

Lucas WJ, Ham B-K, Kim J-Y (2009) Plasmodesmata—bridging the gap between neighboring plant cells. Trends Cell Biol 19:495–503

Malone M (1996) Rapid, long-distance signal transmission in higher plants. Adv Bot Res 22:163–228

McLean BG, Hempel FD, Zambryski PC (1997) Plant intercellular communication via plasmodesmata. Plant Cell 9:1043–1054

Minorsky PV, Spanswick RM (1989) Electrophysiological evidence for a role for calcium in temperature sensing by roots of cucumber seedlings. Plant Cell Environ 12:137–143

Murch SJ (2006) Neurotransmitters, neuroregulators and neurotoxins in plants. In: Baluška F, Mancuso S, Volkmann D (eds) Communication in plants. Springer, Berlin, pp 137–151

Overall RL, Blackman LM (1996) A model of the macromolecular structure of plasmodesmata. Trends Plant Sci 1:307–311

Overall RL, Gunning BES (1982) Intercellular communication in *Azolla* roots: II Electrical coupling. Protoplasma 111:151–160

Overall RL, Wolfe J, Gunning BES (1982) Intercellular communication in *Azolla* roots: I Ultrastructure of plasmodesmata. Protoplasma 111:134–150

Pickard BG (1973) Action potentials in higher plants. Bot Rev 39:172–201

Pickard BG (2007) Delivering force and amplifying signals in plant mechanosensing. Curr Top Membr 58:361–392

Ping Z, Mimura T, Tazawa M (1990) Jumping transmission of action potential between separately placed internodal cells of *Chara corallina*. Plant Cell Physiol 31:299–302

Racusen RH (1976) Phytochrome control of electrical potentials and intercellular coupling in oat coleoptile tissue. Planta 132:25–29

Reid RJ, Overall RL (1992) Intercellular communication in *Chara*: factors affecting transnodal electrical resistance and solute fluxes. Plant Cell Environ 15:507–517

Roy S, Watada AE, Wergin WP (1997) Characterization of the cell wall microdomain surrounding plasmodesmata in apple fruit. Plant Physiol 114:539–547

Schönknecht G, Brown JE, Verchot-Lubicz J (2008) Plasmodesmata transport of GFP alone or fused to potato virus X TGBp1 is diffusion driven. Protoplasma 232:143–152

Shimmen T (2003) Studies on mechano-perception in the Characeae: transduction of pressure signals into electrical signals. Plant Cell Physiol 44:1215–1224

Sibaoka T (1966) Action potentials in plant cells. Symp Soc Exp Biol 20:49–73

Sibaoka T, Tabata T (1981) Electrotonic coupling between adjacent internodal cells of *Chara braunii*: transmission of action potentials beyond the node. Plant Cell Physiol 22:397–411

Skierczynska J (1968) Some of the electrical characteristics of the cell membrane of *Chara australis*. J Exp Bot 19:389–406

Spanswick RM (1972) Electrical coupling between the cells of higher plants: a direct demonstration of intercellular transport. Planta 102:215–227

Spanswick RM (1974) Symplasmic transport in plants. Symp Soc Exp Biol 28:127–137

Spanswick RM (1976) Plasmodesmata and symplasmic transport. Encyclopedia of Plant Physiology, New Series IIB:35–53
Spanswick RM, Costerton JWF (1967) Plasmodesmata in *Nitella translucens*: structure and electrical resistance. J Cell Sci 2:451–464
Stankovic' B, Witters DL, Zawadzki T, Davies E (1998) Action potentials and variation potentials in sunflower: an analysis of their relationships and distinguishing characteristics. Physiol Plant 103:51–58
Stephens N, Qi Z, Spalding EP (2008) Glutamate receptor subtypes evidenced by differences in desensitizatiion and dependence on the *GLR*3.3 and *GLR*3.4 genes. Plant Physiol 146:529–538
Stolarz M, Król E, Dziubińska H, Kurenda A (2010) Glutamate induces series of action potentials and a decrease in circumnutation rate in *Helianthus annuus*. Physiol Plant 138:329–338
Tabata T (1990) Ephaptic transmission and conduction velocity of an action potential in *Chara* internodal cells placed in parallel and in contact with one another. Plant Cell Physiol 31:575–579
Taiz L, Jones RL (1973) Plasmodesmata and an associated cell wall component in barley aleurone tissue. Am J Bot 60:67–75
Tangl E (1879) Über offene Kommunikation zwichen den Zellen des Endosperms einiger Samen. Jahrbücher für wissenschaftliche Botanik 12:170–190
Tucker EB (1982) Translocation in the staminal hairs of *Setcreasea pupurea*. I. A study of cell ultrastructure and cell-to-cell passage of molecular probes. Protoplasma 113:193–201
Tyree MT, Fischer RA, Dainty J (1974) A quantitative investigation of symplasmic transport in *Chara corallina*. II. The symplasmic transport of chloride. Can J Bot 52:1325–1334
Ueki S, Citovsky V (2011) To gate, or not to gate: regulatory mechanisms for intercellular protein transport and virus movement in plants. Mol Plant 4:782–793
van Bel AJE, Knoblauch M, Furch ACU, Hafke JB (2011) (Questions)^n on phloem biology. 1. Electropotential waves, Ca^{2+} fluxes and cellular cascades along the propagation pathway. Plant Sci 181:210–218
van Rijen HVM, Wilders R, Jongsma HJ (1999) Electrical coupling. In: van Bel AJE, van Kesteren WJP (eds) Plasmodesmata. Structure, function, role in cell communication. Springer, Berlin, pp 51–65
Van Sambeek JW, Pickard BG (1976) Modification of rapid electrical, metabolic, transpirational, and photosynthetic changes by factors released from wounds. II. Mediation of the variation potential by Ricca's factor. Can J Bot 54:2651–2661
Veenstra RD (2000) Ion permeation through connexin gap junction channels: effects on conductance and selectivity. Curr Topics Membr 49:95–129
White RG, Barton DA (2011) The cytoskeleton in plasmodesmata: a role in intercellular transport? J Exp Bot 62:5249–5266
Williams SE, Spanswick RM (1976) Propagation of the neuroid action potential of the carnivorous plant *Drosera*. J Comp Physiol A 108:211–223
Wolf S, Deom CM, Beachy RN, Lucas WJ (1989) Movement protein of tobacco mosaic virus modifies plasmodesmatal size exclusion limit. Science 246:377–379
Zambryski PC, Crawford K (2000) Plasmodesmata: gatekeepers for cell-to-cell transport of developmental signals in plants. Annu Rev Cell Dev Biol 16:393–421

Chapter 10
Moon and Cosmos: Plant Growth and Plant Bioelectricity

Peter W. Barlow

Abstract Many of the growth movements of plants (diurnal leaf movements, and perhaps stem dilatation cycles) initiate action potentials which are propagated within the plant body. Action potentials are then able to serve as informational signals that regulate further processes. Some movements appear to be regulated by turning points in the time-courses of the lunisolar tidal accelerative force, when the rate of accelerative change is zero. There are, in addition, other more constitutive bioelectrical phenomena in plants, such as electrical potential differences. These, also, are critically examined in relation to the lunisolar tide. Because of its ever-present nature, it is difficult to analyse experimentally effects of this lunisolar tide on organic processes; nevertheless, it may be possible to take steps towards validating the Moon's effect. This would take advantage of the predictability of the tidal acceleration profile and, hence, experiments could be devised to anticipate possible lunisolar tidal effects on biological events. Certain additional cosmic regulators of bioelectric patterns in plants, such as geomagnetic variations are also discussed, as are the effects of natural seismic events.

> *The world is never quiet, even its silence eternally resounds with the same notes, in vibrations which escape our ears.*
> *Albert Camus.*

P. W. Barlow (✉)
School of Biological Sciences, University of Bristol,
Woodland Road, Bristol, BS8 1UG, UK
e-mail: p.w.barlow@bristol.ac.uk

10.1 Introduction

Even though the adjective 'living' is difficult to define, one may confidently believe that something which is alive shows a capacity not only to increase its mass but also to reproduce. A secure link between these two distinct processes must have come about during the unfolding of life and would have endowed living systems with a potential for development along a time axis. Modulators of development, able to adjust the sizes, forms and cellular composition of organisms, came into existence later. And these modulators, being themselves products of development, became established as intrinsic components of the organism through selectionist, evolutionary processes. In plants, well accepted modulators of growth, regional differentiation and cellular reproduction include the phytohormones. In addition to phytohormones, which are of ancient origin (Sztein et al. 2000; Ross 2010), another class of intrinsic modulators of development and behaviour are the bioelectric impulses which result from ionic exchanges across membranes. Such impulses may be a fundamental feature of living forms and, hence, be even more ancient than the phytohormones (Goldsworthy 1983). These impulses, being channelled within a phytoneural system (Barlow et al. 2009; Barlow 2010), contrive to set up particular spatio-temporal patterns within conglomerates of cells. Much of the early work on plant bioelectricity, particularly in relation to the problem of tissue polarity and the effects thereupon of phytohormones, was summarised by Suzanne Meylan (1971).

Some developmental modulators are extrinsic, however, having their origin in the external environment, beyond the anatomical boundary of the organism. The list of such factors may now be nearly complete. According to classical mechanics, a number of these external influences could be regarded as vibratory, or as having an effective frequency: for example, electromagnetic radiation, sound, mechanical vibration, with units of frequency measured in Hertz (Hz). Their action spectra indicate the biologically effective range of frequencies whose thresholds initiate a growth response that is related to energy flux rate, amplitude and time. The sensing and susception of these frequencies in relation to their biological effects usually depend upon some sort of receptor molecule, or sensitive surface, located within a specialised receptive organ, which, in turn, may be coupled to phytoneural channels of information-flow based on travelling bioelectrical potentials. The latter can be experimentally recorded as time-courses of action potentials (APs), variation potentials and slow wave potentials (Stahlberg et al. 2006).

If it is necessary that information concerning a stimulus should pass, as an AP, along a conductive channel such as a phloem element, then there must be a tremendous scaling-down of energy between the site of reception and the systems which transduce and eventually respond to the AP. For example, following a period in white light, a flux of blue light of 450 nm has been shown to be converted within a young soybean plant into a number of APs (Volkov et al. 2005). This blue-light wavelength has a frequency of 6.66×10^{14} Hz, whereas the APs induced in the plant have a frequency of 3×10^{-4} Hz. The in-coming/out-going scaling factor is therefore 2.22×10^{-18}. Conversely, chemical substances, both natural and synthetic, which themselves seem to have no vibratory capacity except that due to their molecular configuration and atomic orbitals, can also induce periodical bioelectric impulses (Volkov and Mwesigwa 2001). Even water appears to have a small effect on the background bioelectric state: spraying a soybean plant with water was found to lift the resting electropotential difference (EPD) from an amplitude of ~ 0.6 mV (Volkov and Mwesigwa 2001) to one of about 1.5 mV with a frequency of 44 Hz. Similarly, adding water to dry soil surrounding a root system resulted in a sharp change in amplitude of ~ 7 mV with respect to the EPD in the shoot, with the appearance of a slower frequency of 8×10^{-3} Hz superimposed upon the 44 Hz frequency of the resting potential (Volkov and Haacke 1995). It is now considered that the APs induced in response to drought conditions, and perhaps in response to re-hydration also, constitute significant informational signals to which the plant responds with metabolic adjustments of adaptational value (Fromm and Fei 1998).

Clearly, the living plant organism is able to incorporate a huge range of highly energetic influences flowing from the terrestrial environment as well as from the Solar system (e.g., sunlight) and beyond. The hierarchical organisation of plant structure can be regarded as a means by which energy inflows can be transduced and internally redistributed in some other form (e.g. Annila and Kuisamen 2009) by means of similar sets of subsystems at each organisational level (Barlow 1999). But in the absence of a suitable receptor to capture energy and, hence, to buffer the living system, the in-flow of energy can lead to organic instability and even death. Moreover, plants have no buffering receptors for very high energy radiation—for example, X-rays and gamma rays with frequencies of 10^{18}–10^{20} Hz; hence, these particular in-flows, if sustained, are eventually lethal.

The function of sessile plants, both on land and in marine and oceanic environments, is to provide an organic receptive, energy-capturing 'phytodermis' that covers the inorganic crust, or geodermis, of the Earth, and to transduce external energy for the benefit of other life-forms which move upon the surface of the composite phytogeodermis. Plant life is helped to function in this manner by the more gentle, slow vibrations or rhythms in the external environment, such as the cycles of daily light/dark and the lunar tidal forces which bear on the Earth, and which have periods of about 6–12 h, i.e. 1.1–4.6×10^{-5} Hz. Plants respond with cycles of about the same period—for example, the ~ 24 h period of the lunisolar-driven movements of bean leaves (Klein 2007). These responses are completely in tune with their lunisolar gravitational stimulus and, hence, require no scaling up or down. The Sun, Moon and Earth (whose ionosphere has a natural frequency of ~ 8 Hz), as well as its life-forms, constitute a harmonious whole, bound together by gravity. It is in this milieu that the more energetic bioelectrical potentials are then generated to assist the phytogeodermal energy exchange.

Plant bioelectrical features can be simplistically classified as either constitutive or facultative. The former class includes the inherent bioelectrical potential differences which extend throughout the plant body and provide it with an electrical framework; the latter class includes APs which are evoked in response to stimuli, and which travel along phytoneural channels as informative signals. Plants, and life in general, is embraced not only by a constant Earthly gravity but is also held within the grasp of a cyclically varying lunisolar gravity, the solar component being 46% that of the Moon.[1] The present chapter will explore the bioelectric aspects of this plant/lunisolar gravitational relationship.

10.2 The Early Work of Harold Saxton Burr

From a somewhat different perspective to that exposed above, Harold Saxton Burr (1889–1973), E. K. Hunt Professor of Anatomy at Yale University School of Medicine, also grappled with the impact and immensity of influences which reached Earth and affected organic life thereupon. In the 1940s, Burr palpably despaired of world events—both by World War II and by what he saw as the attendant rupture and disintegration of the social fabric—as is revealed in the Introduction to his paper entitled, rather suitably and disturbingly, 'Moon Madness' (Burr 1944). Here, Burr wrote (p. 249): 'In a time of world-wide disruption, the existence of the order [of the Universe] is questioned and man fumbles for some stability'. 'Everyone', he declared, 'sooner or later must determine his relationship to the Universe [and] it becomes especially important... in times like these to re-emphasise not so much the describable order of the Universe but the fact that it has order and that this order can be understood by the mind of man'. Burr goes on (p. 250) to consider how to find a solution: 'the astute procedure is to begin with what lies immediately in front of one... to examine the impact of many environmental forces, both local and universal, on the continued existence of a single living form—a maple tree'.

What was it that Professor Burr hoped the maple tree would tell him? Whatever it was, it led Burr to make numerous observations on the bioelectrical properties of trees, coupling his findings to the passage of the Moon across the skies. The species of 'maple' tree studied by Professor Burr is not named: probably it was *Acer saccharum*, the Sugar Maple. One specimen of this maple tree was growing in his home town of Lyme, Connecticut, USA, a second specimen was located in New Haven CT. Intriguingly, at the same time, Burr was also examining the role of genetic constitution upon the bioelectric potentials of corn caryopses (Burr 1943; Nelson and Burr 1946), and also of their relationship to the morphogenesis of squash fruit (Burr and Sinnott 1944). In this sense, Burr was a pioneer of bioelectromorphogenesis (Levin 2003).

Burr's work with the maple tree consisted of recording the electrical potential difference (EPD) between two silver electrodes inserted within the living bark and

---

[1] The tidal acceleration of the Moon at the Earth's surface is 1.1×10^{-7} g, whereas that of the Sun is 0.46 of this value at 0.52×10^{-7} g. The mass of Earth is 81 times that of the Moon, whereas the mass of the Sun is 3.3×10^5 that of the Earth.

10 Moon and Cosmos: Plant Growth and Plant Bioelectricity 253

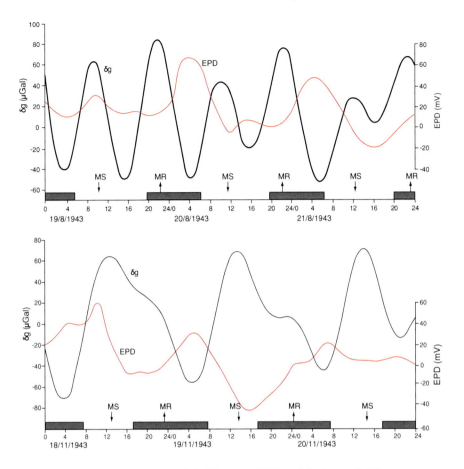

Fig. 10.1 and 10.2 Electrical potential differences (*EPD*, *red line*) recorded in a maple tree by HS Burr in August 1943 at Lyme CT, and in November 1943 at New Haven CT, USA. The corresponding lunisolar-driven gravimetric tidal profiles, δg, are also shown. *Dark bars* on the lower axes indicate night time, as defined by times of sunset and sunrise. MR, MS and associated *short arrows* indicate the times of moonrise and moonset, respectively. Data relating to EPD were taken from Figs. 1 and 2 in Burr (1944)

cambial cells of the trunk, and located 15 and 150 cm above ground level.[2] During a three-day period of observation (Figs. 10.1, 10.2), he found that the EPD, measured in mV, varied rhythmically. This pattern was found in both summer (August 1943) and late autumn (November 1943). At the later date the tree had

---

[2] Electrodes inserted into this location within a tree trunk have the capability of provoking a wound response from the surrounding cellular tissue which can, in turn, isolate the electrode, thereby diminishing the electrical potential reading. This does not seem to have occurred in Burr's experiments, but may account for a diminution of signal in the long-term recordings of Gibert et al. (2006—see their Fig. 6 and results from mid-July 2004 onwards).

presumably lost its leaves, had entered dormancy, and was no longer transpiring, all features which, as we shall see, can regulate the EPD. Burr concluded that the tree was storing electrical energy during each day, the mV values rising from a minimum at sunset to a maximum sometime during the night, and the electricity discharging once a maximum EPD had been reached. The maximum and minimum EPD values nevertheless varied from day to day, another point to which we shall return in relation to a lunar influence upon EPD. Figures 10.1 and 10.2 include the times of sunrise and sunset, as well as of moonrise and moonset, at the two locations.

Another finding which Burr considered significant was that within a 3 month period (mid-June through to mid-September 1943) sharp peaks of EPD occurred on the 1 or 2 days that coincided with New Moon and Full Moon (Burr 1944, his Fig. 3). Four similar sharp peaks of EPD, though obtained from averaged monthly EPD records over a total period of 20 months from a tree of *Quercus cerris*, were also commented by Koppán et al. (2000, 2002). These authors related these EPD peaks (though without any definite evidence) to periods of intense sap flow. Nevertheless, many of Burr's observations alerted him to the possibility that EPD values were subject to lunar modulation, and his later results from 1944 onwards also tend to suggest this possibility. We shall look at these later results of Burr in more detail, together with those from more recent work on EPDs in trees, all within the perspective of a possible lunar influence on the EPD patterns. Moreover, these patterns can be examined in three different time frames, the daily, the monthly and the annual—as Burr had done in his three publications of 1944, 1945 and 1947.

10.3 Methodology

The aids that HS Burr had at his disposal in his investigations of the Moon and its possible influence on EPD were those given by visual inspection of the night sky and almanacs listing the dates of lunar phases. Nowadays, however, lunar influence can be more precisely sought in terms of lunisolar tidal acceleration and other related parameters, including lunar effects upon atmospheric electricity. Because of their basis in physical laws, lunisolar gravitational variations are predictable and, hence, can be investigated retrospectively, allowing the data of Burr and of other workers to be interpreted in the light of these correlates. For example, time-courses of lunisolar tidal acceleration, in the form of a gravimetric tidal profile with units of μGal, can easily be estimated. This parameter relates to the vertical deformation of the Earth's crust due to the gravitational pull of the Sun and Moon. The crust rises and falls as the Moon orbits the Earth, and as the Earth orbits the sun. This moving tidal wave of crustal rise and fall adds or subtracts a certain value, δg, to or from the Earth's local gravitational acceleration and, hence, can be taken as proxy for the lunisolar tidal force exerted upon Earth. For example, a 3 mm elevation or a depression of the Earth's crust, due to the mentioned lunar orbit, corresponds to a local fluctuation of $+1$ μGal or -1 μGal with respect to the Earthly 1 G ($= 9.8 \times 10^9$ μGal).

- Time-courses of δg, referred to here as gravimetric, or lunar, tidal profiles, were estimated using the program 'Etide'. A δg time-course can be produced for any calendar date, or series of consecutive dates, past, present or future, for any particular location upon the Earth's land-surface, given the relevant latitude, longitude, and elevation above sea level.

Because the Etide program estimates δg at the desired location with respect to Universal Time (UTC), it is necessary to adjust the UTC timescale of the δg profile to the local time in order to match the time-course of the recorded biological data. The following Internet site, at the URL mentioned below, provides this adjustment. The site also gives information of whether or not the local time was subject to daylight-saving.

- Local time at places worldwide: http://www.timeanddate.com/worldclock/converter.html

However, it is possible that, in some circumstances, estimates of Solar Time would be provide a more useful and accurate time base.

Other useful data relevant to lunisolar attributes are found at the URLs listed below:

- Dates of lunar phases for any year from 1930 onwards are available from the Moon phase calendar at: http://www.moonconnection.com/moon_phases_calendar.phtml
- Lunar phases for any day of any year are available from: http://tycho.usno.navy.mil/vphase.html
- Lunar distances from Earth at perigee or apogee are available at: http://www.timeanddate.com/astronomy/moon/distance.html
- Times of moonrise and sunrise, and of moonset and sunset are available at: http://aa.usno.navy.mil/data/docs/RS_OneYear.php

Also of interest for a more complete understanding of biological phenomena, especially those of a bioelectrical nature, is the geomagnetic field. There are different indices of geomagnetic activity, and one of the newest and most interesting is the Polar Cap Index, measured near the North and South Poles of the Earth, at Qaanaag (formerly known as Thule) and Vostok, respectively. The polar caps are locations where the Earth's geomagnetic field interacts most intensively with the solar wind. Because organisms respond to magnetic effects (Galland and Pazur 2005), it is of interest to explore the possible relevance of the Polar Cap Index (as recorded at Thule in the Northern hemisphere) on plant bioelectricity. Thus, this Index may be considered biologically relevant.

- Values of the Polar Cap (Thule) Index (no units) are obtained from: ftp://ftp.ngdc.noaa.gov/STP/SOLAR_DATA/RELATED_INDICES/PC_INDEX/THULE/

The daily levels of geomagnetic field disturbance are also estimated by the Disturbance Time Storm (Dst) Index and by the Kp index.

- For the further explanation of the Dst Index, see: http://pluto.space.swri.edu/IMAGE/glossary/dst.html and for values of Dst (units of nanoTesla, nT), see: http://wdc.kugi.kyoto-u.ac.jp/dst_realtime/201108/index.html
- Useful links for Kp index are: http://sunearthday.nasa.gov/swac/tutorials/mag_kp.php; http://www-app3.gfz-potsdam.de/kp_index/description.html; and http://isgi.cetp.ipsl.fr/lesdonne.htm
- All above-mentioned geomagnetic indices and their graphical forms are available on the OMNI2 data web-site at: http://omniweb.gsfc.nasa.gov/form/dx1.html

Data concerning Cosmic-Ray Fluxes at various recording stations can be obtained from:

- ftp://ftp.ngdc.noaa.gov/STP/SOLAR_DATA/COSMIC_RAYS/STATION_DATA/

Regarding the methods by which electrical potentials are measured in botanical material, it seems that each researcher mentioned in this chapter has his/her own particular set up. Therefore, reference should be made to the original articles cited to find out exactly what was done—where the electrodes were placed, and what types of electrodes were used (see also Footnote 2 on p. 4).

10.4 Bioelectricity in the Context of Lunar Parameters

10.4.1 Daily Oscillations of EPD

Now that it is possible to align the gravimetric tidal profile with the EPD profile, we can return to Burr's work and examine these two profiles together, and look for suggestive interrelationships. Burr (1944) recorded EPD values at two-hourly intervals, over two 3 day periods; the contemporaneous δg profiles are constructed of values estimated at 15 min intervals (Figs. 10.1 and 10.2). From past experience with other biological systems (Klein 2007; Barlow et al. 2010; Fisahn et al. 2012), significant regions in the δg profiles are deemed to be their 'turning points'—positions along the data time-course where the rate of change δg is zero. In Figs. 10.1 and 10.2 it can be seen that there are many occasions when there is coincidence between turning points in the δg profile and analogous turning points in the EPD profile. Oscillations of the EPD profiles appear to be the inverse of the oscillations of the δg profiles. [Correspondences could, it is true, be evaluated statistically by means of cross-correlation and local tracking correlation, but for the purposes of the present chapter the visual aspect of the profiles is considered to be sufficiently informative.] Because the profile of δg is regulated by the relative positions of Sun and Moon at the locality in question, some of the δg turning points coincide with sunrise and sunset, and others coincide with moonrise and moonset. Thus, a question still open at present, at least for biological material growing in natural, field conditions, concerns which factor is the more influential upon the EPD pattern—the change from light to dark (or vice versa) or the turning of the lunisolar tide.

Burr (1944) states that the EPD time-courses failed to show any relationship between contemporaneously recorded data for barometric pressure, temperature and humidity. Furthermore, because Burr made measurements during November (see Fig. 10.2), the absence of sap flow during this season of the year would make it unlikely—at least in these particular circumstances—that this variable had any effect on the oscillation of EPD, even though sap flow may have a small effect in other circumstances (see later). It seems, therefore, that in the absence of any other contender, the gravimetric tide is more likely to be the regulator of EPD than the diurnal cycle of night and day.

Could the pattern of EPD reflect an independent diurnal physical oscillation in some structural feature of the tree trunks—their dilatation cycle, for example (Barlow et al. 2010)? Using his maple tree, Burr (1945) recorded the daily change in trunk diameter, δD, in addition to the contemporaneous EPD. Time-courses of δD trace not only a slow increase in diameter due to growth (MacDougal 1921; Cantiani 1978) but also a rhythmic increase and decrease in volume (dilatation) of certain tissues in the bark of the tree (see Zweifel et al. 2000). Daily oscillations were present in time-courses of both EPD and cellular dilatation, as revealed via δD. The EPD increased from a low value at mid-day and reached a maximum either in late afternoon or in the early part of the night, whereupon a fall in EPD commenced. The profile of δD was exactly 12 h out of phase with that of the EPD (Figs. 10.3 and 10.4). Unfortunately, at these later observation times in 1944, the measurements of both variables were not gathered at such close intervals as those gathered for EPD alone in 1943 (see Figs. 10.1 and 10.2)—6 h intervals as opposed to 2 h intervals—thus leaving some scope for interpretation of the exact profiles of EPD and δD. Nevertheless, the smoothed curves in Figs. 10.3 and 10.4 are quite consistent with the given data points. Importantly, the turning points in both the EPD and the δD profiles coincide with turning points in the δg profile. Subsequently, further close correspondences were found between δD and δg in other species of tree (Zürcher et al. 1998; Barlow et al. 2010; Barlow and Fisahn 2012). Now, variation in δD has been shown to be independent of sap flow (transpiration), and even of whether or not the relevant portion of tree is attached to or separated from the whole tree (Cantiani 1978; Cantiani et al. 1994). So, once again it seems likely that the biological variables, EPD and δD, are somehow 'in tune with the Moon'. The initial question of whether δD can affect the EPD, perhaps by some hydrostatic pressure effect, remains open. Both variables, however, may depend upon the solute status of the sapwood and secondary phloem tissues.

Summarising his results, especially with regard to variation in δD, Burr (1945, p. 734) states that: 'It is, therefore, not at all impossible that the lunar cycle produces, in some as yet undiscovered way, tides in the tree'. However, he does add some words of caution regarding the variation of EPD and δD: 'This does not mean necessarily that the moon itself is altering the magnitude of the potential difference. It is quite possible that they are both secondary phenomena, activated by some more basic feature in the Universe' (Burr 1945, p. 733). This point will be discussed in Sect. 10.4.2. However, it is worth remarking that δD is a measure of

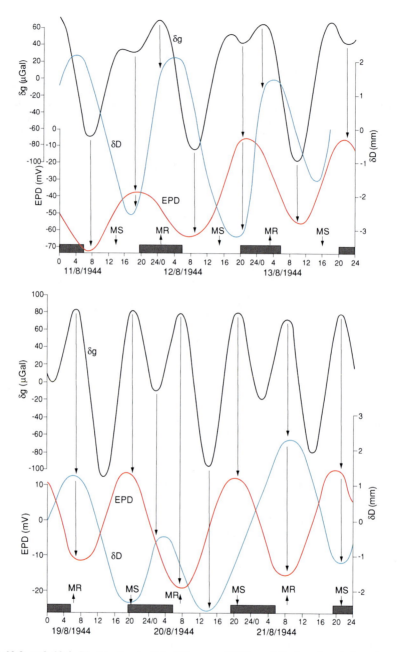

Fig. 10.3 and 10.4 Electrical potential differences (*EPD*, *red line*) and variations in trunk diameter (*δD*, *blue line*), measured with a dendrograph, recorded from a maple tree by HS Burr during two periods in August 1944 at Lyme CT, USA. The corresponding lunisolar-derived gravimetric tidal profiles, *δg*, are also shown. *Long vertical arrows* indicate correspondences between turning points of *δg* and turning points in both the EPD and the *δ*D profiles. Details relating to the time scale on the horizontal axis are as given in the legend to Figs. 10.1 and 10.2. Data relating to EPD and *δ*D are taken from Fig. 1 in Burr (1945)

a growth movement. Other growth movements initiate APs within the phloem (see Sect. 10.8), though whether they could also produce electrical slow wave potentials in response to hydrostatic pressure changes in the xylem due to the lunar-driven variation, δD, is not known. Therefore, on the basis of the inverse relation between oscillations of δD and EPDs, it is tentatively suggested that the cyclic nature of EPD values is dependent upon, or initiated by, the dilatation cycle of the tree stem, and that variations of δg are in some way concerned with this. Logically, the converse is also possible—that EPD affects δD.

In more recent times, commencing with the observations of Morat et al. (1994), time-courses of EPD have shown patterns similar to those described by Burr. For example, Koppán et al. (2000), working with a tree of *Q. cerris*, observed that the decrease in EPD commenced around the time of sunrise, and its increase commenced at around 16:00 h in the afternoon, 3 h before sunset. Here, the turning points in the EPD data of Koppán et al. (2000) are also coincident with those of δg (Fig. 10.5). Similar timings of both δg and electrical potential (EP) can be seen in data of Gibert et al. (2006), whose EP records were obtained from the trunk of a Black Poplar tree, *Populus nigra* (see Fig. 10.6), and also from one of its roots (data not shown). Here, the increase of EP began around the time of sunset, rather than beforehand, as shown in the Fig. 10.5, based on data from Koppán et al. (2000).

Contemporaneous Polar Cap (Thule) Index values, which are measures of the interaction between the solar wind and the magnetosphere (Troshichev et al. 1988), are included in Fig. 10.6. It is interesting to surmise that here there may be some coupling between the profiles of this index and the EPD and δg profiles. The same may be true of some other time-courses of the Polar Cap Index that have been studied in relation to EP profiles in trees of Northern Hemisphere. However, more work needs to be done in this area, especially in the use of indices which are of a global character, such as the Kp and Dst Indices, rather than being of a more local influence [see remarks about the Polar Cap (Thule) Index on p. 7].

A further study of EPs in trees was undertaken by Holzknecht and Zürcher (2006). Using the average output from 12 electrodes inserted into the stem (at intervals over a height of 11 m from ground level) of a specimen of 12-year-old, pot-grown Stone Pine (*Pinus cembra*), these authors recorded an oscillatory diurnal pattern in EP. Additional records were obtained from an 80-year-old Norway Spruce (*Picea abies*). Analyses were over two time periods, one covering nearly a week (19–25 December 2001) for *P. cembra*, the other, for *P. abies*, being between 14 and 18 August 2000. Similar to the results of Burr, Holzknecht and Zürcher (2006) found that an increase of EP values commenced shortly after sunrise and reached a maximum after sunset. Once again, and as other authors had also noted, turning points in the EP profile coincided with turning points in the δg profile (Fig. 10.7). Although no sap flow measurements were undertaken, the time of year (December 2001) suggests that transpiration in these evergreen trees would have been minimal.

The different timings of the rise and fall of EPD or EP values with respect to the times of day or night may be a reflection of either the method used for determining EPDs and EPs or, as seems more likely, that the trees studied were responding to

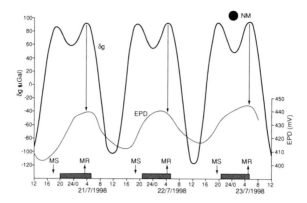

Fig. 10.5 Electrical potential differences (*EPD*) recorded from a tree of *Quercus cerris* during July 1998. New moon (*NM*) occurred on 23 July. The corresponding lunisolar-derived gravimetric tidal profile, δg, is also shown. *Long vertical arrows* indicate correspondences between turning points of δg and turning points in the EPD and δg profiles. Details relating to the time scale on the horizontal axis are as given in the legend to Figs. 10.1 and 10.2. Data for EPD, recorded at location 4 mE on the tree, are redrawn from Fig. 2 in Koppán et al. (2000)

particularly influential turning points in the δg profile, regardless of the time of day. The question then arises as to whether such influential turning points correspond to a peak δg value—as seems to be the case (see Fig. 10.7)—or to a trough in the δg profile. Moreover, slight time differences between profiles of EPD, recorded at different height upon the trunk, and in particular those at the same height but at different positions around the tree-trunk's circumference (Koppán et al. 2000), may be attributable to positional differences in the details of secondary vascular anatomy, which would therefore imply slightly different solute pathways in the vicinities of the different probes (Koppán et al. 2005).

With respect to the oscillations of EPs studied by Holzknecht and Zürcher (2006), Fourier analyses of the EP time-courses showed major periods of 24.7 h and 25 h, respectively. These periods are closer to the duration of the lunar day (24.8 h) than to that of the solar day (24.0 h). However, a Fourier analysis of the EP time-course obtained from *P. cembra* during 11–17 December 2001, was inconclusive: on these dates the authors regarded the tree as being 'not yet "in tune" with the gravimetric tides' due to the Moon. In regard to these latter results, while it is true there may be some slippage of the peaks and troughs of EP in *P. cembra* with respect to δg, the turning points within the EP profile do nevertheless correspond to the peaks of the δg profile, though it is not the same peak which matches up with an EP peak on each successive day. The slippage may simply be a feature of the position in the lunar month when the EP observations were made. This is because the δg profiles around the times of the First and Last Quarters of the Moon are different from those at the times of New or Full Moon (e.g., Fig. 10.8): in the former phases the δg profile is characterised by single peaks, whereas in the latter phases the δg peaks are double. So, any tracking by the

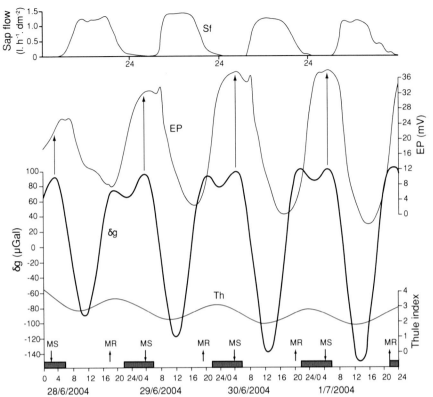

Fig. 10.6 Continuous output traces for electrical potential (*EP*) and rate of sap flow (*Sf*) in a tree of *Populus nigra* during June-July 2004. Also shown are the corresponding lunisolar-derived gravimetric tidal profile, δg, as well as a best-fit curve for the mean hourly values of the Polar Cap (Thule) Index, Th. Long vertical arrows indicate correspondences between turning points of δg and turning points in the EP profile. There is also a general correspondence between the profiles of δg and Th. Details relating to the time scale on the horizontal axis are as given in the legend to Figs. 10.1 and 10.2. Data for EP and Sf are redrawn from Fig. 12 in Gibert et al. (2006)

biological material of the gravimetric δg profile must require some adjustment until it is 'decided' which peak of δg will be followed by the tree. A similar finding was made for δD in relation to δg in some other trees (Barlow et al. 2010). Thus, it remains entirely feasible that turning points in the δg profile provide the regulator of the EP rhythm.

As mentioned above, Burr (1945) hinted that some basic universal feature may influence bioelectric patterns. In a later paper (Burr 1947), Burr suggested that atmospheric electricity and cosmic rays might contribute to the variations in EPD. He was perhaps propelled towards this view by the violent changes in bioelectric patterns recorded from his two experimental maple trees at Lyme and New Haven CT on days in September 1944 when a hurricane passed through this area of New England (Burr 1947). These observations were later confirmed after a subsequent storm event at Lyme CT during 1956 (Burr 1956).

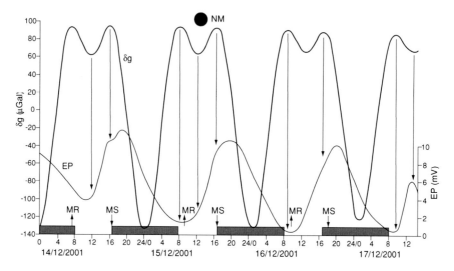

Fig. 10.7 Time-course of electrical potential (*EP*) recorded from a tree of *Pinus cembra* during December 2001. New moon (*NM*) occurs on 15 December. The corresponding lunisolar-derived gravimetric tidal profile, δg, is also shown. *Long vertical arrows* indicate correspondences between turning points of δg and turning points in the EP profile. Because the δg profile is always changing, the various turning points in each profile make different associations during this particular time period, and so the two profiles do not keep exact pace with each other. Data for EP are redrawn from Fig. 4a in Holzknecht and Zürcher (2006). Details relating to the time scale on the horizontal axis are as given in the legend to Figs. 10.1 and 10.2

10.4.2 Monthly Oscillations of EPD

Holzknecht and Zürcher (2006) performed a Fourier analysis of the EP time-course from *P. abies*, using a longer recording period, from 10 November 2000 through to 27 January 2001. The continuously recorded EP time-course revealed an oscillating pattern with a period corresponding to the synodic lunar month (∼29.5 d). The minimal EP values occurred at New Moon, whereas maxima occurred around the time of Full Moon. This is also hinted in the EP results shown in Fig. 10.7. Here, the days with relatively low values of EP coincided with the date of the New Moon.

Results of Gibert et al. (2006) are also pertinent. These authors published a time-course of EP recorded continuously from a single tree of *Populus nigra* (black polar), extending from 18 June 2004 through to 19 July 2004 (Fig. 10.8). EP amplitudes were least just after the time of New Moon and greatest at Full Moon, on July 3. This is similar to what Holzknecht and Zürcher (2006) had found (scc Fig. 10.7), and this New Moon effect on EP amplitude is also evident in the results of Burr (1947, see his Fig. 2) who surveyed EPD values over a period of more than a whole year, between late 1943 and early 1945. Moreover, from the results of Gibert et al. (2006) a statistically valid relationship can be found between

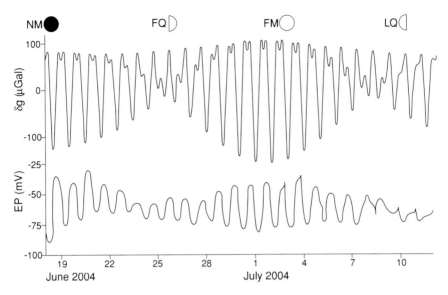

Fig. 10.8 Profiles of electrical potential (*EP, lower panel*) recorded in a tree of *Populus nigra* and of the corresponding lunisolar-derived gravimetric tidal profile (δg, *upper panel*) during 25 days of June–July 2004. Dates of new moon (*NM*), full moon (*FM*) and first and last quarter (*FQ, LQ*) are indicated. New moon occurs on 18 June. Data for EP are redrawn from Fig. 11 of Gibert et al. (2006)

the amplitudes, Δ, of the daily EPs and those of the contemporaneous amplitudes of the daily δg values. Linear regression (Fig. 10.9) yields a correlation coefficient, r = 0.84 (P < 0.001; n = 22).

The readings of EP presented in Fig. 10.8 were taken at location E32 on the poplar tree, about 2.5 m above ground level. Another location, E6, which was 1 m above ground level, showed minimal EPs in March, May, June and July 2004, and perhaps in September and October 2004, also. These minima occurred around the date of the New Moon, as did the minimum EP measured at a location, E0, on a root, during June 2004 (see Fig. 14 in Gibert et al. 2006). Perhaps it is also significant that, during all the months mentioned in Gibert et al. (2006), air temperatures were in the middle of their annual range, and these months were also times when growth of the tree was either increasing (March–June), with renewed cambial and leaf growth, or decreasing (September/October), when cambial and bud dormancy approached. During periods when the air temperature was low (<5°C), EP oscillations vanished or showed distinct down-turns on days when air temperatures were <0°C (see Fig. 10 in Gibert et al. 2006). Other sites of EP measurement higher on the poplar tree (>5 m above ground) did not display such noticeable New Moon/Full Moon effects. There may, therefore, be additional effects, due perhaps to atmospheric conditions, or local microclimatic conditions within the tree's canopy, which mask or over-ride lunar modulation of EP.

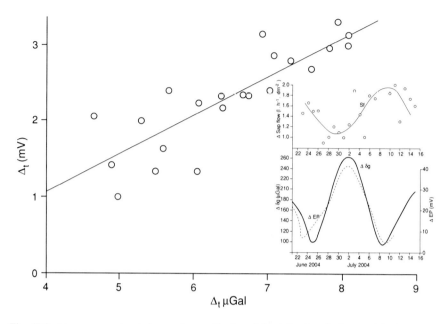

Fig. 10.9 Linear regression between the amplitudes of the daily electrical potential (*EP*) values (in mV) and the corresponding amplitudes of the daily gravimetric tidal δg values (in μGals). In each case, the amplitudes have been transformed as $\Delta_t = \sqrt{x/2}$, where x is the actual amplitude value (as shown in the inset). The regression coefficient, r, is 0.84 (P < 0.001; n = 22). The inset graph shows the original, untransformed amplitude values, Δ, of EP and δg (lower portion of inset), as well as the amplitude, Δ, of the sap flow rates (upper portion of inset). Data for EP and Sf are derived from Fig. 11 in Gibert et al. (2006)

Although the data of Morat et al. (1994) are less clear and not so extensive as those of Holzknecht and Zürcher (2006) and of Gibert et al. (2006), it seems, nevertheless, that in their specimen of an 80-year-old Horse Chestnut tree (*Aesculus hippocastanum*) minimal EPD values also occurred around the time of New Moon. The period of the EPD oscillation was about two weeks, however, indicating that another minimum of EPD occurred at Full Moon. Around the time of both Full and New Moon the amplitudes of δg are maximal, and are related in some complex way to the elapsed time between moonrise and moonset, and also the lunar declination.

How can these effects of lunar phase on EPD be explained? Stolov and Cameron (1964) surveyed values of the geomagnetic index, Kp, over a 31 year period and found a significant (up to 10%) decrease in the mean value (typically about 50 μT) in the 7 days before a Full Moon and a corresponding increase in 7 days after Full Moon. At New Moon, Kp values were also depressed slightly, although Stolov and Cameron (1964) believe the deviations at this time are random. Markson (1971) expanded on these findings and reviewed the modulatory effects of the Moon on atmospheric electricity, as did Mehra (1989). On the basis of a 20-year daily record of atmospheric electric fields at a location in India, the latter author revealed that a peak electric field occurred on days of Full Moon and a steep decline followed during the next 4 days (Mehra 1989). Nishimura and Fukushima (2009) espoused the findings of Stolov and Cameron to support their claim that animals show particular responses at the time of Full Moon. They proposed that magnetoreception could provide the clue to this Full Moon effect in animals.[3] Thus, it is possible that, in some circumstances, variation in atmospheric electric and geomagnetic parameters during the lunar cycle could affect also the bioelectrical pattern of plants. However, if this proposal is generally true, and Kp is the correct index to in relation to biological events, then a case-by-case study should be feasible. However, a preliminary search for such a relationship between Kp values and EPD suggests additional (or other) factors are at work.

10.4.3 Annual Oscillations of EPD

The modulation, by the Moon, of the bioelectric properties of trees over the course of a lunar month, suggests that rhythms of EP (or EPD) with longer cycles might be found. In fact, Burr (1947), when he examined his data concerning the value of EP, recorded at midnight on each day within his maple tree in New Haven CT, believed that he could discern such a prolonged rhythm. His observations took place from mid-October 1943 through to mid-October 1944. He reached the preliminary conclusion that, in addition to monthly cycles of EPD (see also Sect. 10.4.2), the details of which were apparently governed by the lunar phases, there was another, longer cycle of about 6 months. For Burr, this latter pattern did not seem to correspond to any growth feature of the tree, nor to any atmospheric or weather condition. Except for the lunar monthly cycle, he does not suggest any other cycle that could account for the 6 month EP cycle. However, one such cycle becomes evident when the distance, from Earth, of the Moon at perigee is subtracted from its distance at apogee. The temporal pattern of this difference, ε, gives a measure of the constantly changing eccentricity of the lunar orbit. A typical range of ε is $3.5-5.0 \times 10^4$ km. When the values of the monthly maximal EP recorded from Burr's experimental maple tree (Burr 1947) and the monthly values of ε are plotted against time (days), two curves, one the inverse of the other, are obtained (data not shown): a maximum value of ε coincides with a minimum of EP, and vice versa. The period of each cycle is about 220 days ($\sim$7.3 months). The cycle of perigee-apogee distances is constructed of a cycle of the daily values of δg, the amplitudes of which are greatest when the ε values are greatest.

---

[3] Nishimura and Fukushima (2009) considered animal activity in relation to moonlight and linked this with magnetoreception. The light of Full Moon has different polarisation properties compared with the light received from the Moon at other lunar phases.

10.5 Relationship of Bioelectric Potential and Solute Flow

10.5.1 Solute Flow in Secondary Xylem

A number of attempts have been made to clarify the relationship between bioelectrical potential and the flow of solutes within the xylem of plants, as first suggested by Morat et al. (1994). As has already been shown, it is evident, from the time of the early work of HS Burr onwards, that the diurnal pattern of rise and fall of EP is not particularly dependent upon the time of day, as judged by the times of sunrise and sunset. But when times of moonrise and moonset are also considered, then a link between bioelectric potential variation and the lunisolar gravimetric tidal variation, expressed as δg, becomes a possibility. Nor is bioelectric potential a feature peculiar to the actively growing phase of plants since it is present also during their dormant periods. In fact, dormancy should not by any means be regarded as an inactive phase because seed respiration continues and variations in its rate seem correlated with both the barometric pressure (Graviou 1978) and the amplitude of δg (Barlow unpublished). Nevertheless, it is worthwhile to revisit the more recent data on sap flow—by which is meant the stream of solutes through the xylem due to evapotranspiration, and not, in this context, the movement of sucrose solution through the phloem—in the light of bioelectrical potentials, the δg profiles, and other geophysical variables.

Transpirational xylem sap flow has been studied in relation to the already mentioned variation of stem diameter at trees, δD (see Cantiani and Sorbetti Guerri 1989); and although δD is regulated by δg (Barlow et al. 2010), it appears that the variation of δD is independent of sap flow. The evidence comes from trees whose transpiration has been curtailed. During such conditions the dilatation cycle of δD continues as long as the cambium remains alive, and does so even in excised branches (Cantiani and Sorbetti Guerri 1989). Regrettably, no experiments have yet been performed where bio-electrical potential, δD, sap flow, and δg have all been evaluated simultaneously.

Following upon a suggestion (Morat et al. 1994) of a possible link between sap flow and EP, an attempt to examine this was made Koppán et al. (2002). The results of measurements of the two variables, sap flow and EPD, in a specimen of Turkey Oak, *Q. cerris*, growing near Sopron, Hungary, are displayed in Fig. 10.10. Graphed alongside the authors' original data are the contemporaneous values for δg at this geographical location. In the present case, the rise in EPD commenced shortly before sunset and started to fall just after sunrise. (In another study, Holzknecht and Zürcher (2006) found that EP began to decline soon after sunset—Fig. 10.7). What is evident in Fig. 10.10 is that the turning points of EPD, whether from high values to low, or vice versa, coincide with turning points of δg. As for sap flow, its oscillation is approximately the inverse of the oscillation of EPD with respect to the rise and fall of their values (Fig. 10.10). This led Koppán et al. (2002) to remark upon a 'strong correlation' between the two variables, though they qualified this by saying that 'sometimes a 1–2 h phase lag can be observed'.

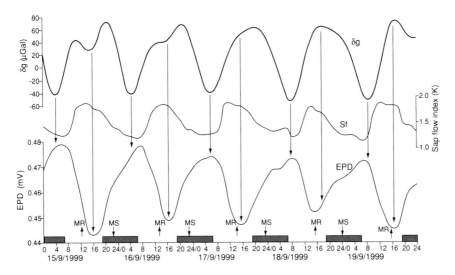

Fig. 10.10 Profiles of electrical potential difference (*EPD*) and sap flow (*Sf*) recorded from a tree of *Quercus cerris* during September 1999, together with the corresponding profile of the gravimetric tide, δg. *Long vertical arrows* indicate correspondences between turning points of δg and turning points in the EP profile. Details relating to the time scale on the horizontal axis are as given in the legend to Figs. 10.1 and 10.2. Data for EPD and Sf, the latter estimated as an index, K, which is linearly related to sap flux density, were redrawn from Fig. 2 of Koppán et al. (2002)

(They do not say whether the 'correlation' was statistically tested, or was, *sensu ampliore*, simply a visual assessment.) These authors also found another 'remarkable correlation' between EPD and the water potential of the air in the neighbourhood of the tree, this latter potential being regulated by air temperature and humidity. With respect to the EP/sap flow relationship, the inset in Fig. 10.9 shows the mean maximal sap flow in the Poplar tree studied by Gibert et al. (2006). In this study, although the relationship between sap flow and EPD looks suggestive, regression of the respective amplitudes shows no significant correlation (r = −0.486; P = 0.05; n = 14), even when one outlying datum on 3 July 2004 is excluded.

Although EPD may be modulated by local humidity and ionic conditions within a tree's canopy, both of which result from transpiration, as suggested by Koppán et al. (2002), there is nevertheless an apparent dependence of EPD variation upon the lunisolar tidal acceleration. Moreover, a further possible factor in this relationship is indicated by observations of the inherent water volume fraction in sapwood. It seems that a certain variable volume of water is always retained within sapwood, the amount of which is not necessarily dependent upon the current rate of sap flow (described by Sparks et al. 2001). So the question is whether this water volume fraction influences EPD. As shown in Fig. 10.11, the daily variation in this water volume fraction appears responsive to variation in δg; hence, this parameter might somehow influence the variation of the EPD.

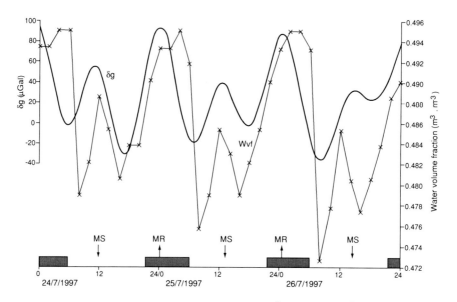

Fig. 10.11 Time-course of water volume fraction (Wvf, m^3 of water per m^3 total volume of sap wood and data points, x) recorded every 4 h over 3 days during July 1997) from a tree of *Pinus contorta*, together with the corresponding time-course of the gravimetric tide, δg. Details relating to the time scale on the horizontal axis are as given in the legend to Figs. 10.1 and 10.2. Original data of Wvf collected by Dr JP Sparks, University of Colorado, USA

In one experiment where 2-year-old avocado (*Persea americana*) trees were placed for 3 days in darkness, transpiration was much reduced but not entirely eliminated (Oyarce and Gurovich 2009). Here, a reduction, but not a complete suppression, of bioelectrical potential variation occurred. The small changes in EP and their timing during each of the 3 days were attributed to conventional circadian rhythms based on molecular and metabolic cycles. However, it cannot be ruled out that a lunisolar tidal influence continued to operate upon this system, and that what was recorded from the trees in these dark conditions was a background rhythm of EP which was influenced by the lunisolar tide. If so, then the more conspicuous changes in EP which are usually seen in illuminated conditions could be a manifestation of a lunisolar-induced EP rhythm that has become amplified following membrane 'irritation' in response to evapo-transpirational water movement. It can be tentatively suggested, therefore, that a background EP or EPD profile is maintained by the lunisolar tide, and that, in keeping with the findings of Polevoi et al. (2003) on the electrical stimulation of sap flow and, together with the early demonstration by Lemström (1901) of an effect of an EPD on water movement in a glass capillary, the prevailing EPD might facilitate the movement of water at the commencement of evapotranspiration during the hours of daylight.

10.5.2 Solute Flow in Phloem

Although important solutes (e.g., sucrose) flow in the sieve elements of the phloem, and phloem is also a transmitter of bioelectrical APs (Fromm and Bauer 1994; Fromm and Lautner 2007) which may regulate the permeability of sieve pores (van Bel et al. 2011), there are no reports of lunisolar gravitation affecting any of these phloem properties. Phloem, it will be recalled, is a channel for phytoneural signals (Barlow 2008). However, there is an interesting report from Beeson and Bhatia (1936) that, in the giant bamboo, *Dendrocalamus stricta* growing in Dehradun, India, during a 12 month period in 1930–1931, the rate of sap flow (estimated from the moisture contents of the bamboo culms) increased during the times between Full Moon and New Moon, and decreased between New Moon and Full Moon. The metabolism of sucrose to starch, which may indicate variations in phloem transport or import/export, was also slightly different between these two phases of the lunar month. In Fig. 10.4 of Beeson and Bhatia (1936), the levels of starch and disaccharides analysed in the culms were higher at Full Moon in 9 months of the 12 month observation period than they were at the next New Moon. These observations might merit re-investigation, together with corresponding measurements of the time-course of phloem bioelectrical potentials. Also, it should be kept in mind that the electric field of the atmosphere varies throughout the lunar month, data for which were collated by Mehra (1989) from recordings made in India over the same period when Beeson and Bhatia (1936) were gathering their observational data. Rather similar findings on carbon allocation, though obtained over only two lunar months in 1996, were reported by Vogt et al. (2002) for three species of shrub growing in the understory of the Luquillo Experimental Forest, Puerto Rico. However, another possibly important factor (in addition to atmospheric electricity) with respect to the findings concerning sugar metabolism in relation to lunar phase is that starch hydrolysis is affected by the proportions of polarised and non-polarised light (Baly and Semmens 1924; Semmens 1947a, b), and that this proportion itself varies according to lunar phase (Gál et al. 2001).

10.6 A Moon-Generated Rhythm that May Initiate Bioelectric Impulses

In the past, many observations were made on the rhythmic movements of bean leaves, and from this and other evidence the idea of a 'physiological clock' was born (Bünning 1956, 1963). Fundamentally, however, these leaf movements are expressions of a 'lunar clock' (Klein 2007): they are initiated upon the turning of the lunar tide. One example will show the idea. The first pair of leaves of the bean, *Canavalia ensiformis*, when they emerge at germination in either continuous light or continuous darkness, make their initial movements at the very times that the

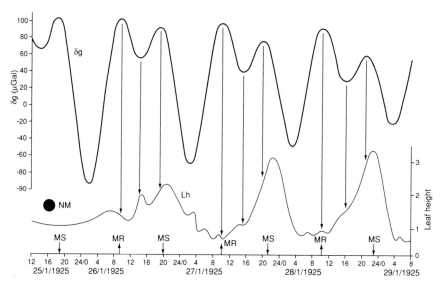

Fig. 10.12 Time-course of the first leaf movements continuously recorded by kymography as leaf height, Lh, of a seedling of *Canavalia ensiformis* germinated and grown in constant light together with the corresponding profile of δg, the gravimetric tide. The scale for Lh is arbitrary. Because of the way in which the kymograph trace is produced, *increasing values* of Lh indicate the descent of the leaf in the vertical plane towards its night-time position, whereas *decreasing values* indicate the ascent of the leaf towards its daytime position. *Long vertical arrows* indicate correspondences between turning points of δg and turning points in the Lh profile. It appears that the δg turning points activate the descent of the leaf. New moon (NM) occurs on 25 January 1925. Around this time no leaf motion was recorded, whereas leaf motion was expressed before and after NM. MS, MR and *small arrows* indicate the time of moonrise and moonset, respectively. Data for Lh, from seedling P19, are redrawn from Fig. 21 in Brouwer (1926)

lunisolar tide is turning, as shown in Fig. 10.12 for one selected case. [Interestingly, the first few movements of this new pair of leaves ceased on days around New Moon, only to recommence on days after this lunar phase (part of these data are shown in Fig. 10.12).] Many other examples from experiments where the rhythms of leaf movements were manipulated by numerous varied and ingenious light/dark regimes, support the hypothesis of a lunar-clock (see Barlow and Fisahn 2012)—and do so more convincingly than interpretations based on a putative endogenous physiological clock.

The reason for mentioning these observations is that diurnal leaf movements can be a source of APs, probably as a result of massive fluxes of ions (e.g., Antkowiak and Englemann 1991; Moshelion et al. 2002) from the cells which sense the passage of lunar time. The link between leaf movements and APs was found most recently in *Chenopodium murale* (Wagner et al. 2006), as well as in many other plants, right from the time of the first discovery 50 years ago by Guhathakurta and Dutt (1961), of the leaf-movement/phytoneuronal-impulse propagation system in *Desmodium gyrans*. Therefore, if leaf movements in general are governed by the lunisolar tidal acceleration, as proposed above, and that these movements also initiate APs, what then can be their possible significance?

Daily leaf movements would be like the rhythmic beating of an animal's heart, though this analogy has already been claimed by Kundt (1998) in connection with the diurnal cycle of root pressure—an autonomous activity which, in its time-course, appears so far to have no obvious relationship to the lunisolar tide (PW Barlow, unpublished). Thus, we may imagine that the output of APs which attend leaf movements, albeit at a slow rate of propagation, could regulate the timing of some event critical in the physiology of the whole plant. It is likely that APs produced by the moving leaf are channelled via the phloem, and the various processes modulated thereby have been mentioned earlier (see also Fromm and Lautner 2007). The beauty of lunisolar tidal generation of bioelectrical signals is that they are perpetuated for as long as the relevant cells remain alive. Moreover, they could persist even when other signals from the external environment are extinguished. The lunar bioelectric clock thus provides a backup for whenever there is a failure of time-keeping based on diurnal solar light and dark periods.

10.7 Other Possible Regulators of Bioelectrical Patterns

Even though in the course of his earlier work HS Burr had considered including atmospheric conditions [including barometric pressure as well as, for him, 'unknown radiations such as cosmic rays' (Burr 1947)] within a proposed 10 year study of the influences of external factors upon bioelectrical patterns, he tended to downplay the possibility that day-to-day weather conditions could be influential on the bioelectric potentials in his experimental maple trees. Now, however, there is geophysical evidence for a lunar influence on the flux of electromagnetic radiation which reaches the Earth from the Sun and from deep space (Stolov and Cameron 1964; Maeda 1968; Markson 1971; Sharma et al. 2010). This type of radiation can be conveniently assessed in terms of a number of indices, including the Polar Cap (Thule) Index (Th), Kp Index and the Disturbance Storm Time Index (Dst). Co-ordinated variations may or may not occur between some or all these geophysical indices and supposedly sensitive biological parameters. On balance, it seems unlikely that either Th or Dst Indices consistently affect day-to-day biological oscillations—those of EPD and δD, for example—to any major degree. Nevertheless, it does seem as though the varying lunisolar tidal acceleration, as expressed through δg, may focus the intensity of the geomagnetic radiations upon Earth; the coordinated oscillations of Th and δg (see Fig. 10.6, for example) may be evidence of such a lunar focussing effect.

Ultra-low frequency (ULF, of <5 Hz) geomagnetic field variations (Fraser-Smith 1993) can be registered by trees (Fraser-Smith 1978), especially the class of Pc1 pulsations in the range 0.2–5 Hz, as well as other classes, Pc2 and Pc3, having frequencies of 0.02–0.2 Hz. The ULF Pc1 variations described by Fraser-Smith

(1978), and which were measured by ground magnetometers, were also detected simultaneously within a reference oak tree, *Quercus lobata*, and took the form of a fluctuating EPD with an amplitude variation of about 0.1 mV. The observed ULF signals were not generated by the tree itself; instead, the oak tree seemed to act like an antenna, capturing the geomagnetic field variations (Fraser-Smith 1978) and transducing them into a bioelectrical impulse. Although Fraser-Smith (1978) suggests that 'electrical storm' conditions, such as witnessed by Burr (1947, 1956) in relation to his maple trees, could have influenced the ULF potentials within and around his reference oak tree, the observed ULF-variations could equally well have occurred during relatively quiet conditions. Fraser-Smith's observations upon his oak tree suggest that all living organisms are continually responding to geomagnetic fluctuations, and not just to the more spectacular effects that accompany hurricanes and electrical storms. A recent and informative appraisal of the physical aspects of ULF activity is due to Kozyreva et al. (2007) ; Khabarova and Dimitrova (2009) consider ULF-variations as one of the most bioeffective parameters of the geomagnetic field. Indeed, in this regard, Fraser-Smith (1978) draws attention to the correspondences between ULF Pc-1 frequencies and the delta waves (frequency 0.5–2.0 Hz) of the human brain, the latter becoming disturbed in certain neurological disorders. However, animals may be able to damp the responses elicited by ULF-variations so that they do not affect their daily life too much, but whether the same is true for plants is not known. Thus, the scaling factor by which incoming impulses are dealt with (see Introduction) may also be regarded, in part, at least, as a measure of a buffering effect.

There has also been speculation (Le Mouël et al. 2010) that passing clouds, laden with electric charge, could, in otherwise constant atmospheric conditions, be the source of 'spontaneous' spikes of EPD readings within trees. Prolonged showers of rain had similar, but more longer lasting, effects (Koppán et al. 1999). Trees may also be responsive to changes in atmospheric barometric pressure. The lunar modulation of these various atmospheric and geophysical parameters has been reviewed by Markson (1971) and Zurbenko and Potrzeba (2010).

The ULF fluctuations considered above are not generated by, for example, fluxes of cosmic rays. However, consideration has been given to these latter fluxes, especially to the biologically effective neutrons to which cosmic rays give rise when these strike the Earth's atmosphere. A question which has emerged is whether cosmic-ray effects (especially those due to neutrons) produce anomalous tree-ring growth during periods of above-average sun-spot numbers (Dengel et al. 2009). Some tree-ring anomalies could be the result of corresponding anomalies in bioelectric potentials which have impinged upon physiological processes in the cambial zone connected with tree-ring development (see Barlow 2005).

Observations of Dodson and Hedeman (1964) showed that, during the period 1952–1963, the count of solar neutrons conformed to a lunar monthly rhythm of 29.5 days. Whether or not this was a "statistical accident", as the authors put it, is perhaps not the point. Rather, it should be asked, firstly, whether this rhythm was present in earlier years and, secondly, has it continued in the years after 1963? And, further, could such long-term cosmic rhythms entrain corresponding bioelectric rhythms in plants? It is conceivable that they could have done so because, as analysed by Markson (1972), 29.5 days was also the period of the bioelectrical potential variation recorded by Burr (1947) in his experimental maple tree during 1943–1944 (see Sect. 10.4.3), and a 30 day period of atmospheric pressure variation was detected by Zurbenko and Potrzeba (2010).

It is well known that cells and organisms are responsive to experimentally applied magnetic fields (Galland and Pazur 2005; and see Cifra 2011 for a recent review). The geomagnetic field of Earth is one that all organisms experience—and to which they may also contribute, especially those trees which live in extensive forest communities. If, as suggested by Nishimura and Fukushima (2009), the quality of light can also affect magnetoperception and response, then even weak light stimuli, such as received from moonlight (see Semmens 1947a; and see also the effects of simulated moon-light, which were investigated by Bünning and Moser 1969), should be investigated for their possible influence upon bioelectric phenomena in plants. Moreover, in an article by Oliver Lodge (1908) reporting upon the work of Lemström (1904), attention is drawn to the luxuriant vegetation in the circumpolar regions of Scandinavia, thriving, so Lemström supposed, under the impress of the Aurora Borealis. Fascinating would be a study of plant bioelectricity in relation to this vivid manifestation of ionisation in the upper atmosphere due to its interaction with particles from the magnetosphere.

In 1991, Toriyama published a summary of the bioelectric disturbances recorded in two silk trees (*Albizzia julibrissima*) during the days and hours immediately preceding a number of earthquakes in Japan (Toriyama 1991). The study period was from 1977 through to 1989. One electrode was placed in sapwood of the recording trees 1.5–3 m above ground level, a second was buried 1 m deep in the soil, 1 m from the tree. The trees were located on the campus of the Tokyo Woman's Christian University. The epicentres of the observed 28 earthquakes with magnitude >7.0 were spread throughout Japan and in off-shore areas. Toriyama classified three types of anomalous bioelectric potential in the trees, which could be related to forthcoming earthquake events. He points out that the geological features of Japan are complex, and that the geophysical and geochemical precursors of earthquakes are varied. This might account for some of the variability of bioelectric responses immediately prior to the earthquakes. Saito (2007) continued Toriyama's work, using a tree of *Ulmus kaeki* as a monitor of bioelectric disturbance. His survey was conducted over the period 2006–2007 and included 18 seismic events. From the intensity of the bioelectric disturbance in the monitor tree, he was able to derive a formula estimating the distance of the recording site from the earthquake's epicentre.

What information could be received by a tree of an impending earthquake? One possibility is that trees perceive subterranean electromagnetic emissions of some sort. Recently, however, geochemical reactions have been suggested as further sources of anticipatory information, at least for animals (Grant et al. 2011). In theory, these reactions could impact on plants also, and induce changes in their bioelectric potentials. Under the effect of tectonic stresses, charge carriers are released from the oxygen matrix of silicates within the Earth's crust. These then interact with ground water, thus creating H_2O_2, as well as enhancing the oxidation of organic materials. Charged ions are also released into the lower atmosphere. The products of these tectonically induced geochemical changes are highly irritable to toads and other amphibious species with wet skins, and no doubt could also be sensed by plants, either through the watery film covering their underground root systems or via their leaves.

An effect of the Moon may be at work in the genesis of Earthquakes. A hypothesis has been developed which proposes that crustal stresses build up as a consequence of long periods of focussed lunisolar tidal acceleration, and that these stresses are later relieved by an earthquake event (Lopes et al. 1990; Kachakhidze et al. 2010, and references therein).

10.8 Discussion

A number of strands of evidence have been brought together which make a circumstantial case for an effect of the Moon on the modulation of bioelectrical potentials in plants. Rather less evidence is available regarding the physiological and developmental processes defined by these potentials and their variations, or about how these potentials might arise, or be stimulated, in the first place. For the moment, let us take as read the proposition that APs are informational, and further, that the daily variations of EPDs are facilitators of some sort of response, or 'event', within plants. However, it may be that many uninformative, or 'passive', signals (bioelectrical 'noise') arise as a consequence of chance perturbations, but that they subsequently decay without eliciting any response simply because the thresholds necessary for their transduction were not attained. In contrast, informative, or 'active', signals are evidently those which modulate physiological events or processes, perhaps by acting as on/off switches. The down-regulation of photosynthesis (Lautner et al. 2005) and the modulation of phloem unloading (Fromm and Bauer 1994) are two instances where active, phloem-borne APs are directed towards a receptive site which thereupon relays further information necessary for a responsive action. Both Volkov and Ranatunga (2006) and Oyarce and Gurovich (2011) have proposed preliminary schemes whereby 'active' bioelectric signals are transduced and subsequently effect new patterns of gene expression appropriate to, for example, a stress response. It is also possible that the exposure of receptive cells to electrical stimuli (APs and so on) could lead to epigenetic modifications of the nuclear chromatin, as has been reported for neural cells in mammals in response to environmental conditions (Sapolosky 2004; Hong et al. 2005). This would be in line with the production of epigenetic modifications, such as have been recorded in plants grown under stressful conditions (e.g., Henderson and Jacobsen 2007), and which perhaps underlie also the electrochemical induction of flowering described by Wagner et al. (2006).

The role of the Moon in the initiation of 'active' APs must remain unresolved for the present. This is largely because lunisolar tidal acceleration is present always and everywhere on Earth, and cannot be switched off or on like a light bulb, as might done in the course of some other type of experiment utilising some other environmental parameter. However, time-courses of the gravimetric tidal acceleration, δg, can be estimated in advance, and so, in theory, pro-active experiments can be planned: anticipated variations in bioelectrical features and biological processes could accordingly be examined in relation to δg at predetermined times.

Turning points in the δg time-course are deemed influential because they are congruent with turning points in the time-course of a physiological process, or in some larger scale growth processes (see, Barlow and Fisahn 2012). Nevertheless, questions remain concerning the significance of the δg turning points, or of lunisolar gravity itself, during the progress of variable biological events. Hypotheses range from the influence of gravitational fields upon rhythmic biochemical processes whose effects manifest as variations in long-term trends of biochemical reactions (Troshichev et al. 2004), to quantum gravity effects mediated by lunar and planetary orbits (Dorda 2010), and which are then expressed in daily variations in biological processes in accordance with δg. Nor is the threshold known for an event induced by a δg turning point—obviously, the zero rate of change of the accelerative force is not reached all at once, but is approached, and then departed from, in a graduated manner (there is a hump in the δg profile, not a spike). Similarly, there may be a period prior to a δg turning point during which there is summation of all the biological information received during the period when the critical rates of change of δg operate.

It goes without saying that, besides the internal milieu of plants, the external biosphere, semiosphere, as well as the magnetosphere, ionosphere and cosmosphere, are exceedingly complex domains, and all of them envelop and penetrate the Earth's phytogeodermal domain. In the areas of human health and physiology, which together comprise arguably the most advanced area of biological research, it is becoming realised that the biospheric and cosmospheric domains impinge upon the workings of the human organism and its social dynamics (Breus et al. 1995; Khabarova and Dimitrova, 2009). This was foreseen by Tchizhevsky (1940), some years before NI Vasil'eva (1998) drew attention to some of the correlations which exist between interacting cycles of solar activity on biological, geophysical and social rhythms. And this interactive cyclicity is also penetrated by the frequencies and phases of the harmonics of the cosmospheric system (Molchanov 1968).

Although such considerations may be aimed at a very long-range view of life and of life processes, an appropriate starting point is nevertheless the day-to-day activities of organisms. However, any set of short-term diurnal events has the possibility of being a pre-condition for amplification over time and, in the long-term, of becoming a set of post-conditional states. Conditional states may be evidenced in those features which, by directed gene mutations or epigenetic modifications, are forwarded into future generations of organisms. The persistent initiation and transmission of bioelectric potentials may be one way in which lunar cycles not only feed into the day-to-day vital processes of plants but also make their effects felt in plant survival strategies and evolution.

Acknowledgments I am grateful to Professor E Klingelé for the gift of the Etide program, to Professor M Mikulecký for analysing the Polar Cap (Thule) Index time-course in Fig. 10.6, to Dr Olga V Khabarova for information about solar wind-magnetosphere interactions, to Professor E Zürcher for helpful remarks, and to Mr Timothy Colborn who expertly prepared the Figures. Data relating to water volume fraction (Fig. 10.11) were kindly provided by Dr JP Sparks and Professor MJ Canny.

References

Annila A, Kuismanen E (2009) Natural hierarchy emerges from energy dispersal. BioSystems 95:227–233
Antkowiak B, Meyer W-E, Engelmann W (1991) Oscillations of the membrane potential of pulvinar motor cells in situ in relation to leaflet movements of *Desmodium motiorum*. J Exp Bot 42:901–910
Baly ECC, Semmens ES (1924) The selective action of polarised light.–I. The hydrolysis of starch. Proc R Soc London, Ser B 97:250–253
Barlow PW (1999) Living plant systems: how robust are they in the absence of gravity? Adv Space Res 23:1975–1986
Barlow PW (2005) From cambium to early cell differentiation within the secondary vascular system. In: Holbrook NM, Zwieniecki MA (eds) Vascular transport in plants. Elsevier Academic Press, Amsterdam, pp 279–306
Barlow PW (2008) Reflections on 'plant neurobiology'. BioSystems 92:132–147
Barlow PW (2010) The origins of life, of living and cognition, and of the phytoneural system. In: Abstracts evolution in communication and neural processing. From first organisms and plants to man and beyond. Modena, 18-19 Nov 2010, pp 5–6
Barlow PW, Fisahn J (2012) Lunisolar tidal force and the growth of plant roots, and some other of its effects on plant movements. Annals of Botany (in press)
Barlow PW, Klingelé E, Mikulecký M (2009) The influence of the lunar-solar tidal acceleration on trees gives a glimpse of how the plant-neurobiological system came into being. In: Abstracts 5th international symposium on plant neurobiology, Firenze, 25–29 May 2009, pp 11–13
Barlow PW, Mikulecký M, Střeštík J (2010) Tree-stem diameter fluctuates with the lunar tides and perhaps with geomagnetic activity. Protoplasma 247:25–43
Beeson CFC, Bhatia BM (1936) On the biology of the Bostrychidae (Coleopt.). Indian For Rec 2:223–323
Breus TK, Cornélisson G, Halberg F, Levitin AE (1995) Temporal associations of life with solar and geophysical activity. Ann Geophys 13:1211–1222
Brouwer G (1926) De Periodieke Bewegingen van de Primaire Bladen bij *Canavalia ensiformis*. HJ Paris, Amsterdam
Bünning E (1956) Die physiologische Uhr. Naturwissenschaftliche Rundschau 9:351–357
Bünning E (1963) Die Physiologische Uhr. Springer, Heidelberg
Bünning E, Moser I (1969) Interference of moonlight with the photoperiodic measurement of time by plants, and their adaptive reaction. Proc Nat Acad Sci USA 62:1018–1022
Burr HS (1943) Electrical correlates of pure and hybrid strains of sweet corn. Proc Nat Acad Sci USA 29(163):166
Burr HS (1944) Moon madness. Yale J Biol Med 16:249–256

Burr HS (1945) Diurnal potentials in the Maple tree. Yale J Biol Med 17:727–735
Burr HS (1947) Tree potentials. Yale J Biol Med 19:311–318
Burr HS (1956) Effect of a severe storm on electric properties of a tree and the earth. Science 124:1204–1205
Burr HS, Sinnott EW (1944) Electrical correlates of form in cucurbit fruits. Am J Bot 31:249–253
Cantiani M (1978) Il ritmo di accrescimento diurno della douglasia del tiglio e del liriodendro a Vallombrosa. L'Italia Forestale e Montana, No 2:57–74
Cantiani M, Sorbetti GF (1989) Transpirazione e ritmo circadiano delle variazzioni reversibli del diametro dei fusti di alcune piante arboree (1^a parte). L'Italia Forestale e Montana, No 5: 341–372
Cantiani M, Cantiani M-G, Guerri Sorbetti F (1994) Rythmes d'accroissement en diameter des arbres forestiers. Révue Forestière Française 46:349–358
Cifra M, Fields JZ, Farhadi A (2011) Electromagnetic cellular interactions. Prog Biophys Mol Biol 105:223–246
Dengel S, Aeby D, Grace J (2009) A relationship between galactic cosmic radiation and tree rings. New Phytol 184:545–551
Dodson HW, Hedeman ER (1964) An unexpected effect in solar cosmic ray data related to 29.5 days. J Geophys Res 69:3965–3971
Dorda G (2010) Quantisierte Zeit und die Vereinheitlichung von Gravitation und Elektromagnetismus. Cuvillier Verlag, Göttingen
Fisahn J, Yazdanbakhsh N, Klingelé E, Barlow P (2012) Sensitivity of developing *Arabidopsis* roots to lunisolar tidal acceleration: a precise backup clock (submitted)
Fraser-Smith AC (1978) ULF tree potential and geomagnetic pulsations. Nature 271:641–642
Fraser-Smith AC (1993) ULF magnetic fields generated by electrical storms and their significance to geomagnetic pulsation generation. Geophys Res Lett 20:467–470
Fromm J, Bauer T (1994) Action potentials in maize sieve tubes change phloem translocation. J Exp Bot 45:463–469
Fromm J, Fei H (1998) Electrical signalling and gas exchange in maize plants in drying soil. Plant Sci 132:203–213
Fromm J, Lautner S (2007) Electrical signals and their physiological function. New Phytol 30:249–257
Gál J, Horváth G, Barta A, Wehner R (2001) Polarization of the moonlit clear night sky measured by full-sky imaging polarimetry at full Moon: comparison of the polarization of moonlit sunlit skies. J Geophys Res 106(D19):22647–22653
Galland P, Pazur A (2005) Magnetoperception in plants. J Plant Res 118:371–389
Gibert D, Le Mouël J-L, Lambs L, Nicollin F, Perrier F (2006) Sap flow and daily electric potential variations in a tree trunk. Plant Sci 171:572–584
Goldsworthy A (1983) The evolution of plant action-potentials. J Theor Biol 103:645–648
Grant RA, Halliday T, Balderer WP, Leuenberger F, Newcomer M, Cyr G, Freund FT (2011) Ground water chemistry changes before major earthquakes and possible effects on animals. Int J Environ Res Pub Health 8:1936–1956
Graviou E (1978) Analogies between rhythms in plant material, in atmospheric pressure, and solar lunar periodicities. Int J Biometeorol 22:103–111
Guhathakurta A, Dutt BK (1961) Electrical correlate of the pulsatory movement of *Desmodium gyrans*. Trans Bose Res Inst 24:73–82
Henderson IR, Jacobsen SE (2007) Epigenetic inheritance in plants. Nature 207:732–734
Holzknecht K, Zürcher E (2006) Tree stems and tides—A new approach and elements of reflexion. Schweizerische Zeitschrift für Forstwesen 157:185–190
Hong EJ, West AE, Greenberg ME (2005) Transcriptional control of cognitive development. Curr Opin Neurobiol 15:21–28
Kachakhidze MK, Kiladze R, Kachakhidze N, Kukhianidze V, Ramishvili G (2010) Connection of large earthquakes occurring moment with the movement of the Sun and the Moon and with the Earth crust tectonic stress character. Nat Hazards Earth Syst Sci 10:1629–1633

Khabarova OV, Dimitrova S (2009) On the nature of people's reaction to space weather and metereological weather changes. Sun Geosph 4:60–71

Klein G (2007) Farewell to the internal clock. A contribution in the field of chronobiology. Springer, New York

Koppán A, Szarka L, Wesztergom V (1999) Temporal variation of electrical signal recorded in a standing tree. Acta Geodaetica Geophysica Hungarica 34:169–180

Koppán A, Fenyvesi A, Szarka L, Wesztergom V (2000) Annual fluctuation in amplitudes of daily variations of electrical signals measured in the trunk of a standing tree. Comptes Rendus de l' Académie des Sciences, Paris, Sciences de la Vie/Life Sciences 323:559–563

Koppán A, Fenyvesi A, Szarka L, Wesztergom V (2002) Measurement of electric potential difference on trees. Acta Biologica Szegediensis 46(3–4):37–38

Koppán A, Szarka L, Wesztergom V (2005) Local variability of electric potential differences on the trunk of *Quercus cerris* L. Acta Silvatica et Lignaria Hungarica 1:73–81

Kozyreva O, Pilipenko V, Engebretson MJ, Yumoto K, Watermann J, Romanova N (2007) In search of a new ULF wave index: comparison of Pc5 power with dynamics of geostationary relativistic electrons. Planet Space Sci 55:755–769

Kundt W (1998) The heart of plants. Curr Sci 75:98–102

Lautner S, Grams TEE, Matyssek R, Fromm J (2005) Characteristics of electrical signals in poplar and responses in photosynthesis. Plant Physiol 138:2200–2209

Le Mouël J-L, Gibert D, Poirier J-P (2010) On transient electric potential variations in a standing tree and atmospheric electricity. CR Geosci 342:95–99

Lemström S (1901) Ueber das Verhalten der Flüssigkeiten in Capillarröhhren unter Einfluss eines elektrischen Luftstromes. Annalen der Physik Series 4 5:729–756

Lemström S (1904) Electricity in agriculture and horticulture. "The Electrician" Printing and Publishing Company Ltd, London

Levin M (2003) Bioelectromagnetics in morphogenesis. Bioelectromagnetics 24:295–315

Lodge O (1908) Electricity in agriculture. Nature 78:331–332

Lopes RMC, Malin SR, Mazzarella A, Palumbo A (1990) Lunar and solar triggering of earthquakes. Phys Earth Planet Inter 59:127–129

MacDougal DT (1921) Growth in trees. Carnegie Institution of Washington Publication No. 307, Washington, p 41

Maeda H (1968) Variations in geomagnetic field. Space Sci Rev 8:555–590

Markson R (1971) Considerations regarding solar and lunar modulation of geophysical parameters, atmospheric electricity and thunderstorms. Pure Appl Geophys 84:161–200

Markson R (1972) Tree potentials and external factors. In: Burr HS (ed) The fields of life. Ballantyne Books, New York, pp 186–206

Mehra P (1989) Lunar phases and atmospheric electric field. Adv Atmos Sci 6:239–246

Meylan S (1971) Bioélectricité. Quelques problèmes. Masson et Cie, Paris

Molchanov AM (1968) The resonant structure of the solar system. The law of planetary distances. Icarus 8:203–215

Morat P, Le Mouël JL, Granier A (1994) Electric potential on a tree. A measurement of the sap flow? Comptes Rendus de l'Académie des Sciences, Paris, Sciences de la Vie/Life Sciences 317:98–101

Moshelion M, Becker D, Czempinski K, Mueller-Roeber B, Attali B, Hedrich R, Moran N (2002) Diurnal and circadian regulation of putative potassium channels in a leaf moving organ. Plant Physiol 128:634–642

Nelson OE, Burr HS (1946) Growth correlates of electromotive forces in maize seeds. Proc Nat Acad Sci USA 32:73–84

Nishimura T, Fukushima M (2009) Why animals respond to the full moon: magnetic hypothesis. Biosci Hypotheses 2:399–401

Oyarce P, Gurovich L (2009) Electrical signals in avocado trees. Responses to light and water availability conditions. Plant Signal Behav 5:34–41

Oyarce P, Gurovich L (2011) Evidence for the transmission of information through electric potentials in injured avocado trees. J Plant Physiol 168:103–108

Polevoi VV, Bilova TE, Shevtsov Yu I (2003) Electroosmotic phenomema in plant tissues. Biology Bulletin (English translation of Izvestiya Akademii Nauk, Seriya Biologicheskaya) 30:133–139

Ross JJ, Reid JB (2010) Evolution of growth-promoting plant hormones. Funct Plant Biol 37:795–805

Saito Y (2007) Preceding phenomena observed by tree bio-electric potential prior to Noto Penninsula off earthquake. Japan Geoscience Union Meeting 2007. 1–14. (http://www.jsedip.jp/English/Papers/070512_JGU%20Meeting%202007-E.pdf)

Sapolsky RM (2004) Mothering style and methylation. Nat Neurosci 7:791–792

Semmens ES (1947a) Chemical effects of moonlight. Nature 159:613

Semmens ES (1947b) Starch hydrolysis induced by polarized light in stomatal guard cells in living plants. Plant Physiol 22:270–278

Sharma S, Dashora N, Galav P, Pandey R (2010) Total solar eclipse of July 22, 2009: its impact on the total electron content and ionospheric electron density in the Indian zone. J Atmos Solar Terr Phys 72:1387–1392

Sparks JP, Campbell GS, Black RA (2001) Water content, hydraulic conductivity, and ice formation in winter stems of *Pinus contorta*: a TDR case study. Oecologia 127:468–475

Stahlberg R, Cleland RE, Van Volkenburgh E (2006) Slow wave potentials—a propagating electrical signal unique to higher plants. In: Baluška F, Mancuso S, Volkmann D (eds) Communication in Plants, Springer-Verlag, Berlin, pp 291–308

Stolov HL, Cameron AGW (1964) Variations of geomagnetic activity with lunar phase. J Geophys Res 69:4975–4981

Sztein AE, Cohen JD, Cooke TJ (2000) Evolutionary patterns in the auxin metabolism of green plants. Int J Plant Sci 161:849–859

Tchizhevsky AL (1940) Cosmobiologie et rythme du milieu extérieur. Acta Medica Scandinavica, Supplement 108:211–226

Toriyama H (1991) Individuality in the anomalous bioelectric potential of silk trees prior to earthquake. Science Report of Tokyo Woman's Christian University 94–95, pp 1067–1077

Troshichev OA, Andrezen VG, Vennerstrøm S, Friis-Christensen E (1988) Magnetic activity in the polar cap—a new index. Planet Space Sci 36:1095–1102

Troshichev OA, Gorshkov ES, Shapovalov SN, Sokolovskii VV, Ivanov VV, Vorobeitchikov VM (2004) Variations of the gravitational field as a motive power for rhythms of biochemical processes. Adv Space Res 34:1619–1624

van Bel AJE, Knoblauch M, Furch ACU, Hafke JB (2011) Questions on phloem biology. 1. Electropotential waves, Ca^{2+} fluxes and cellular cascades along the propagation pathway. Plant Sci 181:210–218

Vasil'eva NI (1998) Correlations between terrestrial and space processes within the framework of universal synchronization. Biophysics 43:694–696

Vogt KA, Beard KH, Hammann S, O'Hara Palmiotto J, Vogt DJ, Scatena FN, Hecht BP (2002) Indigenous knowledge informing management of tropical forests: the link between rhythms in plant secondary chemistry and lunar cycles. Ambio 31:485–490

Volkov AG, Haack RA (1995) Insect-induced bioelectrochemical signals in potato plants. Bioelectrochem Bioenerg 37:55–60

Volkov AG, Mwesigwa J (2001) Electrochemistry of soybean: effects of uncouplers, pollutants, and pesticides. J Electroanal Chem 496:153–157

Volkov AG, Ranatunga DRA (2006) Plants as environmental biosensors. Plant Signal Behav 1:105–115

Volkov AG, Dunkley TC, Labady AJ, Brown CL (2005) Phototropism and electrified interfaces in green plants. Electrochim Acta 50:4241–4247

Wagner E, Lehner L, Normann J, Veit J, Albrechtová J (2006) Hydro-electrochemical integration of the higher plant–Basis for electrogenic flower induction. In: Baluška F, Mancuso S, Volkmann D (eds) Communication in plants. Springer, Berlin, pp 370–389

Zurbenko IG, Potrzeba AL (2010) Tidal waves in the atmosphere and their effects. Acta Geophys 58:356–373

Zürcher E, Cantiani M-G, Sorbetti Guerri F, Michel D (1998) Tree stem diameters fluctuate with tide. Nature 392:665–666

Zweifel R, Item H, Häsler R (2000) Stem radius changes and their relation to stored water in stems of young Norway spruce trees. Tree 15:50–57

Chapter 11
Biosystems Analysis of Plant Development Concerning Photoperiodic Flower Induction by Hydro-Electrochemical Signal Transduction

Edgar Wagner, Lars Lehner, Justyna Veit, Johannes Normann and Jolana T.P. Albrechtová

Abstract In contrast to the classical hypothesis of photoperiodic flower induction involving a flower-inducing hormone 'florigen', it is the goal of this contribution to present the higher plant as a hydraulic-electrochemical signal transducer integrating the plant organs via action potentials in communication with intrinsic and environmental constraints and to replace 'florigen' by a frequency-coded hydro-electrochemical signal pattern. Observation of whole plant behaviour by time lapse photography clearly shows rhythmic integration of the main shoot axis and side branches in rhythmic growth as well as in leaf movements. This was observed with short-day ecotypes of *Chenopodium rubrum* L. and a long-day ecotype of *Chenopodium murale* L. Upon flower induction the phase relationship between rhythmic SER and leaf movements is altered in a very specific way both in the short-day and long-day plant (Wagner et al., Flowering Newslett, 26:62–74, 1998). The experimental set-up for running these investigations is shown in Fig. 11.1. The recording of surface sum action potentials from the surface of Chenopodium plants is achieved with bipolar electrodes and differential amplifiers similar to the equipment used for recording EEGs and ECGs in medicine. Finally, electrophysiograms (EPGs) can be obtained from various phases of plant development (i.e. juvenility, maturity) to be used for applied purposes like plant characterisation and manipulation in nurseries or glasshouse crops. Rhythmic integration of the whole plant possibly involves modulation of turgor pressure via stretch-activated ion channels and concomitant changes in membrane potential. The perception of a flower inducing dark period might lead to a change in electrochemical signalling between leaves and the shoot apical meristem (SAM) and thus represent 'florigen'.

Albrechtová—passed away 29.11.2005

E. Wagner (✉) · L. Lehner · J. Veit · J. Normann · J. T.P.Albrechtová
Institute of Biology II, University of Freiburg, Schänzlestr. 1, 79104 Freiburg, Germany
e-mail: edgar.wagner@biologie.uni-freiburg.de

The switch from the vegetative to the flowering state is a threshold response, systemic in nature and involving not only the apical meristem but also the axillary buds. Thus we believe that the flower inducing signal may be electrical in nature, requiring a holistic biosystem analysis for quantitative ilucidation.

Keywords Action potentials · Aquaporins · Biosystem analysis · Chenopodium · Circadian rhythm · Electrocardiograms · Electrochemical · Electrophysiograms · Facilitation · Floral stimulus · Leaf movements · Meristem · Photoperiod · Plasmodesmata · Symplast

11.1 Introduction: Photoperiodic Flower Induction

The hypothesis that flowering involves a specific stimulus is based upon the demonstration that (a) in photoperiodism the flowering response depends upon the day-length conditions given to the leaves, whereas the response occurs in the apices, and that (b) a floral stimulus can be transmitted via a graft union from an induced partner (donor) to a non-induced one (receptor). Transmission of the floral stimulus by grafting has been demonstrated within various photoperiodic response types, as well as between different photoperiodic response types in interspecific and intergeneric grafts. The physiological evidence for a floral stimulus is clear-cut, but till now the nature of the stimulus has remained obscure (Bernier 1988).

The specific kind of photoperiodic behaviour depends on the exact environmental conditions, as was shown for four different North American ecotypes of *Chenopodium rubrum* (Tsuschiya and Ishiguri 1981). The southern ecotypes display an obligate short-day behaviour under white (W), red (R) and blue (B) light. The most northern ecotype is day neutral in B and W and has an amphiphotoperiodic response in R light. Another northern ecotype has an amphiphotoperiodic response in B and a short-day response in W and R light. The amphiphotoperiodic response in B is modified to day neutral by changing the temperature from 20 to 12°C. These data clearly indicate that photoperiodic behaviour is extremely flexible in adapting to specific environmental conditions.

Irrespective of the flexibility of plants in modifying their photoperiodic behaviour in adapting to specific environmental conditions as mentioned above, the following essentials of the photoperiodic reaction have to be kept in mind as a basis for further considerations on a quantitative analysis of plant photoperiodism (Fukshansky 1981):

- Short-day (SDP) and long-day plants (LDP) show opposite reactions to a given photoperiod.
- Reactions result from coincidence or non-coincidence of light and dark phases of the photoperiod with corresponding phases of an endogenous circadian rhythm and the main photoreceptors are the plant sensory pigment systems, phytochrome and cryptochrome. Circadian rhythm and photoreceptors have the same properties in SDP and LDP.

11 Biosystems Analysis of Plant Development

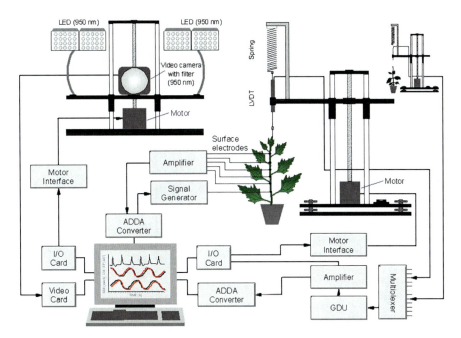

Fig. 11.1 Experimental set-up. Measuring device for long-term recording of changes in electric surface membrane potential, leaf movements (*LM*) and stem elongation rate (*SER*). Video imaging at 950 nm for continuous monitoring of leaf movements in light–dark cycles. Linear voltage differential transformers (*LVDTs*) hooked to the plant stem with a spring loaded constant pull of 1.5 g. Optimal measuring conditions for video imaging and SER registration are maintained via software controlled step motors positioning the devices according to the growth of the plants. Platinum electrodes are used together with a commercial contact gel for measuring and stimulation. Additional sensors monitor electromagnetic noise, temperature, light intensity and humidity. From Normann et al. 2007, Wagner et al. 2006a, b

- Critical photoperiodic induction produces irreversible changes in the leaves of SDP and LDP leading to a common state both in SDP and LDP, as proven by grafting experiments. There is no difference between SDP and LDP in their response towards a common inductor from a grafted leaf from an induced SDP or LDP.

Analysing the kinetics of change at the shoot apical meristem (SAM) during flower initiation can give hints on the mechanism(s) of signal transduction from leaves to SAM during photoperiodic flower induction (Fig. 11.1).

11.2 A Systems Biological Analysis of Development in (the) Higher Plants C. rubrum and C. murale

With the model systems *C. rubrum* (SDP) and *Chenopodium murale* (LDP) growth and behaviour have been studied in response to changes in photo and thermoperiod. With time-laps photography, rhythmic integration of the plant could be demonstrated.

Inter-Organ Communication and Endogenous Rhythm

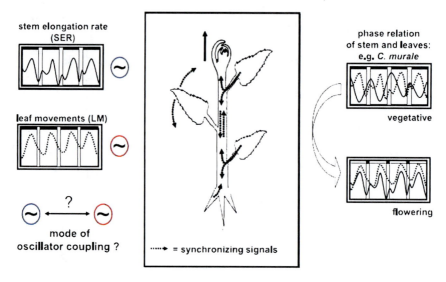

Fig. 11.2 Specific changes in the phasing of SER and LM kinetics appear upon flower induction. Both parameters depend on turgor-mediated growth and their interactions demonstrate rhythmic interorgan communication. LMs and SER are quantified by computer controlled time lapse photography (see Fig. 11.1)

Leaf movements and rhythmic growth were monitored continuously. Upon photoperiodic flower initiation, rhythmic stem extension rate (SER) and leaf movement (LM) change their phase relationship in a specific way (Fig. 11.2). Flower induction correlates to a threshold value for the ratio between integral growth during the dark time span and integral growth during the light time span (Normann et al. 2007).

Both species exhibit a circadian rhythm in the SER with period length of 24.3 ± 0.5 h (*C. rubrum*) and 27.5 ± 0.5 h (*C. murale*). Flowering plants show a significantly shorter period length in the SER of 23.44 ± 0.75 h (*C. rubrum*) and 26.6 ± 0.66 h (*C. murale*). The period lengths of LM mirror the kinetics in the SER, displaying similar increases of frequency in the flower-induced state. While in vegetative plants the kinetics of the SER and LM are 180° out of phase, this phase relationship is shifted after flower induction (Fig. 11.2). Both parameters display clear movement and growth patterns with photoperiod-specific reactions to "light-on" and "light-off" signals. Flower induction correlates to a threshold value of stem growth of 0.6 (*C. rubrum*) and 4.0 (*C. murale*) for the ratio of the integral growth during the dark span over the integral growth in the light span.

The precise output displayed in the growth pattern of the plants (SER and LM) is therefore a systemic reflection of all available environmental inputs into the metabolic networks controlling the timing of communication between organs (Fig. 11.2).

Two hours after the beginning of the critical dark period the pattern of cytoplasmic pH and Ca^{2+} concentration at the apical meristem has changed, possibly indicating the arrival of the flower-inducing signal (Albrechtová et al. 2001; Wagner et al. 2006a; Walczysko et al. 2000).

11.3 The Model System *Chenopodium*: Induction of Flowering from Physiology to Molecular Biology

The model system *Chenopodium spec.* had been established to study photoperiodic control of flowering on the physiological, biochemical and molecular levels.

First, *Chenopodium* was developed as a 'Petri-dish plant' by Cumming (1959) for large-scale screening of photoperiodic flower induction with several latitudinal ecotypes showing short-day, long-day and day-neutral responses (Cumming 1967; Tsuschiya and Ishiguri 1981).

Subsequent studies could demonstrate that phytochrome photoreversibility could not act as an hour glass timer in photoperiodism, but was gated in its light sensitivity by an endogenous (circadian) rhythm presenting photophile and scotophile phases in daily 24 h light–dark cycles (Cumming et al. 1965). Following Bünning's (1942) and Hendricks' (1963) early concepts on metabolic control of timing in photoperiodism, *C. rubrum* has been used to establish an analysis of energy metabolism demonstrating a circadian rhythm in redox state and energy charge as macro parameters timing photoperiodic behaviour (Wagner et al. 1975). Recently, *C. rubrum* was also used in molecular studies on signal transduction in photoperiodic flower induction. An orthologue of *LEAFY*, a transcription factor involved in a signalling cascade leading to flowering in *Arabidopsis* (Nilsson et al. 1998), was identified in *C. rubrum*. Expression kinetics of *LEAFY* orthologue *CrFL* at SAM is related to photoperiod (Veit et al. 2004). Transgenic plants were produced using RNA interference in order to analyse function of *CrFL* in signal transduction in *C. rubrum*.

With studies on the hydro-electrochemical integration of communication in *Chenopodium* plants, we could demonstrate that action potentials precede turgor-mediated LMs and changes in SER (Wagner et al. 2006a, b). Molecular and physiological studies at SAM of *C. rubrum* in transition to flowering presented evidence of changes in turgor and in aquaporin expression (Albrechtová and Wagner 2004; Albrechtová et al. 2004) which are most likely triggered by APs travelling along the stem axis as a line of communication among leaves, roots and SAM. A rapid communication between roots and SAM is inferred from a reduction of O^{15}-water uptake by the roots after cutting off apical meristem (Ohya et al. 2005).

Further research revealed that EPGs are very specific for plants. Mildew attack or dehydration does lead to characteristic EPGs for plants (Lehner 2003; Wagner 2006a, b). It can be assumed that a similar variety of EPGs does exist for plants like Electrocardiograms for humans or animals. From the EPGs and the directionality of

AP propagation it is inferred that APs facilitate flower initiation during transcriptional activation of FLOWERING LOCUS T (FT) in the leaves from where FT seems to travel to the SAM to initiate floral development (Notaguchi et al. 2008; Veit et al. 2004).

The recording of the frequency distribution of spontaneous APs moving basipetal and acropetal on the stem axis resulted in electrophysiograms (EPGs) which could be used to characterise the flowering and vegetative state in *C. rubrum* and *C. murale*. In addition to the characterisation of such phase changes, the information from EPGs has been used for the electrogenic initiation of flowering (Lehner 2002) opening a new field for applications in horticulture, agriculture and silviculture (Wagner et al. 2004).

11.4 Electrophysiology and Plant Behaviour

Chenopodium is very well characterised with respect to the kinetics of photoperiod controlled flower induction in a whole series of latitudinal ecotypes (Cumming 1959, 1967; King 1975). In *C. rubrum* the circadian rhythmic organisation of energy metabolism has been analysed in detail (Wagner et al. 1975, 2004).

In view of demonstrated rhythmic and photoperiod-mediated surface potential phenomena we suggested that the flowering stimulus might be an electrical or electrochemical signal (Wagner et al. 1996, 1997). A similar concept was advanced in relation to systemic effects from wound responses; it was suggested that action potentials could be involved as signalling mechanisms in chilling injury, mechanical perturbation and invasion by pathogens (Davies 1987; Wildon et al. 1989, 1992), as well as for the influence of light and gravity (Davies et al. 1991). Application of direct current, i.e. electrical stimulation, induced proteinase inhibitor II Pin2 gene expression and modulation of photosynthetic activity in the leaves (Herde et al. 1995). Induction of proteinase inhibitor gene expression involved both action potentials and variation potentials in another set of experiments (Stankovic and Davies 1996). Systemic Pin2 gene expression in wild-type and ABA-deficient tomato plants was triggered by mechanical wounding, current application and heat treatment (Herde et al. 1998a, b). Mechanical wounding of *C. rubrum* leaves induced changes in membrane potential, as seen by using a fluorescent probe as indicator (Albrechtová and Wagner 1998).

In sunflower plants electrical activities could be evoked in response to various external stimuli or were generated spontaneously. Action potentials, graded potentials, changes in resting potential and rhythmic electrical activity were observed. Large graded potentials were paralleled by a decrease in growth rate and in the rate of water uptake. Spontaneous action potentials along the stem axis were generated only at night in vegetative plants but they were generated during the day as well as during the night in flowering plants (Davies et al. 1991). Flower induction in *Spinacia* was correlated to the light-stimulated bioelectric response from the leaves (Greppin and Horwitz 1975; Greppin et al. 1973).

As suggested by some previous experiments on the control of flowering (Adamec and Krekule 1989a, b; Adamec et al. 1989), endogenous electrical activities of plants can be manipulated by electric current from outside; specifically in the long-day plant *Spinacia oleracea* (Montavon and Greppin 1983, 1986), as well as in the short-day plant *C. rubrum* (Adamec et al. 1989; Machackova and Krekule 1991; Machackova et al. 1990), photoperiodic flower induction could be inhibited by the application of direct electric current (DC) via felt-tipped electrodes. It was concluded that DC probably interfered with the translocation of a floral stimulus from induced leaves to the SAM (Adamec and Krekule 1989a, b; Adamec et al. 1989).

There is controversial evidence concerning transport of the flower inducing stimulus in the phloem (see Bernier 1988 and references therein), which would be the transport path for the stimulus in florigenic and multicomponent theories. Phloem has been considered as a pathway for signal spreading: but a general symplast/apoplast interaction might be involved as well, as evidenced from studies on fast electric transients induced by electric stimulation in the apoplast of tomatoes (Herde et al. 1998a, b). Temporal changes in plasmodesmal patterning might be involved in modulating electric signal transduction (Liarzi and Epel 2005; Wright and Oparka 1997).

Bearing in mind the circadian rhythm in transcellular current in *Acetabularia* (Novak and Sironval 1976) and the diurnal rhythm in resting membrane potential in *Chenopodium* as shown previously (Wagner et al. 1998), it seems possible that a circadian rhythm in bioelectricity may be of great significance in the circadian rhythmic coordination of the whole plant responsible for sensitivity changes in membrane-bound activities (Vigh et al. 1998). The circadian rhythm in transcellular current probably arises from a circadian rhythm in energy metabolism and rhythmicity of transport processes at the plasma membrane (Albrechtová et al. 2006; Mayer and Fischer 1994; Mills et al. 1994; Pickard 1994).

In *C. rubrum*, a detailed analysis of the occurrence of surface sum APs revealed electric activity in all organs (root, stem, leaves) of the plant. With a series of electrodes along the stem axis the basipetal or acropetal propagation of APs was recorded. Under quite different experimental conditions, so-called reflected APs occasionally could be observed, i.e. APs generated at the SAM were propagating basipetal and then seemed to be reflected acropetal (Fig. 11.3). They have not been studied in detail so far but might be very relevant for analysing the importance of the communication between root meristems and SAM in control of growth and development.

11.5 Circadian Rhythms as Metabolic Bases for Hydro-Electrochemical Signal Transduction

Rhythmicity is one of the characteristics of life which expresses itself at all levels of organisation from unicellular systems to man. Rhythmic phenomena in physiology, development and behaviour of all living systems show period lengths ranging from fractions of a second to hourly, daily and even annual cycles.

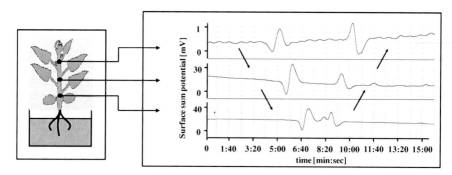

Fig. 11.3 Reflection of an action potential (*AP*) at the basal section of a stem axis of *C. rubrum*. The recordings of the surface sum potential are presented after a light to dark transition of a 5-week-old plant of *C. rubrum* at time 0. The AP moving basipetal from the apical part of the stem axis, is successively passing three bipolar surface electrodes. Below the basal electrodes it is either reflected immediately and is moving acropetal, or, alternatively, the basipetal moving AP might trigger a new AP moving acropetal. Growing conditions: 20°C, 70% rh; photosynthetic active radiation (*PAR*): 120 μmol/m^{2}sec. The mirror image pattern of basipetal, acropetal propagating APs is due to the measuring principle which is using bipolar (+;−) electrodes and a differential amplifier (Lehner 2003) (see also Zawadzki et al. 1991)

The most conspicuous rhythm is the so-called circadian oscillation. The circadian rhythm is an endogenous oscillation of metabolic activity with a period length of exactly 24 h when the organisms are synchronised by the daily light–dark cycle of the earth. In constant conditions, however, its period length is only approximately 24 h, i.e. circadian. In contrast to biological rhythms showing other frequencies, circadian rhythms are temperature compensated and almost unsusceptible to chemical manipulation. It is this stability or homeo-dynamics of period length which qualifies the circadian rhythm as a precise physiological timer and thus is the essence of Bünning's (1973, 1977) theory of the physiological clock.

From an evolutionary point of view, circadian rhythmicity has been considered to be an adaptation of pro- and eukaryotic energy conservation and transformation to optimise energy harvesting by photosynthesis in the daily cycle of energy supply from the environment (Cumming and Wagner 1968; Wagner 1977; Wagner and Cumming 1970; Wagner et al. 1998). It was also assumed that this adaptation is dependent on the division of energy transformation within different compartments of the cell, such as chloroplasts, mitochondria and the glycolytic space involving redox-mediated transcriptional controls (Bauer et al. 1999). In photosynthetic prokaryotes, lacking cell organelles, a metabolic micro-compartmentation allows for a similarly sophisticated regulatory network as in eukaryotes.

High-frequency oscillations in energy-transducing metabolic sequences could give rise to low-frequency oscillations of energy flow in the metabolic network of the whole system, providing a basis for the evolution of a temperature-compensated circadian rhythm (Wagner and Cumming 1970). The clock's periodicity is genetically determined and provides the temporal frame for physiological and behavioural patterns that are necessary for adaptation of organisms and populations to environmental constraints. Thus the circadian rhythmic cell is a hydro-electrochemical oscillator driven or synchronised by the daily dark/light cycle with a temporal compartmentation of metabolism and a network of metabolic sequences to compensate for oxidative stress in adapting to their light environment. This is best shown in the adaptation of photosynthetic machineries from bacteria (Bauer et al. 1999; Joshi and Tabita 1996; Sippola and Aro 2000; Zeilstra-Ryalls et al. 1998) to higher plants (Anderson et al. 1988; Asada 1999), which respond to changing light quality and quantity with coordinated changes of pigmentation, electron transport components, membrane composition,—organisation and— function (Anderson et al. 1988). The acclimation of plants at the cellular level requires interaction among the nucleus, mitochondria and chloroplasts in a regulatory network involving several photoreceptors like phytochromes, blue light receptors and chlorophyll.

The existence of circadian rhythmicity is dependent on the living cell and suggests that metabolic compartmentation spatial and/or temporal is of significance for its generation (Barbier-Brygoo et al. 1997; Flügge 2000). Furthermore, circadian rhythms in photosynthesis, respiration and chloroplast shape suggest that investigation of the metabolic controls may be fruitful for the elucidation of the mechanism of biological rhythms. Interrelations between cellular compartments have already been shown. For example, Könitz (1965) has shown reciprocal changes in the ultrastructure of chloroplasts and mitochondria in daily light–dark cycles. Murakami and Packer (1970) demonstrated that, within the same cell, mitochondria swell and chloroplasts contract upon illumination and the reverse occurs in the dark. The importance of the entire metabolic network for the display of circadian oscillations is underlined by the fact that, in contrast to the temperature-compensated circadian oscillations of the intact system, isolated organelles display high frequency oscillations (Gooch and Packer 1974; Gylkhandanyan et al. 1976) which are temperature dependent. Similarly, in photosynthetic bacteria, cellular signal transduction integrates micro-compartmentation of photosynthesis, carbon dioxide assimilation and nitrogen fixation (Joshi and Tabita 1996). From a detailed analysis of rhythms in enzyme activities involved in compartmental energy metabolism with and without feeding of sugars (cf. Jang and Sheen 1994; Wagner et al. 1983) and on changes of nucleotide pool size levels in the short-day plant *C. rubrum* L., we compiled evidence in favour of circadian rhythmicity in overall energy transduction. Our observations lead us to suggest that circadian rhythmicity, as the timer in photoperiodism, should be based on a circadian rhythm in energy metabolism. This rhythm would be the result of a compensatory control oscillation between glycolysis and oxidative phosphorylation, coupled to photophosphorylation in cyanobacteria (Huang et al. 1990), photosynthetic bacteria and green plants (Wagner 1976a, b, c; Wagner and Cumming 1970; Wagner et al. 1974a, b). This mechanism of circadian rhythmicity could involve energy control of ion transport processes at the membranes of cells and organelles. The membrane's physical state, e.g. modulated by temperature, could control transcription (Vigh et al. 1998) or via frequency coded calcium oscillations lead to differential gene activation and expression (Dolmetsch et al. 1998; Li et al. 1998).

Taking into account the symplastic organisation of higher plants, the concept of compartmental feedback becomes even more attractive in view of bioelectric phenomena. The symplastic organisation of higher plants might be the basis for translocation of electric, photoperiodic and morphogenetic stimuli (Genoud and Métraux 1999). The energetic integration of the entire system could be based on the same symplastic organisation, so that proton translocation (Mitchell 1976) and concomitant ion movements would give rise to a circadian rhythm in electric potential parallelled by circadian LMs and SERs (Aimi and Shibasaki 1975; Wagner et al. 1998).

The circadian rhythm in transcellular current of a single cell, as observed by Novak and Sironval (1976) in *Acetabularia*, probably arises from the compartmental feedback among mitochondria, chloroplasts and glycolysis, as suggested in our concept for a mechanism of circadian rhythmicity. The vacuole could be involved in this control net by acting as a reservoir for metabolites as in the case of oscillations in Crassulacean acid metabolism. A similar concept might hold true for photosynthetic active bacteria in general and their temporal organisation of metabolism, but certainly for circadian rhythmic behaviour of the cyanobacterium *Synechococcus* (Huang et al. 1990; Ishiura et al. 1998).

11.6 Hydraulic-Electrochemical Oscillations as Integrators of Cellular and Organismic Activity

The symplast of higher plants is probably not only the network for rapid electrical integration of metabolic activities but also the route for the translocation of sucrose and the transfer of the flowering stimulus. An observation that may have some bearing on the significance of changes in membranes during signal transduction is the detection of alterations in the distribution of the endoplasmic reticulum in cells of SAM of *Chenopodium album* after photoperiodic stimulation (Gifford and Stewart 1965). In spinach the plasma membrane of apical cells is modified during flower induction (Crèvecoeur et al. 1992; Penel et al. 1988). Changes in pH and Ca^{2+} patterning as measured with fluorescent dyes can be observed between the earliest events of photoperiodically inductive conditions in flower initiation in *C. rubrum* (Albrechtová et al. 2001, 2003; Walczysko et al. 2000). In the case of flower induction, the temporal organisation of development at SAM might involve rhythmic symplastic transport of metabolites and the interaction of rhythmic (bioelectric) signals originating in the leaves (Novak and Greppin 1979) leading to a frequency-coded electrochemical communication between leaves and SAM (Wagner et al. 1998).

Communication by surface membrane action or variation potentials in higher plants has been observed for a series of systemic responses (Wagner et al. 1997).Changes in action potentials triggered by light to dark or dark to light transitions can be related to changes in photosynthetic electron transport (Trebacz and Sievers 1998). Observations on phytochrome action in the moss *Physcomitrella patens* are indicative of activation of plasma membrane anion channels (Ermolayeva et al. 1997) leading to membrane depolarisation as a very first step in signal transduction.

Bearing in mind the circadian rhythm in transcellular current reported in *Acetabularia* (Novak and Sironval 1976), it seems possible that a circadian rhythm in proton flow—of action and variation potentials—may be of great significance in the circadian coordination of the whole plant, and the communication between plant organs like the leaves and SAM in photoperiodic flower induction (Elowitz and Leibler 2000; Smith and Morowitz 2004; Wagner et al. 1998).

The display of circadian rhythms in energy charge and reduction charge favour the concept that the many nonlinear oscillators of cell metabolism are coupled such as to evolve the circadian frequency of the system as a whole (Albrechtová et al. 2006, Smith and Morowitz 2004, Wagner et al. 2000).

An interplay between oscillating enzymatic reactions and contractile elements of the structural proteins of the cell have even been used to design a mechano-chemical model of the biological clock (Sørensen and Castillo 1980), and most significant, with protoplasts from *Phaseolus* pulvinar motor cells a circadian rhythm in volume oscillations could be shown (Mayer and Fischer 1994; See also Norman et al. 2007). This 'electrochemical' view of metabolic control could be very relevant in relation to the mechanism of growth and differentiation. Numerous experimental findings show the involvement of stable electric fields in growing and differentiating cells. The experimental evidence indicates the possibility that the chemical reaction network is controllable by electric fields. This could open up a way to visualise the synchronisation of circadian rhythms by electric and magnetic fields and thus could allow for synchronising inputs from so-called subtle geophysical factors (Olcese 1990).

It is very likely that physiological rhythms are based on the same structural and functional principles underlying Mitchell's (1976) chemiosmotic hypothesis of energy transduction. Allosteric enzymes could function as molecular high frequency oscillators as in glycolysis while compartmental feedback between organelles with vectorially organised metabolic reactions in their enclosing membranes could give rise to a circadian rhythm in energy transduction by proton flow through the entire cell. Thus the coupling between compartmented metabolic sequences is possibly achieved through cycles in nucleotide ratios and ionic balances via transport mechanisms and redox shuttles in the different energy-transducing biomembranes, which could act as coupling elements and frequency transformers. The membranes themselves could act as high frequency oscillators (Morré and Morré 1998; Morré et al. 1999). Such high frequency membrane oscillators could be the basis for perception and transduction of high frequency signals from the environment (Novak and Greppin 1979).

The ratios of coupling nucleotides would be relatively temperature independent and the nucleotides themselves could thus, as rate effectors in compartmental feedback, fulfil the requirements for precise temperature-compensated time keeping. Proton flow and concomitant ion movement through the symplast could be the basis of rhythmic electrochemical integration of the whole plant (Albrechtová et al. 2006; Igamberdiev and Kleczkowski 2003; Stelling et al. 2004; Wagner et al. 1997, 2000).

The circadian pacemaker oscillation expresses itself by temperature-compensated phosphorylation/dephosphorylation or reduction/oxidation cycles as the time standard for the control of transcription and translation controlled ~24 h cycles of protein synthesis and turnover (Albrechtová et al. 2006; Elowitz and Leibler 2000; Richly et al. 2003; Tomita et al. 2005).

11.7 Local Hydraulic Signalling: The Shoot Apex in Transition

Photoperiodic flower induction involves a reorganisation of organogenesis at SAM from vegetative phyllotaxis to floral development. Floral transition includes molecular signalling and physiological and physical changes. Each step in molecular signal transduction is influenced by feedback from sequences of events from other pathways. In this way the 'physiological state' of a plant controls signal transduction and vice versa. Very little is still known on relationships between the molecular and physiological level of the control of organogenesis. To analyse the change from the vegetative to the flowering apex, we studied kinetics of molecular and physiological changes at SAM of *C. rubrum* during floral transition in order to elucidate the kinetic relationships between the events.

Electric and hydraulic long-distance signals are anticipated to be involved in flower induction. The perception of a flower inducing dark span by phytochrome possibly leads to a change in the electro-chemical signalling between leaves and SAM to allow flower initiation to occur (Wagner et al. 1998). Turgor-dependent volume changes, stretch-activated membrane channels and correlated changes in membrane potential might be an essential part of the "hardware" for signal transduction at the cellular and organismic levels. The "software" could involve frequency-coded signals at the cellular, the tissue and organismic levels.

Membrane potentials are ubiquitous in all living cells and provide the energy for the active transport of substances. The depolarisation of the cellular membrane potential can generate APs which in higher plants can be propagated over long distances via the phloem and plasmodesmata. Localised strain of the plasma membrane can elicit electric signals via mechanosensitive ion channels. Such signals can stimulate turgor loss and may even activate genes (Banes et al. 1995; Lang and Waldegger 1997). Due to its mechanisms of action, hydraulic and electric signalling is always coupled.

Physical strain at the surface of SAM was previously suggested to play a key role in the patterning of organogenesis (Green 1994). The distribution of local forces is a result of the upward pressure given by the expanding inner cell layers (corpus) to the sheet on the surface (tunica), and from the local extensibility of cell

11 Biosystems Analysis of Plant Development   293

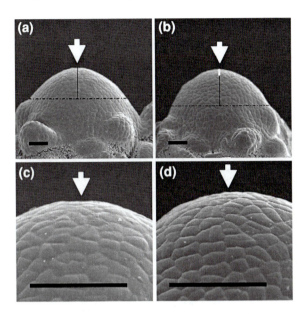

Fig. 11.4 Changes in geometry of SAM during flower inducing treatment. Cryo SEM. **a, c**, control plant, **b, d**, plant at the end of 12 h inductive dark span. The difference is mostly pronounced on the *top* (*arrows*): A subtle depression is visible at SAM of control plant (**a, c**), the SAM of the plant after inductive treatment is well rounded (**b, d**). Both apices have the same diameters of 250 μm (*dotted lines*), but with different heights (full lines, the difference is shown with a *thick white line* in **b**). The ratio height to diameter quantifies the typical difference between both treatments. *Bars* represent 50 μm. From Albrechtová et al. 2004

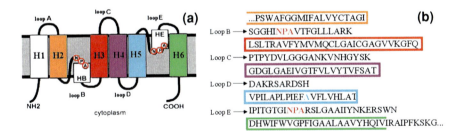

Fig. 11.5 The typical structure of aquaporins (**a**) and the putative protein sequence of a fragment of the aquaporin CrAQP, identified in Chenopodium rubrum (**b**). Helices are indicated by different colours identical in the putative sequence and on the scheme. The typical highly conserved motifs are indicated as red (NPA motif in loops B and E) and blue (alanine in helix H5, typical for plant aquaporins) letters in the sequence. Both C-terminus and N-terminus are missing in the identified fragment

walls at the surface. A signalling network regulates organogenesis, involving molecular, biophysical and biochemical pathways of signal transduction. Based on crosstalk between all pathways, the 'physiological state' of a plant controls signal transduction and vice versa.

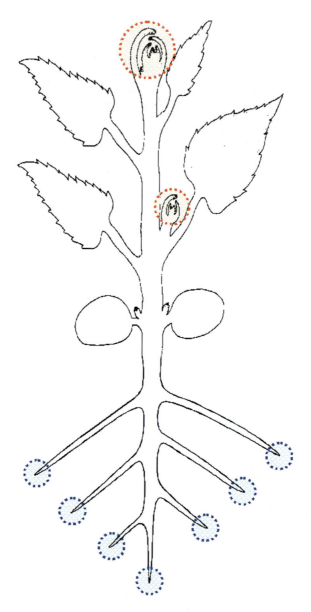

Our studies aim at understanding local water transport and turgor changes as related to changing organogenesis at SAM of *Chenopodium* plants under photoperiodic flower induction. We could show that the size of SAM increases during flower inducing treatment (Fig. 11.4) (Albrechtová et al. 2004). The expansion of the meristem results from cell enlargement rather than from changes in cell division, and therefore is presumably based on water uptake. The phyllotactic

Fig. 11.6 Schematic drawing of a higher plant emphasising elongating meristems as 'centres' for root–shoot coordination. From the data reported so far it is concluded that in daily light–dark cycles the rhythmic activity of the various plant organs [shoot apical meristem(s), leaves, stem (internodes), roots] is synchronised to 24 h with specific phase relationships between organ systems. Hydraulic and electric signals (APs, VPs) and their temporal structure are coordinating the developmental adaptation of the system to endogenous and environmental constraints. APs are predominantly generated at the root and shoot apical meristems. Leaf movements, rhythmic stem elongation (Wagner et al. 1996) and rhythmic root water transport (Lopez et al. 2003) are the main components of the hydraulic system. Changes in turgor are transduced via mechanotransductive ion channels into electric signals (APs). Plant systems can be considered as hydraulic-electrochemical oscillators. They display both circadian and higher frequency oscillations that are obviously taking part in the integration of the plant as a whole in its metabolic and developmental adaptation to the environmental conditions

spiral changes to a circular pattern, visible as local differences in optical properties of cell walls. The results suggest that organogenesis changes long before flower induction is completed. We therefore conclude that the change of organogenesis at SAM during floral transition is initiated by an increased movement of water into SAM leading to its expansion and to the redistribution of the forces at its surface.

A change in water movement could involve aquaporins, which in consequence have been studied in SAM of *C. rubrum*. Aquaporins (AQPs) are highly selective water channels facilitating transport of water across the membrane (Baiges et al. 2002). We identified in *C. rubrum* a gene with high homology to AQPs from other plants, CrAQP (Fig. 11.5). Its expression differs significantly in leaves and in SAM between vegetative- and flower-induced plants (Albrechtová and Wagner 2004). Involvement of AQPs at SAM in flower initiation was proven using application of an inhibitor of aquaporin activity HgCl directly to SAM. HgCl partially inhibited flowering, if applied before or during the dark span.

A comparison of the kinetics of parameters studied revealed that the increase in SAM size is accompanied by an increase in calcium concentration and average pH value at SAM (Albrechtová et al. 2001, 2003; Walczysko et al. 2000). Further studies should reveal whether intracellular pH and calcium concentration can influence water transport by regulating activity of *CrAQP* (Tournaire-Roux et al. 2003). Our further observations confirmed a putative increase of free sucrose at SAM during floral transition, shown in other model plants (Albrechtová and Wagner 2004). The increase in sucrose concentration could lead to an increase in the osmotic pressure in the cells of SAM and thus produce, together with a redistribution of ions, a driving force for water transport.

A central role in the shift of organogenesis during floral transition is thus played by water status, local physical properties of cell walls and distribution of local forces at the surface of the meristem. Altogether, the results support the hypothesis about the involvement of hydraulic signals in organogenesis at SAM. It is anticipated that hydraulic changes at SAM leading to flower initiation are mediated by a specific hydro-electrochemical communication via bidirectional propagation of Aps (Fig. 11.3) between (roots), leaves and SAM.

11.8 Summary and Perspectives: Electrophysiology and Primary Meristems

The hypothesis of a hydraulic electrochemical communication between plant organs prompted studies on the influence of local water transport and turgor changes on organogenesis at SAM of *Chenopodium* plants. Specific changes in shape and size of SAM were found to precede reorganisation of organogenesis under photoperiodic flower induction. Optical properties of cell walls at the surface of SAM were found to precede reorganisation of organogenesis under photoperiodic flower induction. Expression of the aquaporin *CrAQP* increased at SAM during an early phase of flower induction and the application of an inhibitor of aquaporin activity partially inhibited flowering. Changes in ion balance and carbohydrate levels in the cells seem also to be involved in the process. Altogether, the results support a hypothesis on the involvement of hydraulic signals in organogenesis at SAM. It is anticipated that hydraulic changes at SAM leading to flower initiation are mediated by a specific hydro-electrochemical communication among roots, leaves and SAM.

Studies on the uptake of O^{15} labelled water are indicative of a rapid communication between SAM and root meristems (Ohya et al. 2005). Water uptake was strongly reduced after removal of shoot apices, but was not affected, if roots were removed before removal of the apices. The root and shoot primary meristems might be centres of AP generation (Fig. 11.6) as a basis for coordination of physiological processes, such as water uptake and transpiration. The structure and function of both meristems is at present under intensive investigation (Baluska et al. 2004; Bäurle and Laux 2003; Schoof et al. 2000). Their hydro-electrochemical analysis should be most rewarding for an understanding of growth and differentiation.

LMs, rhythmic stem elongation and rhythmic root exudation are components of the hydraulic system. Changes in turgor are transduced via mechanotransductive ion channels into electric signals (APs). Plant systems thus can be considered as hydraulic-electrochemical oscillators displaying both circadian and high frequency oscillations involved in the integration of the plant in its metabolic and developmental adaptation to the environmental conditions.

Further studies will monitor in detail the generation and propagation of APs in parallel to physiological and molecular studies to elucidate the patterns of cooperation between plant organs in response of the plant to seasonal changes in photoperiod as well as biotic and abiotic stress conditions.

The main question of how and where action potentials are generated in plants is still an open one. In plants the main connective element for electric signals are the sieve tubes of phloem vessels and most likely the plasmodesmal bridges between plant cells. The plasmodesmata are dynamic structures which can be changing their position and might be involved in polarity organisation of cells. The function of plasmodesmata as plant "gap junctions" is one of the key questions. The directionality of AP propagation might be involved in the fascilitation of the transmission of chemical signals (Brenner et al. 2006).

References

Adamec L, Krekule J (1989a) Changes in membrane potential in *Chenopodium rubrum* during the course of photoperiodic flower induction. Biol Plant 31:336–343

Adamec L, Krekule J (1989b) Changes in transorgan electric potential in *Chenopodium rubrum* during the course of photoperiodic flower induction. Biol Plant 31:344–353

Adamec L, Machackova I, Krekule J, Novakova M (1989) Electric current inhibits flowering in the short-day plant *Chenopodium rubrum* L. J Plant Physiol 134:43–46

Aimi R, Shibasaki S (1975) Diurnal change in bioelectric potential of *Phaseolus* plant in relation to the leaf movement and light conditions. Plant Cell Physiol 16:1157–1162

Albrechtová JTP, Wagner E (1998) Measurement of membrane potential using a fluorescent probe: 'nerves' in plants? In: Abstract of the 11th international Symposium, workshop on plant membrane biology, Cambridge, p 318

Albrechtová JTP, Wagner E (2004) Mechanisms of changing organogenesis at the apex of *Chenopodium rubrum* during photoperiodic flower induction. Flowering Newslett 38:27–33

Albrechtová JTP, Metzger C, Wagner E (2001) pH-patterning at the shoot apical meristem as related to time of day during different light treatments. Plant Physiol Biochem 39:115–120

Albrechtová JTP, Heilscher S, Leske L, Walczysko P, Wagner E (2003) Calcium and pH patterning at the apical meristem are specifically altered by photoperiodic flower induction in *Chenopodium spp*. Plant Cell Environ 26:1985–1994

Albrechtová JTP, Dueggelin M, Dürrenberger M, Wagner E (2004) Changes in geometry of the apical meristem and concomitant changes in cell wall properties during photoperiodic induction of flowering in *Chenopodium rubrum*. New Phytol 163:263–269

Albrechtová JTP, Vervliet-Scheebaum M, Normann J, Veit J, Wagner E (2006) Metabolic control of transcriptional-translational control loops (TTCL) by circadian oscillation in the redox- and phosporylation state of cells. Biol Rhythm Res 37(4):381–389

Anderson JM, Chow WS, Goodchild DJ (1988) Thylakoid membrane organisation in sun/shade acclimation. Aust J Plant Physiol 15:11–26

Asada K (1999) The water-water cycle in chloroplasts: Scavenging of active oxygens and dissipation of excess photons. Annu Rev Plant Physiol Plant Mol Biol 50:601–639

Baiges I, Schaffner AR, Affenzeller MJ, Mas A (2002) Plant aquaporins. Physiol Plant 115:175–182

Baluska F, Mancuso S, Volkmann D, Barlow PW (2004) Root apices as plant command centres: the unique ''brain-like'' status of the root apex transition zone. Biologia 59(Suppl. 13):7–19

Banes AJ, Tsuzaki M, Yamamoto J, Fischer T, Brigman B, Brown T, Miller L (1995) Mechanoreception at the cellular level: the detection, interpretation, and diversity of responses to mechanical signals. Biochem Cell Biol 73:349–365

Barbier-Brygoo H, Joyard J, Pugin A, Ranjeva R (1997) Intracellular compartmentation and plant cell signalling. Trends in Plant Sci 2:214–222

Bauer CE, Elsen S, Bird TH (1999) Mechanisms for redox control of gene expression. Annu Rev Microbiol 53:495–523

Bäurle I, Laux T (2003) Apical meristems: the plant's fountain of youth. Bio Essays 25:961–970

Bernier G (1988) The control of floral evocation and morphogenesis. Annu Rev Plant Physiol Plant Mol Biol 39:175–219

Brenner ED, Stahlberg R, Mancuso S, Vivanco J, Baluska F, Van Volkenburgh E (2006) Plant Neurobiology: An integrated view of plant signalling. Trends Plant Sci 11:412–419

Bünning E (1942) Untersuchungen über den physiologischen Mechanismus der endogenen Tagesrhythmik bei Pflanzen. Zeitschrift für Botanik 37:433–486

Bünning E (1973) The physiological clock. Springer, Berlin

Bünning E (1977) Die physiologische Uhr. Springer, Berlin

Crèvecoeur M, Crespi P, Lefort F, Greppin H (1992) Sterols and plasmalemma modification in spinach apex during transition to flowering. J Plant Physiol 139:595–599

Cumming BG (1959) Extreme sensitivity of germination and photoperiodic reaction in the genus *Chenopodium* (Tourn.) L. Nature 184:1044–1045

Cumming BG (1967) Early-flowering plants. In: Wilt F, Wessels N (eds) Methods in developmental biology. Thomas Y. Crowell Co., New York, pp 277–299

Cumming BG, Wagner E (1968) Rhythmic processes in plants. Annu Rev Plant Physiol 19:381–416

Cumming BG, Hendricks SB, Borthwick HA (1965) Rhythmic flowering responses and phytochrome changes in a selection of Chenopodium rubrum. Can J Bot 43:825–853

Davies E (1987) Action potentials as multifunctional signals in plants: a unifying hypothesis to explain apparently disparate wound responses. Plant Cell Environ 10:623–631

Davies E, Zawadzki T, Witter JD (1991) Electrical activity and signal transmission in plants: how do plants know? In: Penel C, Greppin H (eds) Plant signalling plasma membrane and change of state. University of Geneva, Geneva, pp 119–137

Dolmetsch RE, Xu K, Lewis RS (1998) Calcium oscillations increase the efficiency and specificity of gene expression. Nature 392:933–936

Elowitz MB, Leibler S (2000) A synthetic oscillatory network of transcriptional regulators. Nature 403:335–338

Ermolayeva E, Sanders D, Johannes E (1997) Ionic mechanism and role of phytochrome-mediated membrane depolarisation in caulonemal side branch initial formation in the moss *Physcomitrella patens*. Planta 201:109–118

Flügge U-I (2000) Transport in and out of plastids: does the outer envelope membrane control the flow? TIPS 5:135–137

Fukshansky L (1981) A quantitative study of timing in plant photoperiodism. Journal of Theoeretical Biology. 63–91

Genoud T, Métraux J-P (1999) Crosstalk in plant cell signaling: structure and function of the genetic network. TIPS 4:503–507

Gifford EM, Stewart KD (1965) Ultrastructure of vegetative and reproductive apices of *Chenopodium album*. Science 149:75–77

Gooch D, Packer L (1974) Oscillatory states of mitochondria. Studies on the oscillatory mechanism of liver and heart mitochondria. Arch Biochem Biophys 163:759–768

Green PB (1994) Connecting gene and hormone action to form, pattern and organogenesis: biophysical transductions. J Exp Bot 45:1775–1788

Greppin H, Horwitz B (1975) Floral induction and the effect of red and far-red preillumination on the light-stimulated bioelectric response of spinach leaves. Z Pflanzenphysiol 75:243–249

Greppin H, Horwitz BA, Horwitz LP (1973) Light-stimulated bioelectric response in spinach leaves and photoperiodic induction. Z Pflanzenphysiol 68:336–345

Gylkhandanyan AV, Evtodienko YV, Zhabotinsky AM, Kondrashova MN (1976) Continuous Sr^{2+}-induced oscillations of the ionic fluxes in mitochondria. FEBS Lett 66:44–47

Hendricks SB (1963) Metabolic control of timing. Science 141:1–7

Herde O, Fuss H, Pena-Cortés H, Fisahn J (1995) Proteinase inhibitor II gene expression induced by electrical stimulation and control of photosynthetic activity in tomato plants. Plant Cell Physiol 36:737–742

Herde O, Pena-Cortés H, Willmitzer L, Fisahn J (1998a) Time-resolved analysis of signals involved in systemic induction of Pin2 gene expression. Bot Acta 111:383–389

Herde O, Pena-Cortés H, Willmitzer L, Fisahn J (1998b) Remote stimulation by heat induces characteristic membrane-potential responses in the veins of wild-type and abscisic acid-deficient tomato plants. Planta 206:146–153

Huang TC, Tu J, Chow TJ, Chen TH (1990) Circadian rhythm of the prokaryote *Synechococcus* Sp. RF-1. Plant Physiol 92:531–533

Igamberdiev AU, Kleczkowski LA (2003) Membrane potential, adenylate levels and Mg^{2+} are interconnected via adenylate kinase equilibrium in plant cells. Biochim Biophys Acta 1607:111–119

Ishiura M, Kutsuna S, Aoki S, Iwasaki H, Andersson CR, Tanabe A, Golden SS, Johnson CH, Kondo T (1998) Expression of a gene cluster *kaiABC* as a circadian feedback process in cyanobacteria. Science 281:1519–1523

Jang J-C, Sheen J (1994) Sugar sensing in higher plants. Plant Cell 6:1665–1679

Joshi HM, Tabita FR (1996) A global two component signal transduction system that integrates the control pf photosynthesis, carbon dioxide assimilation, and nitrogen fixation. Proc Natl Acad Sci USA 93:14515–14520

King RW (1975) Multiple circadian rhythms regulate photoperiodic flowering responses in *Chenopodium rubrum*. Can J Bot 53:2631–2638

Könitz W (1965) Elektronenmikroskopische untersuchungen an *Euglena gracilis* im tagesperiodischen Licht-Dunkel-Wechsel. Planta 66:345–373

Lang F, Waldegger S (1997) Regulating cell volume. Am Sci 85:456–463

Lehner L (2003) Elektrophysiologische Untersuchungen zur Steuerung der Blütenbildung bei Kurz- und Langtagpflanzen. PhD Thesis, University of Freiburg

Li W, Llopis J, Whitney M, Zlokarnik G, Tsien RY (1998) Cell-permeant caged InsP$_3$ ester shows that Ca^{2+} spike frequency can optimize gene expression. Nature 392:936–940

Liarzi O, Epel BL (2005) Development of a quantitative tool for measuring changes in the coefficient of conductivity of plasmodesmata induced by developmental, biotic and abiotic signals. Protoplasma 225:67–76

Lopez F, Bousser A, Sissoëff I, Gaspar M, Lachaise B, Hoarau J, Mahé A (2003) Diurnal regulation of water transport and aquaporin gene expression in maize roots: contribution of PIP2 proteins. Plant Cell Physiol 44:1384–1395

Machackova I, Krekule J (1991) The interaction of direct electric current with endogenous rhythms of flowering in *Chenopodium rubrum*. J Plant Physiol 138:365–369

Machackova I, Pospiskova M, Krekule J (1990) Further studies on the inhibitory action of direct electric current on flowering in the short-day plant *Chenopodium rubrum* L. J Plant Physiol 136:381–384

Mayer W-E, Fischer C (1994) Protoplasts from *Phaseolus occineus* L. pulvinar motor cells show circadian volume oscillations. Chronobiol Int 11:156–164

Mills JW, Lazaro J, Mandel J (1994) Cytoskeletal regulation of membrane transport events. FASEB J 8:1161–1165

Mitchell P (1976) Vectorial chemistry and the molecular mechanics of chemiosmotic coupling: power transmission by proticity. Biochem Soc Trans 4:399–430

Montavon M, Greppin H (1983) Effet sur le développement de l'épinard de l'application d'un potentiel électrique sur le pétiole d'une feuille. Saussurea (Genève) 14:79–85

Montavon M, Greppin H (1986) Développement apical de l'épinard et application d'un potentiel électrique de contrainte. Saussurea (Geneve) 17:85–91

Morré JD, Morré DM (1998) NADH oxidase activity of soybean plasma membranes oscillates with a temperature compensated period of 24 min. Plant J 16:277–284

Morré JD, Morré DM, Penel C, Greppin H (1999) NADH oxidase periodicity of spinach leaves synchronized by light. Int J Plant Sci 160:855–860

Murakami S, Packer L (1970) Light-induced changes in the conformation and configuration of the thylakoid membrane of *Ulva* and *Porphyra* chloroplasts in vivo. Plant Physiol 45:289–299

Nilsson O, Lee I, Blázquez MA, Weigel D (1998) Flowering-time genes modulate the response to *LEAFY* activity. Genetics 149:403–410

Normann J, Vervliet-Scheebaum M, Albrechtova JTP, Wagner E (2007) Rhythmic stem extension grows and leaf movements as markers of plant behaviour: the integral output from endogenous and environmental signals. In: Mancuso S, Shabala S (eds) Rhythms in Plants. Springer, Phenomenology, mechanisms and adaptive significance, pp 199–217

Notaguchi M, Abe M et al (2008) Long-Distance, graft-transmissible action of arabidopsis FLOWERING LOCUS T protein to promote flowering. Oxford J, Life Sci, Plant Cell Physiol 49(11):1645–1658

Novak B, Greppin H (1979) High-frequency oscillations and circadian rhythm of the membrane potential of *Spinach* leaves. Planta 144:235–240

Novak B, Sironval C (1976) Circadian rhythms of the transcellular current in regenerating enucleated posterior stalk segments of *Acetabularia mediterranea*. Plant Sci Lett 6:273–283

Ohya T, Hayashi Y, Tanoi K, Rai H, Suzuki K, Albrechtova JTP, Nakanishi TM, Wagner E (2005) Root-shoot-signalling in *Chenopodium rubrum* L. as studied by ^{15}O labeled water uptake. In: Abstract of the 17th international symposium on botanical congress, Vienna, 17–23 July 2005, p 313

Olcese JM (1990) The neurobiology of magnetic field detection in rodents. Prog Neurobiol 35:325–330

Penel C, Auderset G, Bernardini N, Castillo FJ, Greppin H, Morré J (1988) Compositional changes associated with plasma membrane thickening during floral induction in spinach. Physiol Plant 73:134–146

Pickard BG (1994) Contemplating the plasmalemmal control center model. Protoplasma 182:1–9

Richly E, Dietzmann A, Biehl A, Kurth J, Laloi C, Apel K, Salamini F, Leister D (2003) Covariations in the nuclear chloroplast transcriptome reveal a regulatory master-switch. EMBO Rep 4:491–498

Schoof H, Lenhard M, Haecker A, Mayer KFX, Jürgens G, Laux T (2000) The stem cell population of *Arabidopsis* shoot meristems is maintained by a regulatory loop between the *CLAVATA* and *WUSCHEL* genes. Cell 100:635–644

Sippola K, Aro E-M (2000) Expression of *psbA* genes is regulated at multiple levels in the cyanobacterium *Synechococcus* sp. PCC 7942. Photochem Photobiol 71:706–714

Smith E, Morowitz HJ (2004) Universality in intermediary metabolism. PNAS 101:13168–13173

Sørensen TS, Castillo JL (1980) Spherical drop of cytoplasm with an effective surface tension influenced by oscillating enzymatic reactions. J Colloid Interface Sci 76:399–417

Stankovic B, Davies E (1996) Both action potentials and variation potentials induce proteinase inhibitor gene expression in tomato. FEBS Lett 390:275–279

Stelling J, Gilles ED, Doyle FJ III (2004) Robustness properties of circadian clock architectures. PNAS 101:13210–13215

Tomita J, Nakajima M, Kondo T, Iwasaki H (2005) No transcription-translation feedback in circadian rhythm of KaiC phosphorylation. Science 307:251–254

Tournaire-Roux C, Sutka M, Javot H, Gout E, Gerbeau P, Luu DT, Bligny R, Maurel C (2003) Cytosolic pH regulates root water transport during anoxic stress through gating of aquaporins. Nature 425:393–397

Trebacz K, Sievers A (1998) Action potentials evoked by light in traps of *Dionaea muscipula* Ellis. Plant Cell Physiol 39:369–372

Tsuschiya T, Ishiguri Y (1981) Role of the quality of light in the photoperiodic flowering response in four latitudinal ecotypes of *Chenopodium rubrum* L. Plant Cell Physiol 22:525–532

Veit J, Wagner E, Albrechtová JTP (2004) Isolation of a *FLORICAULA/LEAFY* putative orthologue from *Chenopodium rubrum* and its expression during photoperiodic flower induction. Plant Physiol Biochem 42:573–578

Vigh L, Maresca B, Harwood JL (1998) Does the membrane's physical state control the expression of heat shock and other genes? TIBS 23:369–374

Wagner E (1976a) Endogenous rhythmicity in energy metabolism: Basis for timer-photoreceptor-interactions in photoperiodic control. In: Hastings JW, Schweiger HG (eds) Dahlem konferenzen. Aabkon Verlagsgesellschaft, Berlin, pp 215–238

Wagner E (1976b) Kinetics in metabolic control of time measurement in photoperiodism. J Interdiscipl Cycle Res 7:313–332

Wagner E (1976c) The nature of photoperiodic time measurement: energy transduction and phytochrome action in seedlings of *Chenopodium rubrum*. In: Smith H (ed) Light and plant development. In: Proceedings of the 22nd Nottingham Easter school in agricultural sciences, Butterworth, London

Wagner E (1977) Molecular basis of physiological rhythms. In: Jennings DH (ed) Integration of activity in the higher plant, Society for Experimental Biology, Symposium 31, Cambridge University Press, Cambridge, pp 33–72

Wagner E, Cumming BG (1970) Betacyanine accumulation, chlorophyll content and flower initiation in *Chenopodium rubrum* as related to endogenous rhythmicity and phytochrome action. Can J Bot 48:1–18

Wagner E, Frosch S, Deitzer GF (1974a) Membrane oscillator hypothesis of photoperiodic control. In: De Greef JA, (ed) Proceedings of the annual European symposium on plant photomorphogenesis, Campus of the State University Centre, Antwerpen, pp 15–19

Wagner E, Frosch S, Deitzer GF (1974b) Metabolic control of photoperiodic time measurements. J Interdiscipl Cycle Res 5:240–246

Wagner E, Deitzer GF, Fischer S, Frosch S, Kempf O, Stroebele L (1975) Endogenous oscillations in pathways of energy transduction as related to circadian rhythmicity and photoperiodic control. BioSystems 7:68–76

Wagner E, Haertle U, Kossmann I, Frosch S (1983) Metabolic and developmental adaption of eukaryotic cells as related to endogenous and exogenous control of translocators between subcellular compartments. In: Schenk H, Schwemmler W (eds) Endocytobiology II. Walter de Gruyter & Co., Berlin, pp 341–351

Wagner E, Bonzon M, Normann J, Albrechtová JTP, Greppin H (1996) Signal transduction and metabolic control of timing in photoperiodism. In: Greppin H, Degli Agosti R, Bonzon M (eds) Vistas on biorhythmicity. Geneva University Press, Geneva, pp 3–23

Wagner E, Normann J, Albrechtová JTP, Bonzon M, Greppin H (1997) Photoperiodic control of flowering: electrochemical-hydraulic communication between plant organs—"florigen" a frequency-coded electric signal? In: Greppin H, Penel C, Simon P (eds) Travelling Shot on Plant Development. University of Geneva, Geneva, pp 165–181

Wagner E, Normann J, Albrechtová JTP, Walczysko P, Bonzon M, Greppin H (1998) Electrochemical-hydraulic signalling in photoperiodic control of flowering: Is „florigen" a frequency-coded electric signal? Flowering Newslett 26:62–74

Wagner E, Albrechtová JTP, Normann J, Greppin H (2000) Redox state and phosphorylation potential as macroparameters in rhythmic control of metabolism—a molecular basis for seasonal adaptation of development. In: Vanden Driessche Th et al. (eds) The redox state and circadian rhythms, Kluwer Academic Press, Dordrecht

Wagner E, Lehner L, Normann J, Albrechtová JTP (2004) Electrogenic flower initiation—perspectives for whole plant physiology and for applications in horticulture, agriculture and silviculture. Flowering Newslett 38:3–9

Wagner E, Lehner L, Normann J, Veit J, Albrechtová J (2006a) Hydro-electrochemical integration of the higher plant—basis for electrogenic flower induction. In: Baluska F (ed) Communication in plants. Neuronal aspects of plant life. Springer, Berlin, pp 369–389

Wagner E, Lehner L, Veit J, Normann J, Vervliet-Scheebaum M, Albrechtová JTP (2006b) Control of plant development by hydro-electrochemical signal transduction: a means for understanding photoperiodic flower induction. In: Volkov AG (ed) Plant electrophysiology. Theory and methods. Springer, Berlin, pp 483–501

Walczysko P, Wagner E, Albrechtová JTP (2000) Application of co-loaded Fluo-3 and Fura Red fluorescent indicators for studying the spatial Ca^{2+} distribution in living plant tissue. Cell Calcium 28:23–32

Wildon DC, Doherty HM, Eagles G, Bowles DJ, Thain JF (1989) Systemic responses arising from localized heat stimuli in tomato plants. Ann Bot 64:691–695

Wildon DC, Thain JF, Minchin PEH, Gubb IR, Reilly AJ, Skipper YD, Doherty HM, O'Donnell PJ, Bowles DJ (1992) Electrical signalling and systemic proteinase inhibitor induction in the wounded plant. Nature 360:62–65

Wright KM, Oparka KJ (1997) Metabolic inhibitors induce symplastic movement of solutes from the transport phloem of *Arabidopsis* roots. J Exp Bot 48:1807–1814

Zawadzki T, Davies E, Dziubinska H, Trebacz K (1991) Characteristics of action potentials in *Helianthus annuus*. Physiol Plant 83:601–604

Zeilstra-Ryalls J, Gomelsky M, Eraso JM, Yeliseev A, O'Gara J, Kaplan S (1998) Control of photosystem formation in *Rhodobacter sphaeroides*. J Bacteriol 180:2801–2809

Chapter 12
Actin, Myosin VIII and ABP1 as Central Organizers of Auxin-Secreting Synapses

František Baluška

Abstract In the root apex transition zone, large portion of the polar auxin transport (PAT) is accomplished via endocytic vesicular recycling at F-actin and myosin VIII-enriched cell–cell adhesion domains which are characterized as plant synapses. In these cells, PINs act as vesicular transporters that enrich recycling vesicles and endosomes with auxin, which is then secreted out of cells in a neurotransmitter-like mode. Besides F-actin and myosin VIII, auxin receptor auxin binding protein 1 (ABP1) emerges as critical organizing molecule not only for the plant synapses but also for the whole transition zone. Synaptic auxin transport in root apices is directly linked for sensing environment, and also central for translating these perceptions, via sensory-motoric circuits, into adaptive root tropisms. Finally, PINs acting also as vesicular transportes are suggested to represent transceptors, and the synaptic activity is proposed act as flux sensor for the polar transport of auxin

12.1 Secretion of Auxin at Plant Synapses in Cells of Transition Zone

In the contemporary literature, auxin efflux carriers are considered for the plasma membrane transporters of auxin (Blilou et al. 2005; Wisniewska et al. 2006; Michniewicz et al. 2007; Merks et al. 2007; Kleine-Vehn and Friml 2008). However, in the root apex transition zone, large portion of the polar transcellular

F. Baluška (✉)
IZMB, University of Bonn, Kirschallee 1, 53115 Bonn, Germany
e-mail: baluska@uni-bonn.de

auxin transport is accomplished via endocytic vesicular recycling (Schlicht et al. 2006; Mancuso et al. 2007; Baluška et al. 2008). Although intracellular PINs localized in endocytic vesicles and endosomes have the right polarity to act as vesicular transporters in all plant cells (Merks et al. 2007; Kleine-Vehn and Friml 2008), this scenario was not rigidly tested in recent studies, and PINs are considered to act as auxin transporters only at the plasma membrane. Nevertheless, there are older papers reporting that vesicles isolated from plant cells, both from shoots and roots, are able to accomplish active auxin uptake via presumptive vesicular transporters (Hertel et al. 1983; Lomax et al. 1985; Heyn et al. 1987). Moreover, at least in cells of the root apex transition zone (and maybe also in other plant cells), PINs act also (or predominantly) as vesicular transporters (Schlicht et al. 2006; Mancuso et al. 2007; Baluška et al. 2008) which load endocytic recycling vesicles with auxin which is then secreted out of presynaptic cells via endocytic recyling (Baluška et al. 2003b, 2005, 2009a, b). The vesicular *versus* plasma membrane transporter dilemma of PIN proteins might be solved also via the "transceptor" concept when several amino acid transporters act as both transporter and receptors, depending on their subcellular localization and biological context (Hundal and Taylor 2009; Thevelein and Voordeckers 2009; Gojon et al. 2011; Kriel et al. 2011). This scenario is supported by the fact that many receptor proteins are evolutionarily derived from transporter proteins (Thevelein and Voordeckers 2009) and there are data which indicate that PINs act both as transporters and receptors in plants (Zhao et al. 2002; Hössel et al. 2005).

12.2 Secretion of Auxin is Linked to Polar Transport of Calcium

The currently dominant concept of the polar auxin transport (PAT) accomplished solely via the plasma membrane-associated auxin transporters is not compatible also with some older data reporting that there is polar transport of calcium which is closely interlinked with the polar transport of auxin (Evans 1964; Goswami and Audus 1976; dela Fuente 1984; deGuzman and dela Fuente 1984; Lee et al. 1983, 1984; Lee and Evans 1985). Moreover, there are additional features of the PAT which also implicate the vesicular secretion of auxin, rather than simple diffusion via the cytoplasm which is then followed by the export via plasma membrane carriers (Box 1). Importantly, the polar transport of auxin is strictly oxygen dependent (Goldsmith 1967a, b), strongly implying that the polar transport is not only energetically demanding active process but that it also requires continous supply of energy via respiration. Interestingly in this respect, peaks of auxin fluxes in the root apex transition zone are associated with the peak of oxygen influx (Mancuso et al. 2005, 2007; McLamore et al. 2010a, b) and the auxin and oxygen peaks are interdependent (McLamore et al. 2010a; Mugnai et al. 2012).

Box 1: Features of the PAT Implicating its Synaptic Secretory Mode

a. Cell–cell adhesion polarity domains (but not actively growing lateral domains) in root apices are enriched with F-actin.
b. Cell–cell adhesion polarity domains (but not actively growing lateral domains) in root apices are very active in endocytosis and endocytic vesicle recycling.
c. Auxin efflux transporter of the PIN family accumulate within cell–cell adhesion polarity domains via the polarized endocytic vesicle recycling.
d. Not only PINs and several other plasma membrane proteins, but also cell wall pectins are recycling at these cell–cell adhesion polarity domains of root apices.
e. Polarity of PINs requires intact F-actin, endocytic vesicle recycling, and cell–cell adhesion.
f. PAT is tightly linked with the active endocytic vesicle recycling of PINs.
g. PINs have the right polarity to transport auxin into endosomes and recycling vesicles and our auxin immunolocalization studies reported that auxin is enriched within these cell–cell adhesion domains.
h. Importantly, no study showed that the PINs are NOT active in their auxin transport when inserted into endosomes or vesicles. Moreover, vesicles isolated from plant cellsare able to accomplish active auxin uptake via presumptive vesicular transporters.
i. Cells in the root apices show highest vesicle recycling (synaptic activity) in the transition zone which is the most active one also with respect of the PAT and electric spiking activity.
j. Plant synaptotagmins localize to cell–cell adhesion domains in root apices. These plant-specific synaptotagmins were also implicated in plant endocytosis and vesicle trafficking.
k. Neurotransmitters glutamate and serotonin control root growth and root system architecture Plant glutamate receptors are relevant for this root control and glutamate has recently been shown to be released into the apoplast via vesicular transport.
l. Immunological synapses have been reported also for plants (Kwon et al. 2008). This again strongly supports general usefulness of the synaptic concept in plants.
m. Finally, also our original concept of the symbiotic plant synapse (Baluška et al. 2005a, b) was highlighted recently by the group of Jose Feijo (see the Fig. 2 in Lima et al. 2009).

12.3 Plant Synapses are Organized by F-actin, Endocytosis, and Endocytic Vesicular Recycling

Cells of plant organs, especially of root apices, are aligned into regular cell files due to tight cell–cell adhesions at their end poles (Baluška et al. 2003a). In the last decade, these cell–cell adhesion domains have been characterized as plant synaptic domains specialized for endocytois and vesicle recycling based cell–cell transport of auxin (Baluška et al. 2003b, c; Baluška et al. 2005a, 2009a; Barlow et al. 2004; Wojtaszek et al. 2004). These domains are characterized by dynamic fluid cell walls, when already 2 h of brefeldin A exposure results in internalization of almost all pectins and xyloglucans from cell walls into the endocytic BFA-induced compartments (Baluška et al. 2002, 2005b; Hause et al. 2006; Dhonukshe et al. 2006). In contrast to other cell walls, especially longitudinal cell walls, synaptic domains maintain their juvenile nature based predominantly on high amount of pectins/callose and low amount of cellulose (Volkmann and Baluška 2006). This allows to keep their fluid synaptic nature necessary for effective cell–cell transport of auxin. Transported auxin and recycling cell wall components localize to the same recycling vesicles (Šamaj et al. 2004, 2005), allowing tight integration of sensory perceptions and the PAT with the plant specific synaptic plasticity (for root phototropism see Wan et al. 2012). Besides their enrichment with F-actin and myosin VIII (Baluška et al. 1997, 2000), one of the characteristic features of plant synapses is that their composition mimicks that of endocytic BFA-induced compartments (Šamaj et al. 2004, 2005). In other words, trans-Golgi network (TGN) and early endosomes (EE), which get trapped within BFA-induced compartments (Baluška et al. 2002, 2005b; Dettmer et al. 2006; Hause et al. 2006; Dhonukshe et al. 2006; Šamaj et al. 2004, 2005) during their recycling (due to the block in the secretory branch of endocytic vesicular recycling), represent primary building blocks of plant synapses. Importantly in this respect, TGN/EE compartments serve as the most important building blocks also for the cytokinetic cell plates (Baluška et al. 2002, 2005b; Dettmer et al. 2006; Hause et al. 2006; Dhonukshe et al. 2006; Chow et al. 2008; Lam et al. 2008; Zhang et al. 2011b) which represent the structural precursor of plant synapse (Hause et al. 2006; Dhonukshe et al. 2006). After cytokinetic cell plate matures into young cell wall, the synaptic nature of this wall depends on the amount of cell–cell transport of auxin across this cell wall which keeps the cell wall in juvenile fluid state. The more active PAT is accomplished, based on synaptic vesicle recycling (Baluška et al. 2003c, 2005b; Schlicht et al. 2006; Mancuso et al. 2007; Baluška et al. 2008), the more abundant is the synaptic F-actin, and the more prominent is also the fluid nature of synaptic cell walls (Baluška et al. 2003b, 2005b; Wojtaszek et al. 2004). Root tissues with high PAT, such as cells of endodermis, pericycle, epidermis, and outer stele have also much more abundant F-actin, both at the synapses, as well as longitudinal bundles of F-actin connecting the opposite synaptic poles of these cells (Volkmann and Baluška 1999; Baluška et al. 1997, 2000; Baluška and Hlavacka 2005; Kasprowicz et al. 2009).

12.4 Myosin VIII as Endocytic Plant Myosin

Plants are unique with respect of their myosins, as they have only two plant-specific classes XI and VIII, but lack any myosins from general (present in all other eukaryotes) classes I and II. In Arabidopsis, there are four myosins of the class VIII and similar myosins have be found in all analyzed plant genomes (Reddy and Day 2001; Thompson and Langford 2002; Bezanilla et al. 2003; Peremyslov et al. 2011). In contrast to myosins of the class XI (evolutionarily similar to myosins of the class V), that are localized to diverse organelles and underly their motilities (Peremyslov et al. 2011), myosins of the class VIII are localized also to the plasma membrane and represent the only plant myosins relevant for plant endocytosis (Baluška et al. 2003b, 2004; Volkmann et al. 2003; Wojtaszek et al. 2007; Golomb et al. 2008; Sattarzadeh et al. 2008), as well as for the architecture and gating of plasmodesmata (Baluška et al. 2001, 2003b, 2004; Volkmann et al. 2003; Wojtaszek et al. 2007; White and Barton 2011). As plasmodesmata emerge to be organized via synaptic principle of membrane adhesion domains (Baluška et al. 2001, 2003b; Mongrand et al. 2010; Maule et al. 2011; Tilsner et al. 2011), and are assembled/maintained via polarized endocytosis and endocytic vesicle recycling (Baluška et al. 2005b), the roles of myosin VIII in plasmodesmata and synapses are very similar (Baluška et al. 2003b, 2005b, 2009a; Barlow et al. 2004; Wojtaszek et al. 2004, 2007).

12.5 PIN Polarity is Dependent on Plasma Membrane: Cell Wall and Cell-to-Cell Adhesions

Synaptic nature of PIN polarity is obvious from its dependence on both the cell–cell adhesion, as well as the plasma membrane—cell wall adhesion (Baluška et al. 2003b; Boutté et al. 2006; Wojtaszek et al. 2007; Pinosa 2010; Feraru et al. 2011). The plasma membrane—cell wall adhesions are based on the actin cytoskeleton and myosin VIII (Baluška et al. 2000, 2001, 2003b, 2004; Volkmann et al. 2003; White and Barton 2011). Central role in synaptic cell–cell adhesion is played by pectins recycling via endocytosis and endocytic vesicle recycling through TGN/EE compartments representing specific synaptic organelle of synaptically active plant cells (Baluška et al. 2002, 2005b; Hause et al. 2006; Dhonukshe et al. 2006; Chow et al. 2008; Lam et al. 2008; Zhang et al. 2011b). Synaptic cell–cell adhesion is upstream of endogenous polarity determinants. As soon as, cell–cell adhesion is lost, also the PIN polarity vanishes (Boutté et al. 2006; Pinosa 2010; Feraru et al. 2011). Due to their mechanically weaker cell walls, having fluid nature allowing endocytosis of cell wall pectins, (Baluška et al. 2002, 2003b, 2005b), plant synapses emerge to act as mechanosensitive subcellular domains (Baluška et al. 2003b, 2005b; Baluška and Volkmann 2011). This feature can explain not only why does PAT follow tightly the gravity vector (Baluška et al. 2005b, 2009a; Baluška and Volkmann 2011), but also why the polarized vesicle recycling is positioned below mechanically weakened end poles, imposing on them their synaptic nature.

12.6 Plasmodesmata as Electrical Synapses

Plasmodesmata have several other features which indicate that they act as plant-specific electrical synapses. Although differing significantly from animal electrical synapses generated by connexins, plasmodesmata allow electric coupling between adjacent cells (Spanswick 1972; Drake et al. 1978; Lew 1994). In root apices, the density of plasmodesmata/electrical synapses is known to be much higher at cross walls/chemical synapses than at the side walls (Juniper and Barlow 1969; Seagull 1983; Zhu et al. 1998a, b). Plasmodesmata density gets lower with tissue maturation or cell–cell separation (Zhu et al. 1998a, b; Zhu and Rost 2000). Moreover, plasmodesmata structure and gating are very sensitive to environmental stimuli such as light (Epel and Erlanger 1991) and gravity (Šamaj et al. 2006). Although longitudinal cell wall have much lower plasmodesmata densities, here plasmodesmata are grouped into large pit fields which can be considered for a "mini-synapses," being also focus of active endocytosis and endocytic vesicle recycling (Baluška et al. 2004; Šamaj et al. 2006).

12.7 ABP1 as Auxin Receptor for Electrical Responses

Auxin binding protein 1 (ABP1) is auxin receptor at the plasma membrane with its auxin-binding domain exposed to the extracellular matrix. Auxin binding to the plasma membrane associated ABP1 induces membrane hyperpolarization and often initiates action potentials. These older data obtained mostly with studies performed on isolated protoplasts (Shen et al. 1990; Barbier-Brygoo et al. 1989; Rück et al. 1993; Maurel et al. 1994). Interestingly, these electrical responses to auxin, mediated at the plasma membrane via its receptor APB1, induce also physico-chemical changes to the plasma membrane as evidenced by the loss of fluorescence (Dahlke et al. 2010) by the endocytic tracer synaptoRed reagent (known also as FM4-64) name of which is derived from the fact that it is marker for neuronal synapses and their recycling vesicles. In intact root apices, similarly as in neuronal synapses, FM4-64 accumulates in their synapses as well as in their recycling vesicles (Baluška et al. 2003b, 2005a, 2009a, b, 2010; Schlicht et al. 2006; Mancuso et al. 2007).

12.8 ABP1 as Auxin Receptor for Endocytosis Feeding into Synaptic Organelle TGN/EE

Rather surprisingly, ABP1 has been shown to act as auxin receptor underlying high rates of clathrin-mediated endocytosis (Robert et al. 2010) at plant synapses in roots. This feature seems to be responsible for the permanent character of TGNs in root apex cells with active synapses, as active endocytosis feeds back into the TGN heavily, causing also the early endosome nature of TGN in these root cells (Šamaj et al. 2005; Dettmer et al. 2006; Viotti et al. 2010). In fact, this feature of root cells is characteristic also to animal/human neuronal cells having active synapses (Baluška 2010). Interestingly, root apex cells engaged in secretion, such as secretory root cap cells or elongating root cells, inactivate their synapses visualized through brefeldin A-induced compartments (Baluška et al. 2002, 2005a, 2010) and also lose their TGNs as independent organelles (Kang 2011). These cells are very active in secretion, but less so in endocytosis, so that their TGNs are quickly used up via secretion. This is the reason why secretory root cap cells and rapidly elongating cells do not generate large BFA-induced compartments (Baluška et al. 2002). In the *abp1* mutant lines, there is general inhibition of endocytosis, which blocks feeding from the plasma membrane into TGNs (Robert et al. 2010), and root apex cells of this mutant also use up most of their TGNs. Thus, BFA is not inducing formation of large BFA-induced compartments in root apex cells of *abp1* mutant line. Besides underlying high rates of endocytosis, synaptic ABP1 transmits signals from neurotransmitter auxin released by the adjacent synaptic cell partner and traversing the synaptic cleft. Binding of auxin to ABP1 on the plasma membrane of adjacent cell has two fundamental effects: it inhibits the ABP1-mediated endocytosis (Robert et al. 2010) and it induces rapid (within 30 s) activation of plant Rho GTPases Rop2 (Xu et al. 2010).

12.9 Evolution of Plant Synapses: From ABP1 to Synaptic Endocytosis and Vesicle Recycling

ABP1 is acting as auxin receptor at the plasma membrane, but most of ABP1 is trapped within endoplasmic reticulum (ER) via the KDEL sequence which retrieves most of the ER exported ABP1 from the cis Golgi back to ER (Napier et al. 2002). *Physcomitrella patens* ABP1 lacks this KDEL-based ABP1 retrieval (Panigrahi et al. 2009). This implicates that ABP1 was not ER localized in early plants (Tromas et al. 2010). Auxin-secreting plant synapses evolved together with vascular system and PAT machinery based on PIN proteins (Friml 2003; Paponov et al. 2005; Teale et al. 2006; Krecek et al. 2009) which recycle between the plasma membrane and TGN/EE compartments. This PIN-based transport is essential for plant development, sensory perceptions, as well as for sensory-motor circuits underlying plant tropisms (Baluška et al. 2005b, 2009a, b, 2010).

Polar cell–cell transport of auxin evolved presumably from the ancient auxin detoxification mechanisms which included exporting of auxin out of cells or storing of auxin within ER and vacuoles. This scenario is supported by the role of ABC transporters in auxin cell–cell transport as these transporters are essential for cellular homeostasis of eukaryotic cells (Tarr et al. 2009). At the transition from early land plants to vascular plants, key innovations evolved and all these are closely linked with, and based on, PAT across cell–cell adhesion domains which progressively evolved into plant synapses of recent higher plants. These key innovations allowed formation of continuous vascular elements, integrated into networks, as well as evolution of roots and their sensoric root apices searching for water and plant nutrition. Somewhere during this evolutionary cross road, plant cells evolved plasma membrane PIN transporters from original ER located PINs (Mravec et al. 2009; Krecek et al. 2009), as well as ABP1 equiped with the KDEL peptide (Panigrahi et al. 2009) retrieving it from the early secretory pathway back to ER (Tromas et al. 2010). This allowed only limited and highly regulated access for ABP1 to the plasma membrane. The plasma membrane ABP1 integrated with the clathrin-based endocytosis machinery. In addition, there were further important innovations related to the endocytosis and vesicle recycling via TGN/EE which allowed evolution of plant synapses. Here we can mention ADP-ribosylation factor GTPases (ARFs), as well as their ARF GEFs and ARF GAPs, which are active only within the endomembrane trafficking in eukaryotic cells. In early vascular plants, some of them such as ARF1 (Baluška et al. 2000; Xu and Scheres 2005) and ARF GEF GNOM gained new functions at the plasma membrane related to endocytosis and endocytic vesicle recycling (Xu and Scheres 2005; Teh and Moore 2007; Richter et al. 2007, 2010).

12.10 Evolution of Plant Synapses: Expansion of Synaptic PINs During Plant Evolution

Key evolutionary innovations of vascular plants—the formation of vascular system and true roots was associated with the emergence of PINs localized in recycling vesicles and associated with the plasma membrane (Mravec et al. 2009; Krecek et al. 2009; Tromas et al. 2010). In cells of the transition zone, PINs obviously act as vesicular transporters (Schlicht et al. 2006; Mancuso et al. 2007; Baluška et al. 2008). This new feature allowed synaptic communication via signal-mediated release of auxin into the synaptic space between two adjacent root cells connected via synaptic cell–cell adhesion domain. Besides increasing the number of synaptic PINs, being higher in more evolved monocot species such as rice and Sorghum in comparison with dicot species such as Arabidopsis (Krcek et al. 2009; Wang et al. 2009; Shen et al. 2010), the highest number of synaptic PINs is active in root apices where two inverted fountains of the PAT streams determine both formation and maintenance of the transition zone (Blilou et al. 2005; Baluška et al. 2005b, 2009a, b, 2010). The monocot-specific PINs of the classes 9 and 10 are expressed in roots and emerge to be involved in formation and development of adventitious roots (Wang et al. 2009; Shen et al. 2010). It is important to note that the root complexity is much higher in monocots than in dicots (Hochholdinger and Zimmermann 2008), implicating that roots are further evolving rapidly. It might be that the very tight human—crop plant coevolution, which started some 15 and 10 thousand years ago, and which is so successful with several monocot species, is based on more evolved root synapses of these highly evolved plant species.

12.11 Are Fungal Infections Related to the Opposite (Shootward) Polarity of PIN2?

In root cells with increased auxin, calcium and IP3 levels, the rootward PIN polarity switches to the opposite shootward polarity (Zhang et al. 2011a). Interestingly in this respect, fungal invasions increase auxin level in root apices and also drive lateral root primordial initiation and formation (Splivallo et al. 2009; Felten et al. 2009, 2010). PIN2 is unique among other PINs as it shows rootward polarity in the cortex cells of the meristem, but then it switches into the shootward polarity in the transition zone (Chen and Masson 2005; Rahman et al. 2010), as it is the case also in the all epidermis cells. One possible evolutionary scenario why PIN2 has the opposite cellular polarity is that repeated fungal invasions of roots resulted in increased auxin levels in epidermis and cortex cells, switching the rootward PIN polarity into the PIN2-specific shootward polarity.

12.12 Did ABP1 Activity Result in Formation of the Transition Zone?

Although the original transition zone was elaborated using maize roots, it proved to be valid also for Arabidopsis roots. In Arabidopsis, which is currently in the focus of plant biology, it has been shown that this root apex zone has unique roles in hormonal control of root growth (Blilou et al. 2005; Dello Ioio et al. 2007; Wolters and Jürgens 2009). In fact, the transition zone acts as a zone which integrates several sensory and hormonal systems into adaptive root motoric responses—also known as root tropisms (Baluška et al. 2009a, b, 2010). Cells of this zone are transporting auxin via two inverted fountains, complexity which is based not only of PINs driving mostly axial auxin transport but also on PIN3 and PIN2 which drive lateral transport (Blilou et al. 2005; reviewed in Baluška et al. 2010). Importantly, the auxin transport in root apices is directly linked for sensing environment, and also for translating these perceptions, via sensory-motor circuits, into adaptive root tropisms (Baluška et al. 2003c, 2005b, 2009a, b, 2010). Intriguingly, recent reports revealed that ABP1 activity is important for formation of the transition zone in root apices of Arabidopsis. ABP1 mediates formation of the transition zone, which is lost in roots of *abp*1 mutant (Tromas et al. 2009, 2010). To perform this unique role, ABP1 activity defines zone of competence for PLETHORA transcription factor to control cell fate at the basal border of the apical meristem when cells enter the transition zone (Tromas et al. 2009). Importantly, ABP1 activity is also important for the basipetal (shootward) but not for the acropetal (rootward) PAT (Effendi et al. 2011; Scherer 2011). This highlights the importance of PIN2, which is the only endogenous transporter of the PIN family responsible for the basipetal auxin transport in roots, for both the formation and maintenance of the transition zone (reviewed in Baluška et al. 2010). Finally, ABP1 and PIN2 are essential for the early auxin-up-regulated gene expression (Effendi et al. 2011; Scherer 2011), which is downstream of auxin signaling at the root synapse.

12.13 Plant Synaptic Activity Emerge as Elusive Flux Sensor for the Polar Transport of Auxin

Polar transport of auxin, based on the canalization principle of Tsvi Sachs (Sachs 1969, 1991), requires hypothetical flux sensor which was elusive ever since this model has been proposed (Merks et al. 2007; Stoma et al. 2008). The synaptic nature of transcellular PAT, based on synaptic secretion recycling vesicles having presumably quantal sizes as it is the case of neuronal synaptic vesicles, solves very elegantly this mystery. Another flux sensor can be envisioned at the external leaflet of the plasma membrane where synaptically released auxin activates ABP1. Moreover, if PINs act as receptors for auxin at the plasma membrane and transporters for auxin in endosomes and endocytic vesicles, then this transceptor PIN concept can be further elaborated for purposes of the auxin flux sensor. If the synaptic activity acts as a elusive flux sensor, and PINs are vesicular transportes and transceptors, then recent mathematical and computer models (Kramer 2009; Garnett et al. 2010; Krupinski and Jönsson 2010; Wabnik et al. 2010, 2011; Jönsson et al. 2011) are not taking into account the biological complexity behind the PAT. Emerging candidate for flux sensor like molecule sensing the synaptic activity (rate of endocytic recycling) is Brevix Radix (BRX) protein which recycles via clathrin-mediated endocytosis at plant synapses (Santuari et al. 2011) and performs nuclear translocation mediated via auxin and its polar transport (Scacchi et al. 2010; Santuari et al. 2011). Synaptic activity can be tightly linked with nuclear gene expression activity via synapto-nuclear shuttling of BRX in the dependency of synaptic activity. BRX recycling and endocytosis is high in the transition zone where BRX obviously accomplishes plant synapse—nucleus integration (Scacchi et al. 2009, 2010; Depuydt and Hardtke 2011; Santuari et al. 2011).

BRX action is integrated with BRI1, which is endosomal receptor for brassinosteroids (Geldner et al. 2007). Interestingly, BRI1 has recently been reported to control levels of PIN2 and PIN4 auxin efflux carriers in root apices (Hacham et al. 2011).

12.14 Importance of Active Plant Synapses in the Transition Zone for Tropisms and Organogenesis: From Ionic and Electric Oscillations Towards Gene Expression Oscillations

Synaptic activity-based polar transport of auxin is essential for all kind of plant tropisms as well as for plant organogenesis, both in roots, and shoots. In roots, all tropisms are essentially linked also with the shootward transport of auxin mediated via PIN2 recycling (Chen and Masson 2005; Schlicht et al. 2008; Shen et al. 2008). Inhibitors of PAT, such as NPA and TIBA, as well as general inhibitor of endocytic vesicle recycling BFA, block not only all plant tropisms, but also root and shoot organogenesis. This fact places the auxin secreting plant synapses in the central position of plant organogenesis and behavior (Baluška et al. 2009b; Trewavas 2009; Hodge 2009). The highest synaptic activity has been scored in the transition zone of the root apex which acts as sensory-motoric nexus, representing plant command centers in the sense of the "root-brain" hypothesis of Charles and Francis Darwin (Baluška et al. 2009a, b). Importantly in this respect, the PAT is not only showing clear peak in the transition zone (Mancuso et al. 2005, 2007; McLamore et al. 2010a; Wan et al. 2012), but also show highly synchronized oscillations in its fluxes at this unique root apex zone (Mancuso et al. 2005; McLamore et al. 2010a). Both auxin efflux and influx oscillate but their amplitude peaks are out of phase (McLamore et al. 2010a). In addition to auxin, also other transported molecules, including oxygen, nictric oxide (NO), potassium, glutamate, calcium, H^+ ions show oscillations in their fluxes at the transition zone (Shabala et al. 1997, 2006; Mancuso and Boselli 2002; McLamore et al. 2010b; Mugnai et al. 2012). These oscillatory transport activities result in oscillations of the root apex electric field (Scott 1957; Jenkinson and Scott 1961) and of electric spikes, especially at the transition zone (Masi et al. 2009) which also shows the most prominent electric field from the whole root apex (Collings et al. 1992).

These ionic/electric oscillations are associated with gene expression oscillations in root apices of Arabidopsis (Moreno-Risueno et al. 2010). Cells in the meristem are not showing this feature but when they enter the transition zone, they start to show rhythmic pulsation with a period about 6 h and show this phenomenon along the elongation region. Progression of this oscillatory wave of gene expression seems to require active synapses because it starts in region reaching the highest synaptic activity and it is terminated at the basal limit of elongation zone, where also high synaptic activity ceases (Baluška et al. 2002, 2009a, 2010). As discussed above, BRX is shuttling between active root synapses and nuclei where it controls gene expression (Depuydt and Hardtke 2011; Santuari et al. 2011). As BRX is needed for expression of auxin-responsive genes, this protein might, together with some other proteins shuttling between active synapses and nuclei, integrate synaptic activities with gene expression patterns. This oscillatory network at the root apices represents an endogenous developmental mechanism regulating two processes: bending of root apices and periodic root branching via control of the competence for lateral root formation (Moreno-Risueno et al. 2010; Traas and Vernoux 2010; Van Norman et al. 2011; Moreno-Risueno and Benfey 2011). Intriguingly, there are two large groups of genes oscillate out of phase—one set of 2,084 genes was shown to oscillate in phase with DR5, whereas the second set of 1409 genes oscillated in antiphase to DR5 expression (Moreno-Risueno et al. 2010). Expression patterns of these two sets of genes behave as propagating waves over the longitudinal root axis, closely resembling similar waves of oscillating gene expression along the dorsal spine axis during the vertebrate segmentation (Moreno-Risueno et al. 2010; Traas and Vernoux 2010; Van Norman et al. 2011; Moreno-Risueno and Benfey 2011). This surprising similarity between the plant root clock and animal segmentation clock provides another example of convergent evolution between higher plants, especially their roots, and animals (Baluška 2010; Baluška et al. 2009b; Baluška and Mancuso 2009).

12.15 Conclusion

Polar synaptic auxin transport integrates complex and immobile plant bodies with their ever changing environments, allowing their phenotypic plasticity. The PAT serves as both sensing and executing system allowing effective translation of complex streams of sensory perceptions into motoric and phenotypic adaptive responses. This sensory and cognitive aspect of plant biology is changing our understanding of higher plants. In the root apex transition zone, the large portion of transported auxin is secreted across plant synapses. Here, PIN proteins act as vesicular transporters and might be viewed at the plasma membrane as transceptors, integrating both transporter and receptor features. This would allow, together with the inherently quantal nature of auxin-enriched secretory vesicles, to implement the long sought flux sensor into the PAT system. Fluid nature of synaptic cell walls, when especially pectins recycle together with PIN proteins via the auxin-enriched vesicles, allows tight integration of the cell wall, vesicles, and cytoskeleton at the auxin secreting synapses. Besides auxin, also glutamate may be enriched within vesicles via putative vesicular glutamate transporters. Recently, glutamate has been shown to be released into the apoplast of plant cells via brefeldin A-sensitive vesicular transport, triggering then calcium influx into these cells (Vatsa et al. 2011)

References

Baluška F (2010) Recent surprising similarities between plant cells and neurons. Plant Signal Behav 5:87–89
Baluška F, Hlavacka A (2005) Plant formins come to age: something special about cross-walls. New Phytol 168:499–503
Baluška F, Mancuso S (2009) Plants and animals: convergent evolution in action? In: Baluška F (ed) Plant-environment interactions from sensory plant biology to active plant behavior. Springer Verlag, Berlin, pp 285–301
Baluška F, Volkmann D (2011) Mechanical aspects of gravity-controlled growth, development and morphogenesis. In: Wojtaszek P (ed) Mechanical integration of plant cells and plants. Springer Verlag, Berlin, pp 195–222
Baluška F, Vitha S, Barlow PW, Volkmann D (1997) Rearrangements of F-actin arrays in growing cells of intact maize root apex tissues: a major developmental switch occurs in the postmitotic transition region. Eur J Cell Biol 72:113–121
Baluška F, Barlow PW, Volkmann D (2000) Actin and myosin VIII in developing root cells. In: Staiger CJ, Baluška F, Volkmann D, Barlow PW (eds) Actin–a dynamic framework for multiple plant cell functions. Kluwer Academic Publishers, Dordrecht, pp 457–476
Baluška F, Cvrčková F, Kendrick-Jones J, Volkmann D (2001) Sink plasmodesmata as gateways for phloem unloading. Myosin VIII and calreticulin as molecular determinants of sink strength? Plant Physiol 126:39–46
Baluška F, Hlavačka A, Šamaj J, Palme K, Robinson DG, Matoh T, McCurdy DW, Menzel D, Volkmann D (2002) F-actin-dependent endocytosis of cell wall pectins in meristematic root cells: insights from brefeldin a-induced compartments. Plant Physiol 130:422–431
Baluška F, Wojtaszek P, Volkmann D, Barlow PW (2003a) The architecture of polarized cell growth: the unique status of elongating plant cells. BioEssays 25:569–576
Baluška F, Šamaj J, Wojtaszek P, Volkmann D, Menzel D (2003b) Cytoskeleton–plasma membrane—cell wall continuum in plants: emerging links revisited. Plant Physiol 133: 482–491
Baluška F, Šamaj J, Menzel D (2003c) Polar transport of auxin: carrier-mediated flux across the plasma membrane or neurotransmitter-like secretion? Trends Cell Biol 13:282–285
Baluška F, Šamaj J, Hlavačka A, Kendrick-Jones J, Volkmann D (2004) Myosin VIII and F-actin enriched plasmodesmata in maize root inner cortex cells accomplish fluid-phase endocytosis via an actomyosin-dependent process. J Exp Bot 55:463–473
Baluška F, Liners F, Hlavačka A, Schlicht M, Van Cutsem P, McCurdy D, Menzel D (2005a) Cell wall pectins and xyloglucans are internalized into dividing root cells and accumulate within cell plates during cytokinesis. Protoplasma 225:141–155
Baluška F, Volkmann D, Menzel D (2005b) Plant synapses: actin-based adhesion domains for cell-to-cell communication. Trends Plant Sci 10:106–111
Baluška F, Schlicht M, Volkmann D, Mancuso S (2008) Vesicular secretion of auxin: evidences and implications. Plant Signal Behav 3:254–256
Baluška F, Schlicht M, Wan Y-L, Burbach C, Volkmann D (2009a) Intracellular domains and polarity in root apices: from synaptic domains to plant neurobiology. Nova Acta Leopold 96:103–122
Baluška F, Mancuso S, Volkmann D, Barlow PW (2009b) The 'root-brain' hypothesis of Charles and Francis Darwin: revival after more than 125 years. Plant Signal Behav 4(1121):1127
Baluška F, Mancuso S, Volkmann D, Barlow PW (2010) Root apex transition zone: a signalling-response nexus in the root. Trends Plant Sci 15:402–408
Barbier-Brygoo H, Ephritikhine G, Klämbt D, Ghislain M, Guern J (1989) Functional evidence for an auxin receptor at the plasmalemma of tobacco mesophyll protoplasts. Proc Natl Acad Sci U S A 86:891–895
Barlow PW, Volkmann D, Baluška F (2004) Polarity in roots. In: Lindsey K (ed) Polarity in plants. Blackwell Publishing, pp 192–241

Bezanilla M, Horton AC, Sevener HC, Quatrano RS (2003) Phylogenetic analysis of new plant myosin sequences. J Mol Evol 57:229–239

Blilou I, Xu J, Wildwater M, Willemsen V, Paponov I, Friml J, Heidstra R, Aida M, Palme K, Scheres B (2005) The PIN auxin efflux facilitator network controls growth and patterning in Arabidopsis roots. Nature 433:39–44

Boutté Y, Crosnier MT, Carraro N, Traas J, Satiat-Jeunemaitre B (2006) The plasma membrane recycling pathway and cell polarity in plants: studies on PIN proteins. J Cell Sci 119: 1255–1265

Chen R, Masson PH (2005) Auxin transport and recycling of PIN proteins in plants. In: Samaj J, Baluska F, Menzel D (eds) Plant endocytosis. Springer Verlag, pp 139–157

Chow CM, Neto H, Foucart C, Moore I (2008) Rab-A2 and Rab-A3 GTPases define a trans-golgi endosomal membrane domain in Arabidopsis that contributes substantially to the cell plate. Plant Cell 20:101–123

Collings DA, White RG, Overall RL (1992) Ionic current changes associated with the gravity-induced bending response in roots of *Zea mays* L. Plant Physiol 100:1417–1426

Dahlke RI, Luethen H, Steffens B (2010) ABP1: an auxin receptor for fast responses at the plasma membrane. Plant Signal Behav 5:1–3

deGuzman CC, dela Fuente RK (1984) Polar calcium flux in sunflower hypocotyl segments, I. The effect of auxin. Plant Physiol 76:347–352

Dela Fuente RK (1984) Role of calcium in the polar secretion of indoleacetic acid. Plant Physiol 76:334–342

Dello Ioio R, Linhares FS, Scacchi E, Casamitjana-Martinez E, Heidstra R, Costantino P, Sabatini S (2007) Cytokinins determine Arabidopsis root-meristem size by controlling cell differentiation. Curr Biol 17:678–682

Depuydt S, Hardtke CS (2011) Hormone signalling crosstalk in plant growth regulation. Curr Biol 21:R365–R373

Dettmer J, Hong-Hermesdorf A, Stierhof YD, Schumacher K (2006) Vacuolar H^+-ATPase activity is required for endocytic and secretory trafficking in Arabidopsis. Plant Cell 18: 715–730

Dhonukshe P, Baluška F, Schlicht M, Hlavačka A, Šamaj J, Friml J, Gadella TWJ Jr (2006) Endocytosis of cell surface material mediates cell plate formation during plant cytokinesis. Dev Cell 10:137–150

Drake GA, Carr DJ, Anderson WP (1978) Plasmolysis, plasmodesmata, and the electrical coupling of oat coleoptile cells. J Exp Bot 29:1205–1214

Effendi Y, Rietz S, Fischer U, Scherer GF (2011) The heterozygous abp1/ABP1 insertional mutant has defects in functions requiring polar auxin transport and in regulation of early auxin-regulated genes. Plant J 65:282–294

Epel BL, Erlanger MA (1991) Light regulates symplastic communication in etiolated corn seedlings. Physiol Plant 83:149–153

Evans EC III (1964) Polar transport of calcium in the primary root of *Zea mays*. Science 144:174–177

Felten J, Kohler A, Morin E, Bhalerao RP, Palme K, Martin F, Ditengou FA, Legué V (2009) The ectomycorrhizal fungus *Laccaria bicolor* stimulates lateral root formation in poplar and Arabidopsis through auxin transport and signaling. Plant Physiol 151:1991–2005

Felten J, Legué V, Ditengou FA (2010) Lateral root stimulation in the early interaction between *Arabidopsis thaliana* and the ectomycorrhizal fungus *Laccaria bicolor*: is fungal auxin the trigger? Plant Signal Behav 5:864–867

Feraru E, Feraru MI, Kleine-Vehn J, Martinière A, Mouille G, Vanneste S, Vernhettes S, Runions J, Friml J (2011) PIN polarity maintenance by the cell wall in Arabidopsis. Curr Biol 21:338–343

Friml J (2003) Auxin transport–shaping the plant. Curr Opin Plant Biol 6:7–12

Garnett P, Steinacher A, Stepney S, Clayton R, Leyser O (2010) Computer stimulation: the imaginary friend of auxin transport biology. BioEssays 32:828–835

Geldner N, Hyman DL, Wang X, Schumacher K, Chory J (2007) Endosomal signaling of plant steroid receptor kinase BRI1. Genes Dev 21:1598–1602

Gojon A, Krouk G, Perrine-Walker F, Laugier E (2011) Nitrate transceptor(s) in plants. J Exp Bot 62:2299–2308

Goldsmith MHM (1967a) Movement of pulses of labeled auxin in corn coleoptiles. Plant Physiol 42:258–263

Goldsmith MHM (1967b) Separation of transit of auxin from uptake: average velocity and reversible inhibition by anaerobic conditions. Science 156:661–663

Golomb L, Abu-Abied M, Belausov E, Sadot E (2008) Different subcellular localizations and functions of Arabidopsis myosin VIII. BMC Plant Biol 8:3

Goswami KKA, Audus LJ (1976) Distribution of calcium, potassium and phosphorous in *Helianthus anuus* hypocotyls and *Zea mays* coleoptiles in relation to tropic stimuli and curvatures. Ann Bot 40:49–64

Hacham Y, Sela A, Friedlander L, Savaldi-Goldstein S (2011) BRI1 activity in the root meristem involves post-transcriptional regulation of PIN auxin efflux carriers. Plant Signal Behav 7:68–70

Hause G, Šamaj J, Menzel D, Baluška F (2006) Fine structural analysis of brefeldin a-induced compartment formation after high-pressure freeze fixation of maize root epidermis: compound exocytosis resembling cell plate formation during cytokinesis. Plant Signal Behav 1:134–139

Hertel R, Lomax TL, Briggs WR (1983) Auxin transport in membrane vesicles from *Cucurbita pepo* L. Planta 157:193–201

Heyn A, Hoffmann S, Hertel R (1987) In vitro auxin transport in membrane vesicles from maize coleoptiles. Planta 172:285–287

Hochholdinger F, Zimmermann R (2008) Conserved and diverse mechanisms in root development. Curr Opin Plant Biol 11:70–74

Hodge A (2009) Root decisions. Plant Cell Environ 32:628–640

Hössel D, Schmeiser C, Hertel R (2005) Specificity patterns indicate that auxin exporters and receptors are the same proteins. Plant Biol 7:41–48

Hundal HS, Taylor PM (2009) Amino acid transceptors: gate keepers of nutrient exchange and regulators of nutrient signaling. Am J Physiol Endocrinol Metab 296:E603–E613

Jenkinson IS, Scott BIH (1961) Bioelectric oscillations of bean roots: further evidence for a feedback oscillator. I. Extracellular response to oscillations in osmotic pressure and auxin. Aust J Biol Sci 14:231–247

Jönsson H, Gruel J, Krupinski P, Troein C (2011) On evaluating models in computational morphodynamics. Curr Opin Plant Biol 15:103–110

Juniper BE, Barlow PW (1969) The distribution of plasmodesmata in the root tip of maize. Planta 89:352–360

Kang BH (2011) Shrinkage and fragmentation of the trans-Golgi network in non-meristematic plant cells. Plant Signal Behav 6:884–886

Kasprowicz A, Szuba A, Volkmann D, Baluška F, Wojtaszek F (2009) Nitric oxide modulates dynamic actin cytoskeleton and vesicle trafficking in a cell type-specific manner in root apices. J Exp Bot 60:1605–1617

Kleine-Vehn J, Friml J (2008) Polar targeting and endocytic recycling in auxin-dependent plant development. Annu Rev Cell Dev Biol 24:447–473

Kramer EM (2009) Auxin-regulated cell polarity: an inside job? Trends Plant Sci 14:242–247

Krecek P, Skupa P, Libus J, Naramoto S, Tejos R, Friml J, Zazímalová E (2009) The PIN-FORMED (PIN) protein family of auxin transporters. Genome Biol 10:249

Kriel J, Haesendonckx S, Rubio-Texeira M, Van Zeebroeck G, Thevelein JM (2011) From transporter to transceptor: signaling from transporters provokes re-evaluation of complex trafficking and regulatory controls: endocytic internalization and intracellular trafficking of nutrient transceptors may, at least in part, be governed by their signaling function. BioEssays 33:870–879

Krupinski P, Jönsson H (2010) Modeling auxin-regulated development. Cold Spring Harb Perspect Biol 2:a001560

Kwon C, Panstruga R, Schulze-Lefert P (2008) Les liaisons dangereuses: immunological synapse formation in animals and plants. Trends Immunol 29:159–166

Lam SK, Cai Y, Hillmer S, Robinson DG, Jiang L (2008) SCAMPs highlight the developing cell plate during cytokinesis in tobacco BY-2 cells. Plant Physiol 147:1637–1645

Lee JS, Evans ML (1985) Polar transport of calcium across elongation zone of gravistimulated roots. Plant Cell Physiol 26:1587–1595

Lee JS, Mulkey TJ, Evans ML (1983) Gravity-induced polar transport of calcium across root tips of maize. Plant Physiol 73:874–876

Lee JS, Mulkey TJ, Evans ML (1984) Inhibition of polar calcium movement and gravitropism in roots treated with auxin-transport inhibitors. Planta 160:536–543

Lew RR (1994) Regulation of electrical coupling between Arabidopsis root hairs. Planta 193:67–73

Lima PT, Faria VG, Patraquim P, Ramos AC, Feijó JA, Sucena E (2009) Plant-microbe symbioses: new insights into common roots. Bioessays 31:1233–1244

Lomax TL, Mehlhorn RJ, Briggs WR (1985) Active auxin uptake by zucchini membrane vesicles: quantitation using ESR volume and ApH determinations. Proc Natl Acad Sci U S A 82:6541–6545

Mancuso S, Boselli M (2002) Characterisation of the oxygen fluxes in the division, elongation and mature zones of vitis roots: influence of oxygen availability. Planta 214:767–774

Mancuso S, Marras AM, Magnus V, Baluska F (2005) Noninvasive and continuous recordings of auxin fluxes in intact root apex with a carbon nanotube-modified and self-referencing microelectrode. Anal Biochem 341:344–351

Mancuso S, Marras AM, Mugnai S, Schlicht M, Zarsky V, Li G, Song L, Hue HW, Baluška F (2007) Phospholipase Dζ2 drives vesicular secretion of auxin for its polar cell–cell transport in the transition zone of the root apex. Plant Signal Behav 2:240–244

Masi E, Ciszak M, Stefano G, Renna L, Azzarello E, Pandolfi C, Mugnai S, Baluska F, Arecchi FT, Mancuso S (2009) Spatiotemporal dynamics of the electrical network activity in the root apex. Proc Natl Acad Sci U S A 106:4048–4053

Maule AJ, Benitez-Alfonso Y, Faulkner C (2011) Plasmodesmata—membrane tunnels with attitude. Curr Opin Plant Biol 14:1–8

Maurel C, Leblanc N, Barbier-Brygoo H, Perrot-Rechenmann C, Bouvier-Durand M, Guern J (1994) Alterations of auxin perception in rolB-transformed tobacco protoplasts. Time course of rolB mRNA expression and increase in auxin sensitivity reveal multiple control by auxin. Plant Physiol 105:1209–1215

McLamore ES, Diggs A, Calvo Marzal P, Shi J, Blakeslee JJ, Peer WA, Murphy AS, Porterfield DM (2010a) Non-invasive quantification of endogenous root auxin transport using an integrated flux microsensor technique. Plant J 63:1004–1016

McLamore ES, Jaroch D, Rameez Chatni M, Porterfield DM (2010b) Self-referencing optrodes for measuring spatially resolved, real-time metabolic oxygen flux in plant systems. Planta 232:1087–1099

Merks RMH, de Peer YV, Inze D, Beemster GTS (2007) Canalization without flux sensors: a traveling-wave hypothesis. Trends Plant Sci 12:384–390

Michniewicz M, Zago MK, Abas L, Weijers D, Schweighofer A, Meskiene I, Heisler MG, Ohno C, Zhang J, Huang F, Schwab R, Weigel D, Meyerowitz EM, Luschnig C, Offringa R, Friml J (2007) Antagonistic regulation of PIN phosphorylation by PP2A and PINOID directs auxin flux. Cell 130:1044–1056

Mongrand S, Stanislas T, Bayer EMF, Lherminier J, Simon-Plas F (2010) Membrane rafts in plant cells. Trends Plant Sci 15:656–663

Moreno-Risueno MA, Benfey PN (2011) Time-based patterning in development: the role of oscillating gene expression. Transcription 2:124–129

Moreno-Risueno MA, Van Norman JM, Moreno A, Zhang J, Ahnert SE, Benfey PN (2010) Oscillating gene expression determines competence for periodic Arabidopsis root branching. Science 329:1306–1311

Mravec J, Skůpa P, Bailly A, Hoyerová K, Krecek P, Bielach A, Petrásek J, Zhang J, Gaykova V, Stierhof YD, Dobrev PI, Schwarzerová K, Rolcík J, Seifertová D, Luschnig C, Benková E, Zazímalová E, Geisler M, Friml J (2009) Subcellular homeostasis of phytohormone auxin is mediated by the ER-localized PIN5 transporter. Nature 459:1136–1140

Mugnai S, Azzarello E, Baluška F, Mancuso S (2012) Local root apex hypoxia at the transition zone induces NO-mediated hypoxic acclimation of the whole root. Plant Cell Physiology [Accepted]

Napier RM, David KM, Perrot-Rechenmann C (2002) A short history of auxin-binding proteins. Plant Mol Biol 49:339–348

Panigrahi KC, Panigrahy M, Vervliet-Scheebaum M, Lang D, Reski R, Johri MM (2009) Auxin-binding proteins without KDEL sequence in the moss *Funaria hygrometrica*. Plant Cell Rep 28:1747–1758

Paponov IA, Teale WD, Trebar M, Blilou I, Palme K (2005) The PIN auxin efflux facilitators: evolutionary and functional perspectives. Trends Plant Sci 10:170–177

Peremyslov VV, Mockler TC, Filichkin SA, Fox SE, Jaiswal P, Makarova KS, Koonin EV, Dolja VV (2011) Expression, splicing, and evolution of the myosin gene family in plants. Plant Physiol 155:1191–1204

Pinosa F (2010) Analysis of PIN2 polarity regulation and *Mob1* function in Arabidopsis root development. PhD Thesis, Albert-Ludwig-Universität Freiburg im Breisgau, Germany

Rahman A, Takahashi M, Shibasaki K, Wu S, Inaba T, Tsurumi S, Baskin TI (2010) Gravitropism of *Arabidopsis thaliana* roots requires the polarization of PIN2 toward the root tip in meristematic cortical cells. Plant Cell 22:1762–1776

Reddy ASN, Day IS (2001) Analysis of the myosins encoded in the recently completed *Arabidopsis thaliana* genome sequence. Genome Biol 2:0024.1–0024.17

Richter S, Geldner N, Schrader J, Wolters H, Stierhof YD, Rios G, Koncz C, Robinson DG, Jürgens G (2007) Functional diversification of closely related ARF–GEFs in protein secretion and recycling. Nature 448:488–492

Richter S, Anders N, Wolters H, Beckmann H, Thomann A, Heinrich R, Schrader J, Singh MK, Geldner N, Mayer U, Jürgens G (2010) Role of the GNOM gene in Arabidopsis apical-basal patterning–from mutant phenotype to cellular mechanism of protein action. Eur J Cell Biol 89:138–144

Robert S, Kleine-Vehn J, Barbez E, Sauer M, Paciorek T, Baster P, Vanneste S, Zhang J, Simon S, Čovanová M, Hayashi K, Dhonukshe P, Yang Z, Bednarek SY, Jones AM, Luschnig C, Aniento F, Zažímalová E, Friml J (2010) ABP1 mediates auxin inhibition of clathrin-dependent endocytosis in Arabidopsis. Cell 143:111–121

Rück A, Palme K, Venis MA, Napier RM, Felle H (1993) Patch-clamp analysis establishes a role for an auxin binding protein in the auxin stimulation of plasma membrane current in *Zea mays* protoplasts. Plant J 4:41–46

Sachs T (1969) Polarity and the induction of organized vascular tissues. Ann Bot 33:263–275

Sachs T (1991) Cell polarity and tissue patterning in plants. Development 91(Suppl 1):83–93

Santuari L, Scacchi E, Rodriguez-Villalon A, Salinas P, Dohmann EM, Brunoud G, Vernoux T, Smith RS, Hardtke CS (2011) Positional information by differential endocytosis splits auxin response to drive Arabidopsis root meristem growth. Curr Biol 21:1918–1923

Šamaj J, Baluška F, Voigt B, Schlicht M, Volkmann D, Menzel D (2004) Endocytosis, actin cytoskeleton and signalling. Plant Physiol 135:1150–1161

Šamaj J, Read ND, Volkmann D, Menzel D, Baluška F (2005) The endocytic network in plants. Trends Cell Biol 15:425–433

Samaj J, Chaffey NJ, Tirlapur U, Jasik J, Volkmann D, Menzel D, Baluška F (2006) Actin and myosin VIII in plasmodesmata cell–cell channels. In: Baluška F et al (eds) Cell–cell channels. Landes bioscience, pp 119–134

Sattarzadeh A, Franzen R, Schmelzer E (2008) The Arabidopsis class VIII myosin ATM2 is involved in endocytosis. Cell Motil Cytoskel 65:457–468

Scacchi E, Osmont KS, Beuchat J, Salinas P, Navarrete-Gómez M, Trigueros M, Ferrándiz C, Hardtke CS (2009) Dynamic, auxin-responsive plasma membrane-to-nucleus movement of Arabidopsis BRX. Development 136:2059–2067

Scacchi E, Salinas P, Gujas B, Santuari L, Krogan N, Ragni L, Berleth T, Hardtke CS (2010) Spatio-temporal sequence of cross-regulatory events in root meristem growth. Proc Natl Acad Sci U S A 107:22734–22739

Scherer GF (2011) AUXIN-BINDING-PROTEIN1, the second auxin receptor: what is the significance of a two-receptor concept in plant signal transduction? J Exp Bot 62:3339–3357

Schlicht M, Strnad M, Scanlon MJ, Mancuso S, Hochholdinger F, Palme K, Volkmann D, Menzel D, Baluška F (2006) Auxin immunolocalization implicates vesicular neurotransmitter-like mode of polar auxin transport in root apices. Plant Signal Behav 1:122–133

Schlicht M, Samajová O, Schachtschabel D, Mancuso S, Menzel D, Boland W, Baluska F (2008) D'orenone blocks polarized tip growth of root hairs by interfering with the PIN2-mediated auxin transport network in the root apex. Plant J 55:709–717

Scott BIH (1957) Electric oscillations generated by plant roots and a possible feedback mechanism responsible for them. Aust J Biol Sci 10:164–179

Seagull RW (1983) Differences in the frequency and disposition of plasmodesmata resulting from root cell elongation. Planta 159:497–504

Shabala S, Shabala L, Gradmann D, Chen Z, Newman I, Mancuso S (2006) Oscillations in plant membrane transport: model predictions, experimental validation, and physiological implications. J Exp Bot 57:171–184

Shen WH, Davioud E, David C, Barbier-Brygoo H, Tempé J, Guern J (1990) High sensitivity to auxin is a common feature of hairy root. Plant Physiol 94:554–560

Shen H, Hou NY, Schlicht M, Wan Y, Mancuso S, Baluška F (2008) Aluminium toxicity targets PIN2 in Arabidopsis root apices: effects on PIN2 endocytosis, vesicular recycling, and polar auxin transport. Chin Sci Bull 53:2480–2487

Shen C, Bai Y, Wang S, Zhang S, Wu Y, Chen M, Jiang D, Qi Y (2010) Expression profile of PIN, AUX/LAX and PGP auxin transporter gene families in *Sorghum bicolor* under phytohormone and abiotic stress. FEBS J 277:2954–2969

Spanswick RM (1972) Electrical coupling between cells of higher plants: A direct demonstration of intercellular communication. Planta 102:215–227

Splivallo R, Fischer U, Göbel C, Feussner I, Karlovsky P (2009) Truffles regulate plant root morphogenesis via the production of auxin and ethylene. Plant Physiol 150:2018–2029

Stoma S, Lucas M, Chopard J, Schaedel M, Traas J, Godin C (2008) Flux-based transport enhancement as a plausible unifying mechanism for auxin transport in meristem development. PLoS Comput Biol 4:e1000207

Tarr PT, Tarling EJ, Bojanic DD, Edwards PA, Baldán Á (2009) Emerging new paradigms for ABCG transporters. Biochim Biophys Acta 1791:584–593

Teale WD, Paponov IA, Palme K (2006) Auxin in action: signalling, transport and the control of plant growth and development. Nat Rev Mol Cell Biol 7:847–859

Teh OK, Moore I (2007) An ARF–GEF acting at the Golgi and in selective endocytosis in polarized plant cells. Nature 448:493–496

Thevelein JM, Voordeckers K (2009) Functioning and evolutionary significance of nutrient transceptors. Mol Biol Evol 26:2407–2414

Thompson RF, Langford GM (2002) Myosin superfamily evolutionary history. Anat Rec 268:276–289

Tilsner J, Amari K, Torrance L (2011) Plasmodesmata viewed as specialised membrane adhesion sites. Protoplasma 248:39–60

Traas J, Vernoux T (2010) Oscillating roots. Science 329:1290–1291

Trewavas A (2009) What is plant behaviour? Plant Cell Environ 32:606–616

Tromas A, Braun N, Muller P, Khodus T, Paponov IA, Palme K, Ljung K, Lee JY, Benfey P, Murray JA, Scheres B, Perrot-Rechenmann C (2009) The AUXIN BINDING PROTEIN 1 is required for differential auxin responses mediating root growth. PLoS ONE 4:e6648

Tromas A, Paponov I, Perrot-Rechenmann C (2010) AUXIN BINDING PROTEIN 1: functional and evolutionary aspects. Trends Plant Sci 15:436–446

Van Norman JM, Breakfield NW, Benfey PN (2011) Intercellular communication during plant development. Plant Cell 23:855–864

Vatsa P, Chiltz A, Bourque S, Wendehenne D, Garcia-Brugger A, Pugin A (2011) Involvement of putative glutamate receptors in plant defence signaling and NO production. Biochimie 93:2095–2101

Viotti C, Bubeck J, Stierhof YD, Krebs M, Langhans M, van den Berg W, van Dongen W, Richter S, Geldner N, Takano J, Jürgens G, de Vries SC, Robinson DG, Schumacher K (2010) Endocytic and secretory traffic in Arabidopsis merge in the trans-Golgi network/early endosome, an independent and highly dynamic organelle. Plant Cell 22:1344–1357

Volkmann D, Baluška F (1999) The actin cytoskeleton in plants: from transport networks to signaling networks. Microsc Res Tech 47:135–154

Volkmann D, Baluška F (2006) Gravity: one of the driving forces of evolution. Protoplasma 229:143–148

Volkmann D, Mori T, Tirlapur UK, König K, Fujiwara T, Kendrick-Jones J, Baluška F (2003) Unconventional myosins of the plant-specific class VIII: endocytosis, cytokinesis, plasmodesmata–pit-fields, and cell-to-cell coupling. Cell Biol Int 27:289–291

Wabnik K, Govaerts W, Friml J, Kleine-Vehn J (2011) Feedback models for polarized auxin transport: an emerging trend. Mol BioSyst 7:2352–2359

Wabnik K, Kleine-Vehn J, Balla J, Sauer M, Naramoto S, Reinöhl V, Merks RM, Govaerts W, Friml J (2010) Emergence of tissue polarization from synergy of intracellular and extracellular auxin signaling. Mol Syst Biol 6:447

Wan Y-L, Jasik J, Wang L, Hao H, Volkmann D, Menzel D, Mancuso S, Baluška F, Lin J-X (2012) The signal transducer NPH3 integrates the phototropin1 photosensor with PIN2-based polar auxin transport in arabidopsis root phototropism. Plant Cell [In press]

Wang JR, Hu H, Wang GH, Li J, Chen JY, Wu P (2009) Expression of PIN genes in rice (*Oryza sativa* L.): tissue specificity and regulation by hormones. FEBS J 277:2954–2969

White RG, Barton DA (2011) The cytoskeleton in plasmodesmata: a role in intercellular transport? J Exp Bot 62:5249–5266

Wisniewska J, Xu J, Seifertová D, Brewer PB, Ruzicka K, Blilou I, Rouquié D, Benková E, Scheres B, Friml J (2006) Polar PIN localization directs auxin flow in plants. Science 312:883

Wojtaszek P, Volkmann D, Baluška F (2004) Polarity and cell walls. In: Lindsey K (ed) Polarity in Plants. Blackwell Publishing, pp 72–121

Wojtaszek P, Baluska F, Kasprowicz A, Luczak M, Volkmann D (2007) Domain-specific mechanosensory transmission of osmotic and enzymatic cell wall disturbances to the actin cytoskeleton. Protoplasma 230:217–230

Wolters H, Jürgens G (2009) Survival of the flexible: hormonal growth control and adaptation in plant development. Nat Rev Genet 10:305–317

Xu J, Scheres B (2005) Dissection of Arabidopsis ADP-RIBOSYLATION FACTOR 1 function in epidermal cell polarity. Plant Cell 17:525–536

Xu T, Wen M, Nagawa S, Fu Y, Chen JG, Wu MJ, Perrot-Rechenmann C, Friml J, Jones AM, Yang Z (2010) Cell surface- and rho GTPase-based auxin signaling controls cellular interdigitation in Arabidopsis. Cell 143:99–110

Zhang J, Vanneste S, Brewer PB, Michniewicz M, Grones P, Kleine-Vehn J, Löfke C, Teichmann T, Bielach A, Cannoot B, Hoyerová K, Chen X, Xue HW, Benková E, Zažímalová E, Friml J (2011a) Inositol trisphosphate-induced Ca^{2+} signaling modulates auxin transport and PIN polarity. Dev Cell 20:855–866

Zhang L, Zhang H, Liu P, Hao H, Jin JB, Lin J (2011b) Arabidopsis R-SNARE proteins VAMP721 and VAMP722 are required for cell plate formation. PLoS ONE 6:e26129

Zhao H, Hertel R, Ishikawa H, Evans ML (2002) Species differences in ligand specificity of auxin-controlled elongation and auxin transport: comparing Zea and Vigna. Planta 216:293–301

Zhu T, Lucas WJ, Rost TL (1998a) Directional cell-to-cell communication, in the Arabidopsis root apical meristem. I. An ultrastructural, and functional analysis. Protoplasma 203:35–47

Zhu T, O'Quinn RL, Lucas WJ, Rost TL (1998b) Directional cell-to-cell communication in Arabidopsis root apical meristem. II. Dynamics of plasmodesmatal formation. Protoplasma 204:84–93

Zhu T, Rost TL (2000) Directional cell-to-cell communication, in the Arabidopsis root apical meristem. III. Plasmodesmata turnover and apoptosis in meristem and root cap cells during four weeks after germination. Protoplasma 213:99–107

Chapter 13
Ion Currents Associated with Membrane Receptors

J. Theo M. Elzenga

Abstract The coordination of cellular physiology, organ development, life cycle phases and symbiotic interaction, as well as the triggering of a response to changes is the environment, in plants depends on the exchange of molecules that function as messengers. Binding of these messenger molecules to receptor proteins, is transduced through a network of second messengers into a response. Electrophysiological processes, like ion channel activation, regulation of ion pumps and membrane potential changes, have been shown to be involved in signal perception, transduction and in the cellular response. Examples are the perception and response to pathogen invasion, the signal exchange between plants and microbial symbionts, the systemic response to herbivore attack and the interaction between pollen tube and pistil tissue. In this chapter an overview of the coupling between ligands binding to a receptor protein and subsequent ion flux changes is given.

13.1 Introduction

Plants are exposed to a plethora of environmental perturbations that potentially can affect their fitness. The relevant environmental factors are both abiotic (wind and other mechanical stress, temperature, light, drought, salt, flooding, pollutants like toxic metals, nutrient availability) and biotic (herbivory by various animals like insects, nematodes or vertebrates, pathogenic bacteria, viruses or fungi, roots of competing plants, etcetera). For plant survival adequate responses in physiology

J. T. M. Elzenga (✉)
Plant Ecophysiology, University of Groningen, Nijenborgh 7,
9747 AG Groningen, The Netherlands
e-mail: j.t.m.elzenga@rug.nl

(closure of stomata, synthesis of defense compounds), morphology (reduction of leaf area, initiation of new roots) or anatomy (forming barriers against pathogen spreading, transition from sun to shade leaves, development of aerenchyma in the root and stem) are essential. Timely initiation of these responses require an 'early warning system' and plants indeed are equipped with sensory systems that in their sensitivity rivals those of animals. It is therefore no surprise that it has been proposed to enslave plants, connect them to electronic equipment and use them as environmental biosensors (Volkov and Ranatunga 2006).

The elements of sensory systems can be divided into three main components: the sensor, the transducer and the effector. In the plant responses that involve a developmental shift, the effector elements consist of gene promoter activation, leading to gene transcription. In the responses that require no developmental changes, but are based on cell- or tissue-autonomous responses, the effector element can be enzymes that are activated (for instance, pigment biosynthesis under high light exposure) or activation of ion transport (for instance, closing and opening of stomata). The transduction of an environmental signal can be a relatively complex process: intracellular activation of second messenger systems, long distance signaling by plant hormones, iRNA, metabolic intermediates or electrical signals. In plants, as in animals, the variety of environmental signals is sensed by an equally wide variety of sensor types. Sensors can be pigment molecules (for light), ion channels (for mechanical stress), soluble cytoplasmic receptor protein (hormones), membrane bound proteins (pathogen recognition), etc.

Noteworthy is the potential role of ion transport processes in all three elements of a signal-to-response chain of events. This can be illustrated with these examples: in guard cells ion fluxes across the membrane are part of the *effector* system, root hairs responding to chemical or mechanical stimuli use Ca^{2+} channels and Ca^{2+} pumps in the modulation of cytoplasmic $[Ca^{2+}]$, a second messenger, *transducing* the signal and stretch-activated ion channels are likely to be involved as *sensor* of mechanical perturbation (in guard cells (Zhang et al. 2007), in pollen tube tips (Dutta and Robinson 2004)). But mostly the perception of the signal and the response at the cell, tissue, organ or plant level are coupled through a signal transduction cascade. A nice example of such a cascade is the perception of light by phototropin leading through activation of a signal transduction cascade to drastic changes in the activity of several transporter proteins and the closure of stomates. Phototropin is a membrane-bound, blue and UV-A sensitive, photoreceptor, using a light oxygen voltage (LOV) photosensory domain (Demarsy and Fankhauser 2009), involved in a range of plant responses including phototropic bending of stems and hypocotyls, chloroplast migration in response to light intensity changes and stomatal opening. Phototropins represent a class of receptor kinases that appear exclusively in plants (Lin 2000; Briggs and Christie 2002). The signal transduction pathway, involving autophosphorylation of the Serine residues in the activation loop, following excitation of the photoreceptor is still not fully elucidated. Phototropins form a complex with several proteins, among which is the AHA2 H^+-ATPase. End targets of the signal transduction pathway(s) in guard cells are a malate and Cl^- permeable anion channel (inhibited) and H^+-ATPase (stimulated) (see review in Inoue et al. 2010).

In this chapter the focus will be on ion fluxes that are directly coupled to ligand-membrane receptor binding events.

13.2 Role of Electrical Signals in Plant Development

Electrical signals have been known to play an important role in signal transduction for over a century. The presence of action potentials, resembling the electrical signals in nerve cells, in plants was realized already in 1873 (see review by Fromm and Lautner 2007). The early studies of electrical properties of excitable cells were done on the giant cells of *Chara* and *Nitella* and only later on the squid giant axon (Johnson et al. 2002). Long distance signaling along the phloem strands is instrumental in movement of plant organs upon mechanical triggering as is the case in carnivorous Venus fly trap plants, in the leaf and petiole folding of *Mimosa pudica* and the tentacle movement to wrap around a captured insect in *Drosera* species. The long distance electrical signals being transmitted that have action potential-like characteristics and have been found, apart from the examples already mentioned, in *Zea mays* (triggered by re-irrigation and by cold shock), *Luffa* (by cooling and electrical triggering) and *Lycopersicon* (by electrical triggering). While transmitted action potentials consist of a purely electrical, propagated depolarization wave along the phloem strands or other electrically coupled cells, a second long distance, so-called variation potential or VP, depends on a hydraulic wave transmitted through the xylem, locally inducing electrical responses. These VP's are triggered by heat or wounding and have been found in *Vicia*, *Solanum*, *Helianthus*, *Mimosa*, *Populus* (all by heating) and *Pisum* (by wounding). The physiological responses range from regulation of leaf movement, to decreased elongation growth and modification of gass exchange (Fromm and Lautner 2007; Yan et al. 2009).

Whereas long distance signaling depends on, either the transport of hormones and other signaling molecules between organs the plants through the mass flow conduits (phloem and xylem), or on electrical signaling, the short distance integration between cells of developmental and physiological processes relies mainly on ligand-receptor interactions (De Smet et al. 2009). It is interesting to note that ion transport can be involved in both and might serve as a connection between long and short distance signaling.

13.3 Ligand-Binding Receptors

13.3.1 Types of Membrane Bound Receptors

Plasma membrane localized receptor proteins play an important role in the perception and transduction of signals that originate from outside the cell into the cell. There are different types of receptor proteins such as ion channel-linked receptors

(Davenport 2002), G protein-linked receptors (Grill and Christmann 2007) and receptor protein kinases (RPKs, Kacperska 2004).

A general mechanism of signaling cascades in animals and plants is the reversible phosphorylation/dephosphorylation-dependent regulation of protein activity by protein kinases and phosphatases. RPKs catalyze the phosphate transfer from ATP onto histidine, serine, threonine or tyrosine in the protein backbone. According to their kinase activities, RPKs are classified into three main groups: (1) the receptor tyrosine kinases (RTKs), phosphorylating tyrosine residues, (2) the serine/threonine kinase receptors (STKRs), phosphorylating serine and threonine residues, and (3) receptor histidine kinases (RHKs) phosphorylating histidine residues (Becraft 2002).

In animals the majority of RPKs are receptor tyrosine kinases (RTKs). In plants, RHKs and STKRs are the main types of receptor protein kinases. Examples of RHKs are the ethylene receptor ETR1, and the cytokinin receptor CRE1.

Based on homology searches in completely and nearly completely sequenced genomes, plants appear to lack proteins with the RTK signatures, instead they contain structurally and functionally similar receptors called receptor like kinases (RLKs) which have the serine/threonine kinase signature (Castells and Casacuberta 2007). Similar to RTKs, four domains can structurally be distinguished in RLKs: a targeting peptide (signal peptide), an extracellular receptor domain, a trans-membrane domain and an intracellular kinase domain (Fig. 13.1, Shiu and Bleecker 2001). The four domains work together in order to transduce the signal. In short, the receptor domain binds a ligand, an event that triggers a conformational change in the protein. Subsequently, the oligomerization state of the RLK changes and by a still unknown mechanism the signal is intramolecularly transduced across the plasma membrane to the kinase domain. The kinase domain is activated and intracellular signaling transduction cascades are triggered. The genome of *Arabidopsis thaliana* contains more than 600 homologues of RLKs. Of most of these RLKs the ligand is unknown, as is their function. The few RLKs for which we do know its function and/or ligand, indicate that these receptors are involved in a wide range of processes, such as symbiosis with arbuscular mycorrhiza forming fungi, pathogen recognition and thus plant resistance, cell growth regulation, brassinosteroid signaling, priming of innate immune response and organ development (De Smet et al. 2009).

13.4 Small Signaling Peptides

Small peptides have recently been given a more prominent place, next to phytohormones, in intercellular signaling in processes as diverse as growth and development, biotic and abiotic stress responses and self and non-self recognition (Butenko et al. 2009; Katsir et al. 2011; Germain et al. 2006; Fukuda and Higashiyama 2011). The evolutionary advantage of using small peptides for signaling purposes is the flexibility in regulating responses to peptide signaling

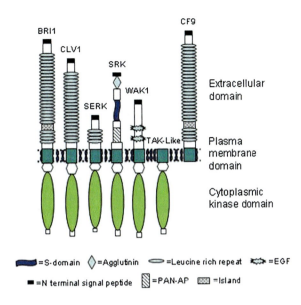

Fig. 13.1 Examples of protein domain configurations of membrane-bound receptor-like kinases, indicating the general structure of RLKs and illustrating the diversity in extracellular domains of RLKs. *Cf9* represents the RLPs that do not have a kinase domain. *TAK-like* represents the kinases that don't have an extracellular domain. The other RLKs represent the RLKs classes that have extracellular, trans-membrane and kinase domain

molecules through modular combinations of receptors (Wheeler and Irving 2010). Peptide signaling molecules are secreted into the apoplast, often undergo a proteolytic cleavage, and act on receptors on nearby or adjacent cells. In case a receptor has been identified, they are all receptor-like kinases that form oligomers upon ligand binding and then undergo mutual phosphorylation, which triggers a down stream signaling cascade. The small peptides that have been demonstrated to be active in various plant processes, belong to large gene families (CLE/ESR, RALF-like, DVL/ROT, IDA, Egg apparatus 1-like and HT) of which the function still largely remains to be elucidated and for which the receptor is unknown. Of the small peptides, the RALF family, named after the rapid alkalization it induces in cultured cells, is known to interfere with ion fluxes across the membrane. Since the effect of RALF on ion fluxes resembles those of systemin (see below) a signaling peptide with a clear function in plant pathogen defense, the other RALF-like peptides are suspected to have equally important functions (Bedinger et al. 2010). In the next paragraphs examples of plant developmental or defensive processes, like symbiosis, pollen tube self-incompatibility, and pathogen and herbivory detection, will be discussed.

13.4.1 Legume-Rhizobium Symbiosis

In the establishment of the Rhizobium-legume symbiosis, leading to the macroscopically visible nodules on the roots of the plants, several signals are exchanged between plant and bacterium. The compounds that are used for these signals are oligosaccharides (Nod factors), surface polysaccharides produced by the bacteria, phytohormones and, more recently discovered, plant-encoded small peptides. One of the genes that is up-regulated in *Medicago trunculata* upon the addition of the lipochito-oligosaccharides, known as Nod factors, that trigger the developmental cascade leading to nodule formation, is Rapid Alkanisation factor 1 (RALF1) (Batut et al. 2011), a 49 amino acid peptide capable of arresting root and pollen tube growth. The addition of RALF to cultured cells leads to an alkalinisation of the media, due to blockage of the proton pumping ATPase (Pearce et al. 2010). Although the specific role of RALF1 in the nodule formation still needs to be elucidated, the almost ubiquitous presence of RALF, and its proposed role in many developmental processes, hints at an important role in symbioses interaction. The interaction between the legume *Lotus japonicus* and the bacterial symbiont *Mesorhizobium loti*, requires two LysM-type serine/threonine receptor kinases in the plant, NFR1 and NFR5, that enable the recognition of the bacterium. One of the earliest cellular responses when exposed to the Nod factor, lipochitin-oligosaccharide signal is a rapid membrane depolarization and concommittant alkalinization of the external medium. Mutant plants that carry a mutated NFR5 allele do not have any detectable response (Radutolu et al. 2003).

13.4.2 Pollen Tube Growth and Guidance

Pollen tube growth exhibits polar tip growth, which is a specialized form of growth found also in root hairs, fungal hyphae and algal rhizoids (Campanoni and Blatt 2007). Polar increase in cytoplasmic [Ca^{2+}] is considered a signal transduction step in the the regulation of tip growth, while the role of other ion channels (K^+, Cl^- and H^+) is assumed (Michard et al. 2009). Calcium might enter the cell through hyperpolarization-activated channels (Qu et al. 2007) or through a stretch activated Ca^{2+} channel (Dutta and Robinson 2004). Pollen tube growth is controlled by a specialized mechanism that in higher plants prevents self-fertilization. Interaction between a pollen tube and a pistil that both contain the same allele of a so-called S-locus, will result in rejection of incompatible or 'self' pollen. The S-determinants of the pistil in Papaver (PrsS) are small, secreted, cysteine-rich proteins. The S-determinants of the Papaver pollen, PrpS, is a trans-membrane protein. When pollen tubes are exposed to the 'incompatible' PrsS the cytoplasmic [Ca^{2+}] instantaneously is raised, followed by inhibition of pollen tube growth and programmed cell death (reviewed by Bosch et al. 2008). In whole cell patch clamp experiments addition of PrsS leads to activation of a ligand gated Ca^{2+} channel, permeable to both Ca^{2+} and Ba^{2+} and completely blocked by La^{3+} (Fig. 13.2) (Wu et al. 2011). Tubes of compatible pollen will continue to grow and be directed to the female gametophyte to deliver the non-motile sperm cells and enable fertilization of the egg cell. Directional growth of the pollen tube is partly mechanical guidance and partly chemotrophic (Takeuchi and Higashiyama 2011). Straight growth in the style is mechanically guided, but once the tip of the pollen tube leaves the style, the pollen tube is exposed to ovular factors and is attracted to the

Fig. 13.2 Stepped voltage protocol reveals PrsS-stimulated, Ca^{2+}-permeable conductance in Papaver rhoas. **a** Cartoon of the voltage step protocol. **b** family of currents obtained from an untreated pollen protoplast using the stepped voltage mode protocol (see **a**). **c** A family of currents recorded from the same cell as in **b**, subjected to the voltage step protocol after exposure to incompatible PrsS (After Wu et al. 2011)

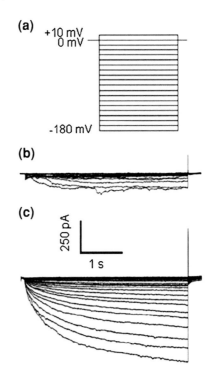

embryo sac. Interestingly Arabidopsis cation/H^+ exchanger mutants (CHX21 and CHX23) do not grow in the direction of the ovules, but remain inside the transmitting tract in het ovary. It is very likely that the interaction between ovary tissue and pollen tube involves extensive exchange of signals. Of the 620 RLK genes present in Arabidopsis 43 are expressed in germinating pollen tubes. In the still putative schedule of the tip of the growing pollen tube, approaching the ovules two plasma membrane Ca^{2+} transporters (CNGC18, regulated by cyclic nucleotides produced by still unknown cyclases and GLR (Ca^{2+}-permeable, glutamate receptor channel, regulated by D-serine produced by the sporophytic cells of the ovules) feature prominently.

13.4.3 Natriuretic Peptide and Salt Stress

Natriuretic peptides were first described in animals where they are involved in salt and osmotic balance. The regulatory pathway involves changes in the intracellular cyclic GMP. In plants natriuretic peptides occur in Arabidopsis and many other plants and induces protoplast swelling and stomatal opening (Wang et al. 2007). Plant natriuretic peptides (AtPNP-A and AtPNP-B) in Arabidopsis are excreted into the apoplast and expressed at elevated levels when cells are exposed to heat, hyperosmotic solutions and salt (Wang et al. 2011). Apart from modulating the water permeability of the membrane through regulation of PIP activity, natriuretic peptide affect K^+, Na^+, and H^+ fluxes in *Z. mays* roots (Pharmawati et al. 1999) and in Arabidopsis roots (Ludidi et al. 2004) and modulates the proton gradient across plasma membrane vesicles derived from potato leaf vesicles. (Maryani et al. 2000).

13.4.4 Pathogen and Herbivory Recognition

The paradigm for plant defense reactions involves recognition of pathogens through specific receptors that bind pathogen associated patterns (PAMPs), a mechanism termed PAMP-triggered immunity (Nicaise et al. 2009). Molecules that trigger responses carry patterns that are not present in the host plant proteins or oligosaccharides, examples of which are flagellin, chitin, peptidoglycans and lipopolysaccharides and harpin protein HrpZ (Zimaro et al. 2011). Plants can also respond to endogenous molecules, released by the pathogen invading the plant and derived from damaged cell wall or cuticula. These elicitors are called danger associated molecular patterns (DAMPS) (Boller and Felix 2009; Toer et al. 2009). A third class of elicitors is formed by the pathogen virulence molecules, called effectors, which are aimed at suppressing the defense response of the host (Dodds and Rathjen 2010).

In an earlier study pathogen associated elicitors, like fragments of cell wall proteins, were shown to induce opening of large conductance, Ca^{2+} permeable channels (Zimmermann et al. 1997). Small peptides have been shown to be involved in the innate immune response. They bind to the membrane of suspension cells and trigger an alkalinisation in the cell suspension media in Arabidopsis (Pearce et al. 2008), Soybean (Pearce et al. 2010a, 2010b). In Arabidopsis a similar response induced by the peptide AtPep1 depends on a receptor with guanylyl cyclase activity and the subsequent activation of a Ca^{2+} channel by cGMP (Qi et al. 2010; Krol et al. 2010). Since the influx of Ca^{2+} and the subsequent rise in intracellular $[Ca^{2+}]$ is found for a variety of stimuli, including triggering by MAMPs, such as flg22 (flagellin, see below), elf18 and Pep1, the elicitor-induced transient $[Ca^{2+}]_{cyt}$ rise could be successfully used to screen for signaling mutants in Arabidopsis (Ranf et al. 2011). In the response of Arabidopsis root cells to flagellin and elf18, the other downstream events, like changes in K^+, anion and H^+ fluxes and membrane depolarization, seem to depend strictly on this rise of cytoplasmic $[Ca^{2+}]$.

Upon herbivore attack a number of changes can be observed at the cellular level in the plant: phosphorylation of proteins, increase in cytoplasmic calcium concentration, depolarization of the membrane, changes in ion fluxes across the membrane, formation of reactive oxygen species and upregulation of defense associated genes. Changes in membrane potential and ion fluxes are amongst the fastest responses to herbivore attack. A herbivory-induced membrane potential change is followed by a fast electrical signal that can travel through the plant (Maffei et al. 2007a, b). The change in membrane potential can originate through a herbivore triggered increase in hydrogenperoxide, which has a strong membrane depolarizing effect, or the release of the peptide Pep13. In solanaceae herbivores trigger the hydrolytic cleavage of prosystemin to form the oligopeptide systemin, which affects ion transport and the membrane potential upon binding to its receptor SR160 (see the next paragraph). In other plant species the signaling molecules that allow herbivore-specific recognition are not as well known. Oral secretion from feeding insects could contain potential elicitors. Different proteins, for instance a beta-glucosidase from caterpillars of the White Cabbage Butterfly, do trigger defense responses. However, none of the compounds that induce defense responses have been tested for the induction of ion flux changes (Ferry et al. 2004; Mithoefer and Boland 2008; Heil 2009). Until now systemin analogues have not emerged.

13.4.5 Specific Signaling Molecule Perception: Two Case Studies, Systemin and Flagellin

Systemin

To deter herbivores a wide array of sophisticated defense mechanisms has evolved in plants. Induction of these mechanisms is triggered by feeding of the herbivore, not only near the bite or puncture made by the animal, but also in plant parts at considerable distance from this site. This systemic induction of defense is attributed to the plant hormone jasmonic acid or its derivative methyl-jasmonate (Farmer 2003; Lorenzo 2005). In solaneous plants the jasmonic acid signaling system is fine-tuned by the introduction of the peptide hormones, systemin and hydroxyproline-rich systemin, as signaling intermediates (Ryan 2003).

In tomato the 18-amino acid peptide, systemin, is released upon feeding and locally induces the defense response. Initially, it was assumed that this peptide was the signal molecule responsible for the systemic activation of defense, however, the currently preferred view on the systemic dispersal of the wound defense response is that not systemin, but jasmonic acid (JA) is the systemic signal molecule. Systemin, then, functions locally by inducing and amplifying JA and its own production along the vascular tissues and strengthens the systemic response (Schilmiller 2005).

Exposing plants to systemin, either indirectly through wounding or directly by application of a systemin-containing solution, results in a well-defined response. Two phases can be discriminated in this response: an early phase, which consists of changes in ion-fluxes across the plasma membrane, simultaneous changes in the cytoplasmic Ca^{2+} concentrations (Felix 1995; Meindl 1998; Moyen 1996; Moyen 1998; Schaller 1999) and an increase in the production of reactive oxygen species (Orozco-Cardenas 2001), and a late phase, which consists of changes in gene expression. All Solanaceae that are closely related to tomato, the subtribe solaneae (tomato, potato, black nightshade and pepper), respond to exposure to tomato systemin, while more distantly related species did not.

A few years ago the receptor for systemin SR160 was isolated from *Solanum esculentum* suspension-cultured cells and identified as an RLK (Scheer 2002), with a high similarity to the brassinosteroid-receptor BRI1. SR160 is expressed in a

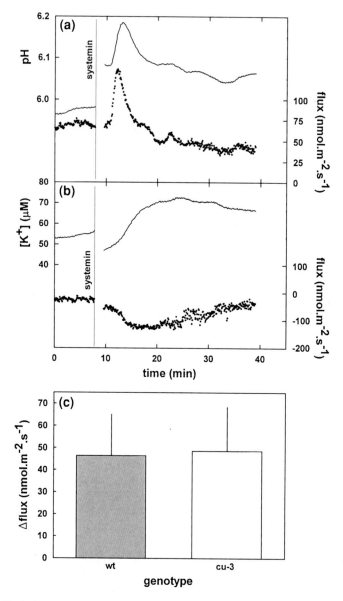

Fig. 13.3 Typical systemin-induced changes in proton and potassium fluxes from cu3 *S. pimpinellifolium* leaf tissue. **a** pH of the unstirred layer and proton flux. **b** K$^+$ concentration of the unstirred layer and K$^+$ flux. *Continuous lines*, pH or K$^+$ concentration at 10 and 50 µm from the cell wall (*left* y axis). *Dotted lines*, Calculated H$^+$ or K$^+$ flux based on the pH and [K$^+$] at 10 and 50 µm (*right* y axis). A *vertical line* indicates the moment of the addition of systemin. When the flux becomes more positive, this means either an increase of the influx or a reduction of the efflux. **c** The amplitude of the first oscillation of the systemin-induced proton-flux change in wild-type and cu3 *S. pimpinellifolium* plants

tissue specific way, highly expressed in vascular tissue and absent in mesophyll cells (Hind et al. 2010). Functional expression of the *SR160* gene in tobacco, a plant species that does not responds to systemin, resulted in an systemin-induced alkalinization of the medium similar to the one observed in tomato suspension cells (Scheer 2003). Systemin has a physiological effect by inhibition through a receptor mediated signal transpduction pathway of the plasma membrane ATP-ases, thereby causing an alkalinization of the medium. When ligand-receptor binding is prevented by suramin, the alkalinization is blocked (Strattmann et al. 2000).

The structural similarity between SR160 and BRI1 triggered speculation on the involvement of BRI1 in the binding of systemin. However, it was later demonstrated that the Bri-1 protein is not required for wound-signaling (Holton et al. 2007; Malinowski 2009). Lanfermeijer et al. (2008) confirmed with the MIFE (a ion selective vibrating probe system) that systemin elicited changes in H^+ and K^+ fluxes from tomato-leaf tissues in situ, even in the cu-3 mutant, which does not have a functional BRI1 (Fig. 13.3).

Flagellin

The presence of microbes is perceived by plants by receptors that bind highly conserved molecular signatures that are not present in the host, but are microbe-associated. These signatures are collectively known as micobe-associated molecular patterns or MAMPS. The Arabidopsis pattern recognition receptor FLAGELLIN-SENSING 2 (FLS2) is genetically well defined and mediates defense responses against bacterial pathogens. Stimulation of this receptor through its ligand, the bacterial protein flagellin, leads to a defense response and ultimately to increased resistance. In Arabidopsis flagellin induced a rapid, dose-dependent membrane potential depolarization in both mesophyll and root hair cells. Using ion-selective microelectrodes, pronounced anion, K^+ and H^+ current and concentration changes were recorded upon application of flagellin, indicating that the signaling cascade targets plasma membrane ion channels. The depolarization coincides with an increase in cytosolic calcium and requires phosphorylation, as K-252a prevents the flagellin-induced potential change. The flagellin-induced membrane depolarization depends on the activity of the FLS2-associated receptor-like kinase BAK1, suggesting that activation of FLS2 leads to BAK1-dependent, calcium-associated plasma membrane anion channel opening as an initial step in the pathogen defense pathway (Jeworuztzki et al. 2010).

13.5 Conclusion

Considering the number of genes present in the genomes of species that have been sequence and the instances where receptor-ligand binding events lead to changes in membrane potential, ion fluxes and extra-cellular and intra-cellular ion concentrations, it can be expected that electrical signaling and involvement of ion transport processes is widely utilized in plant responses and plant development processes. Since many of the small peptides have not been annotated in the database (Wheeler and Irving 2010), the search for receptors, functions and specific down-stream signaling networks will continue.

References

Batut J, Mergaert P, Masson-Bovin C (2011) Peptide signaling in the rhizobium-legume symbiosis. Curr Opin Microbiol 14:181–187
Becraft PW (2002) Receptor kinase signaling in plant development. Annu Rev Cell Dev Biol 18:163–192
Bedinger PA, Pearce G, Covey PA (2010) RALFS peptide regulators of plant growth. Plant Signal Behav 5:1342–1346
Boller T, Felix G (2009) A renaissance of elicitors: perception of microbe-associated molecular patterns and danger signals by pattern-recognition receptors. Annu Rev Plant Biol 60:379–406
Bosch M, Poulter NS, Vatovec S, Franklin VE (2008) Initiation of programmed cell death in self-incompatibility: role for cytoskeleton modifications and several caspase-like activities. Mol Plant 1:879–887
Briggs WR, Christie JM (2002) Phototropins 1 and 2: versatile plant blue-light receptors. Trends Plant Sci 7:204–210
Butenko MA, Vie AK, Brembu T, Aalen RB, Bones AM (2009) Plant peptides in signaling: looking for new partners. Trends Plant Sci 14:255–263
Campanoni P, Blatt MR (2007) Membrane trafficking and polar growth in root hairs and pollen tubes. J Exp Bot 58:65–74
Castells E, Casacuberta JM (2007) Signaling through kinase-defective domains: the prevalence of atypical receptor-like kinases in plants. J Exp Bot 58(13):3503–3511
Davenport R (2002) Glutamate receptors in plants. Ann Bot 90:549–557
De Smet I, Voss U, Juergens G, Beeckman T (2009) Receptor-like kinases shape the plant. Nature Cell Biol 11:1166–1173
Demarsy E, Fankhauser C (2009) Higher plants use LOV to perceive blue light. Curr Opin Plant Biol 12:69–74
Dodds PN, Rathjen JP (2010) Plant immunity: towards an integrated view of plant-pathogen interactions. Nat Rev Genet 11:539–548
Dutta R, Robinson KR (2004) Identification and characterization of stretch-activated ion channels in pollen protoplasts. Plant Physiol 135:1398–1406
Farmer EE, Almeras E, Krishnamurthy V (2003) Jasmonates and related oxylipins in plant responses to pathogenesis and herbivory. Curr Opin Plant Biol 6:372–378
Felix G, Boller T (1995) Systemin induces rapid ion fluxes and ethylene biosynthesis in *Lycopersicon peruvianum* cells. Plant J 7:381–389
Ferry N, Edwards MG, Gatehouse JA, Gatehouse AMR (2004) Plant-insect interactions: molecular approaches to insect resistance. Curr Opin Biotechnol 15:155–161
Fromm J, Lautner S (2007) Electrical signals and their physiological significance in plants. Plant Cell Env 30:249–257
Fukuda H, Higashiyama T (2011) Diverse functions of palnt peptides: entering a new phase. Plant Cell Physiol 52:1–4
Germain H, Chevalier E, Matton DP (2006) Plant bioactive peptides: an expanding class of signaling molecules. Can J Bot 84:1–19
Grill E, Christmann A (2007) A plant receptor with a big family. Şcience 315:1676–1677
Heil M (2009) Damaged-self recognition in plant herbivore defense. Trends Plant Sci 14:356–363
Hind SR, Malinowski R, Yalamanchili R, Stratmann JW (2010) Tissue-type specific systemin perception and the elusive systemin receptor. Plant Signal and Behav 5:42–44

Holton N, Cano-Delgado A, Harrison K, Montoya T, Chory J, Bishop GJ (2007) Tomato brassinosteroid insensitive1 is required for systemin-induced root elongation in solanum pimpinellifolium but is not essential for wound signaling. Plant Cell 19:1709–1717

Inoue S, Takemiya A, Shimazaki K (2010) Phototropin signaling and stomatal opening as a model cas. Curr Opin Plant Biol 13:587–593

Jeworutski E, Roelfsema MRG, Anschuetz U, Krol E, Elzenga JTM, Felix G, Boller T, Hedrich R, Becker D (2010) Early signaling through the Arabidopsis pattern recognition receptors FLS2 and EFR involves Ca^{2+}-associated opening of plasma membrane anion channels. Plant J 62:367–378

Johnson BR, Wyttenbach RA, Wayne R, Hoy RR (2002) Action potentials in a giant algal cell: a comparative approach to mechanisms and evolution of excitability. J Undergraduate Neurosci Educ 1:A23–A27

Kacperska A (2004) Sensor types in signal transduction pathways in plant cells responding to abiotic stressors: do they depend on stress intensity? Physol Plant 122:159–168

Katsir L, Davies KA (2011) Peptide signaling in plant development. Curr Biol 21:356–365

Krol E, Mentzel T, Chinchilla D, Boller T, Felix G, Kemmerling B, Postel S, Arents M, Jeworutzki E, Al-Rasheid KAS, Becker D, Hedrich R (2010) Perception of the Arabidopsis danger signal peptide 1 involves the pattern recognition receptor AtPEPR1 and its close homologue AtPEPR2. J Biol Chem 285:13471–13479

Lanfermeijer FC, Staal M, Malinowski R, Stratmann JW, Elzenga JTM (2008) Micro-electrode flux estimation confirms that the Solanum pimpinellifolium cu3 mutant still responds to systemin. Plant Physiol 146:129–139

Lin C (2000) Plant blue-light receptors. Trends Plant Sci 5:337–342

Lorenzo O, Solano R (2005) Molecular players regulating the jasmonate signaling network. Curr Opin Plant Biol 8:532–540

Ludidi N, Morse M, Sayed M, Wherrett T, Shabala S, Gehring C (2004) A recombinant plant natriuretic peptide causes rapid and spatially differentiated K+, Na+ and H+ flux changes in Arabidopsis thaliana roots. Plant Cell Physiol 45:1093–1098

Maffei ME, Mithoefer A, Boland W (2007a) Before gene expression: early events in plant—insect interaction. Trends Plant Sci 12:310–316

Maffei ME, Mithoefer A, Boland W (2007b) Insect feeding on plants: rapid signals and responses preceding the induction of phytochemical release. Phytochemistry 68:2946–2959

Malinowski R, Higgins R, Luo Y, Piper L, Nazir A, Bajwa VS, Clouse SD, Thompson PR, Stratmann JW (2009) The tomato brassinosteroid receptor BRI1 increases binding of systemin to tobacco plasma membranes, but is not involved in systemin signaling. Plant Mol Biol 70:603–616

Maryani MM, Shabala S, Gehring CA (2000) Plant natriuretic peptide immunoreactants modulate plasma-membrane H+ gradients in Solanum tuberosum L. leaf tissue vesicles. Arch Biochem Biophys 376:456–458

Meindl T, Boller T, Felix G (1998) The plant wound hormone systemin binds with the N-terminal part to its receptor but needs the C-terminal part to activate it. Plant Cell 10:1561–1570

Michard E, Alves F, Feijo JA (2009) The role of ion fluxes in polarized cell growth and morphogenesis: the pollen tube as an experimental paradigm. Int J Dev Biol 53:1609–1622

Mithoefer A, Boland W (2008) Recognition of herbivory-associated molecular patterns. Plant Physiol 146:825–831

Moyen C, Hammond-Kosack KE, Jones J, Knight MR, Johannes E (1998) Systemin triggers and increase of cytoplasmic calcium in tomato mesophyll cells: Ca^{2+} mobilization from intra-and extracellular compartments. Plant Cell Envir 21:1101–1111

Moyen C, Johannes E (1996) Systemin transiently depolarizes the tomato mesophyll cell membrane and antagonizes fusicoccin-induced extracellular acidification of mesophyll tissue. Plant Cell Environ 19:464–470

Nicaise V, Roux M, Zipfel C (2009) Recent advances in PAMP-triggered immunity against bacteria: pattern recognition receptors watch over and raise the alarm. Plant Physiol 150: 1638–1647

Orozco-Cárdenas ML, Narváez-Vásquez J, Ryan CA (2001) Hydrogen peroxide acts as a second messenger for the induction of defense genes in tomato plants in response to wounding, systemin, and methyl jasmonate. Plant Cell 13:179–191

Pearce G, Munske G, Yamaguchi Y, Ryan CA (2010a) Structure-activity studies of GmSubPep, a soybean peptide defense signal derived from an extracellular protease. Peptides 31:2159–2164

Pearce G, Yamaguchi Y, Barona G, Ryan CA (2010b) A subtilisin-like protein from soybean contains an embedded cryptic signal that activates defense related genes. Proc Nat Acad Sci USA 107:14921–14925

Pearce G, Yamaguchi Y, Munske G, Ryan CA (2008) Structure-activity studies of AtPep1, a plant peptide signal involved in the innate immune response. Peptides 29:2083–2089

Pearce G, Yamaguchi Y, Munske G, Ryan CA (2010c) Structure-activity studies of RALF, rapid alkalinization factor, reveal an essential—YISY—motif. Peptides 31:1973–1977

Pharmawati M, Shabala S, Newman IA, Gehring CA (1999) Natriuretic peptides and cGMP modulate K+, Na+, and H+ fluxes in Zea mays roots. Mol Cell Biol Res Commun 2:53–57

Qi Z, Verma R, Gehring C, Yamaguchi Y, Zgao Y, Ryan CA, Berkowitz GA (2010) Ca^{2+} signaling by plant Arabidopsis thaliana Pep peptides depends on AtPepR1, a receptor with guanylyl cyclase activity, and cGMP-activated Ca^{2+} channels. Proc Nat Acad Sci USA 107:21193–21198

Qu H-Y, Shang Z-L, Zhang S-L, Liu L-M, Wu J-Y (2007) Identification of hyperpolarization-activated calcium channels in apical pollen tubes of Pyrus pyrifolia. New Phytol 174:524–536

Radutolu S, Madsen LH, Madsen EB, Felle HH, Umehara Y, Groenlund M, Sato S, Nakamura Y, Tabata S, Sandal N, Stougaard J (2003) Plant recognition of symbiotic bacteria requires two LysM receptor-like kinases. Nature 425:585–592

Ranf S, Grimmer J, Poeschl Y, Pechner P, Chinchilla D, Scheel D, Lee J (2011) Defense-related calcium signaling mutants uncovered via a quantitative high-throughput screen in arabidopsis thaliana. Mol Plant. doi:10.1093/mp/ssr064

Ryan CA, Pearce G (2003) Systemins: a functionally defined family of peptide signals that regulate defensive genes in Solanaceae species. Proc Natl Acad Sci USA 100:14577–14580

Schaller A (1999) Oligopeptide signaling and the action of systemin. Plant Mol Biol 40:763–769

Scheer J, Ryan C (2002) The systemin receptor SR160 from Lycopersicon peruvianum is a member of the LRR receptor kinase family. Proc Natl Acad Sci USA 99:9585–9590

Scheer JM, Pearce G, Ryan CA (2003) Generation of systemin signaling in tobacco by transformation with the tomato systemin receptor kinase gene. Proc Natl Acad Sci USA 100:10114–10117

Schilmiller AL, Howe GA (2005) Systemic signaling in the wound response. Curr Opin Plant Biol 8:369–377

Shiu SH, Bleecker AB (2001) Plant receptor-like kinase gene family: diversity, function, and signaling. Sci STKE 113: re22

Strattmann J, Scheer J, Ryan CS (2000) Suramin inhibits initiation of defense signaling by systemin, chitosan, and a β-glucan elicitor in suspension-cultured Lycopersicon peruvianum cells. Proc Natl Acad Sci USA 97:8862–8867

Takeuchi H, Higashiyama T (2011) Attraction of tip-growing pollen tubes by the female gametophyte. Curr Opin Plant Biol 14:614–621

Toer M, Lotze MT, Holton N (2009) Receptor-mediated signaling in plants: molecular patterns and programmes. J Exp Bot 60:3645–3654

Volkov A, Ranatunga DFA (2006) Plants as environmental biosensors. Plant Signal Behavior 1:105–115

Wang YH, Gehring C, Gahill DM, Irving HR (2007) Plant natriuretic peptide active site determination and effects on cGMP and cell volume regulation. Funct Plant Biol 34:645–653

Wang YH, Gehring C, Irving HR (2011) Plant natriuretic peptides are apoplastic and paracrine stress response molecules. Plant Cell Physiol 52:837–850

Wheeler JI, Irving HR (2010) Evolutionary advantages of secreted peptide signaling molecules in plants. Funct Plant Biol 27:382–394

Wu J, Wang S, Gu Y, Zhang S, Publicover SJ, Franklin-Tong VE (2011) Self-incompatibility in Papaver rhoeas activates nonspecific cation conductance permeable to Ca^{2+} and K^+. Plant Physiol 155:963–973

Yan X, Wang Z, Huang L, Wang C, Hou R, Xu Z, Qiao X (2009) Research progress on electrical signals in higher plants. Prog Nat Sci 19:531–541

Zhang W, Fan L-M, Wu W-H (2007) Osmo-sensitive and stretch-activated Calcium-permeable channels in *Vicia faba* guard cells are regulated by actin dynamics. Plant Physiol 143:1140–1151

Zimaro T, Gottig N, Garavaglia BS, Gehring C, Ottado J (2011) Unraveling plant responses to bacterial pathogens through proteomics. J Biomed Biotechnol. doi:10.1155/2011/354801

Zimmerman S, Nuernberger T, Frachisse J-M, Wirtz W, Guern J, Hedrich R, Scheel D (1997) Receptor-mediated activation of a plant Ca^{2+}-permeable ion channel involved in pathogen defense. Proc Nat Acad Sci USA 94:2751–2755

Chapter 14
Characterisation of Root Plasma Membrane Ca^{2+}-Permeable Cation Channels: Techniques and Basic Concepts

Vadim Demidchik

Abstract Calcium is an important macronutrient, which is required for structural and regulatory needs. Elevation of the cytosolic free Ca^{2+} is widely accepted as a major signalling mechanism in plants, which encodes the information about a multitude of exogenous and endogenous stimuli. Ca^{2+} uptake by roots from the soil is a passive process, which is catalysed by specialised plasma membrane proteinaceous pores that are called cation channels. These channels demonstrate different biophysical and physiological characteristics and may belong to a number of gene families. This chapter summarises data on physiological techniques and basic concepts for investigation of Ca^{2+}-permeable cation channels in plant root cells. The main focus of the chapter is on patch-clamp technique, ion-selective vibration microelectrodes and Ca^{2+}-aequorin luminometry. In the past two decades, a combination of these techniques allowed to investigate Ca^{2+}-permeable channels in higher plants and establish their crucial roles in plant cell physiology.

14.1 Introduction

Plants consist of two 'principal' parts that are non-photosynthesising roots and photosynthesising shoots. For a number of plant species, fully developed leaves appear only for a short period of time (4–5 months), while roots are always present. Leaves are highly specialised in photosynthesis. They are covered by wax cuticula that prevents an exchange with the environment (exception is stomata).

V. Demidchik (✉)
Department of Physiology and Biochemistry of Plants, Biological Faculty
Belarusian State University, Independence Ave 4, 220030 Minsk, Belarus
e-mail: dzemidchyk@bsu.by

In contrast, roots are predominantly 'open system' and directly contact the soil solution, biota, water and minerals. Roots are very diverse anatomically, they can grow rapidly and 'explore' the soil, searching for water, macro- and micronutrients. Roots play the role of 'mechanical anchoring systems', holding plant in the soil, and interact with a number of stresses (salinity, pathogens, heavy metals, drought, etc.). Roots, like animals or fungi, are heterotrophs and rely on organic substances synthesised in leaves. Some researchers have recently hypothesised that root tips, which consist of many small non-differentiated cells and abundantly express synaptotagmins (proteins that function at synapses), may have brain-like activities functioning as 'command centres' (Baluška et al. 2010). This hypothesis might be a bit 'too brave' but root tips are indeed unique in their structure, function and behaviour. Intriguingly, the density of ion currents and fluxes, which is very high in synaptic membranes, is 3–4 times higher in root tips than in mature root cells (Demidchik et al. 2002b, 2003a) and 7–10 times higher than in leaf mesophyll (Shabala et al. 2006).

Roots have sophisticated ion transport and signal recognition systems. Among those systems, Ca^{2+}-permeable cation channels are particularly important. These channels play a number of critical physiological roles. First, they are a route for uptake of Ca^{2+}, Mg^{2+}, Zn^{2+} and a number of monovalent cations. Second, they regulate the gradient of electrochemical potential across the membrane (together with K^+ and Cl^- channels), which is a driving force for transport of ions. Third, they are responsible for the generation of so-called 'Ca^{2+} signals', which are the transient elevations of Ca^{2+} activity due to Ca^{2+} influx through the channels. Ca^{2+} signals seem to be necessary for regulation, signalling and coordination of cell functions (Hetherington and Brownlee 2004). Transient elevation of cytosolic Ca^{2+} activity ($[Ca^{2+}]_{cyt}$) causes dramatic modifications of catalytic capacities of enzymes, cytoskeleton, vesicular transport and gene expression. It is widely accepted that Ca^{2+} signals are likely to be generated in any plant tissue, and that this signalling event is central to all regulatory processes in plants (Hetherington and Brownlee 2004). Fourth, Ca^{2+}-permeable cation channels mediate 'polar Ca^{2+} influx' in growing cells, such as root hairs or cells of root elongation zone (Demidchik et al. 2002b, 2007; Foreman et al. 2003). This is required for stimulation of exocytosis and delivery of new membrane and cell wall material in the growing part of the cell (reviewed by Datta et al. 2011). This chapter summarises and discusses (in simple terms) key aspects of study of Ca^{2+}-permeable cation channels in plants.

14.2 Cation Channels in Plants

Plant cell is surrounded by polysaccharide cell wall with nanometer pores (usually 1–10 nm) (Fleischer et al. 1999) and the plasma membrane with 'subnanometer' pores (0.2–0.5 nm) (Carbone 2004). This allows cell to regulate substance, energy and information exchange with the environment and other cells. Plant membrane pores are formed by transmembrane proteins that can be divided into two main categories: ion channels (selective for cations or anions) and aquaporins (transporting water). Some of ion channels are permeable to Ca^{2+} and called the Ca^{2+}-permeable cation channels. Plants do not have homologs of classical animal Ca^{2+} channels, which are, by several orders of magnitude, more selective for Ca^{2+} than to K^+ and Na^+ (Demidchik et al. 2002a). Therefore plant Ca^{2+}-transporting channels are often called Ca^{2+}-permeable cation channels or Ca^{2+}-permeable nonselective cation channels (NSCCs) (Tester 1990; Gelli and Blumwald 1997; Pei et al. 2000; Demidchik et al. 2002a, b, 2003a; Demidchik and Maathuis 2007; Jammes et al. 2011).

There are two major approaches for investigation of Ca^{2+}-permeable cation channels, which are as follows: (A) molecular approach (identification of genes, expression patterns, molecular structure, studying key functional moieties and their relation with function) and (B) physiological approach (characterisation of biophysical and pharmacological properties, such as voltage dependence, selectivity, kinetics, parameters of single channels, sensitivity to natural and synthetic modulators, and examination of functions at the cellular, tissue, organ, organismal or population levels). While molecular approaches for study of cation channels are similar to those that are routinely used in research on any plant gene, the physiological techniques are 'highly specialised' and require specific biophysical knowledge and practical skills in classical plant physiology.

In higher plants, Ca^{2+}-permeable cation channels/NSCCs are likely to belong to the following four gene families: (1) cyclic nucleotide gated channels (CNGCs; 20 genes in arabidopsis, 16 in rice); (2) ionotropic glutamate receptors (abbreviations: IGRs, iGluRs, GLRs; 20 genes in arabidopsis, 13 in rice, 61 in poplar); (3) mechanosensitive cation channels (abbreviations: MSCs, MSCCs; 10 genes in arabidopsis); (4) two-pore channel (TPC1; 1 gene in arabidopsis). Schematic structure of these channels is shown in Fig. 14.1a. Plant ion channels have not yet been crystallised; therefore their 3-D structure and gating mechanisms are unknown. They probably form oligomeric complexes of same or different subunits, similar to their animal counterparts (4 subunits for CNGCs, 4–5 for GLRs, 4–6 for MSCCs and 2 for TPC1) (Demidchik et al. 2002a; Davenport 2002; Demidchik and Maathuis 2007; Haswell et al. 2011). Data on their expression pattern are available online on 'genevestigator.com' and have recently been summarised by Jammes et al. (2011). Apart from AtCNGC18 and AtCNGC2, all tested channels have demonstrated a significant level of expression in roots. Selective filters of plant NSCCs, with exception for K^+-selective moieties in few CNGCs, are not studied in enough details; however, this has been suggested that they are Ca^{2+}-permeable (Demidchik et al. 2002a; Jammes et al. 2011).

Since the development of first plant electrophysiological techniques, cation transport has been attracting a great deal of attention from physiologists (Osterhout 1908, 1958; Osterhout et al. 1927; Yurin 1969). Although the importance of Ca^{2+} as a macronutrient and regulatory agent has been known for a very long time (Yurin 1969), mechanisms of Ca^{2+} transport in higher plants have only been elucidated in the past 20 years (see for review Schroeder and Thuleau 1991;

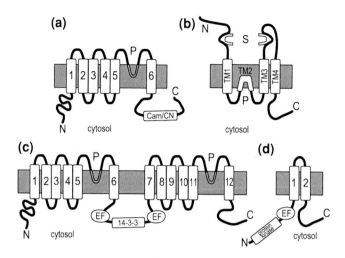

Fig. 14.1 Transmembrane topology and major functional domains of putative Ca^{2+}-permeable cation channels in plant membranes. **a** Cyclic nucleotide gated channels: shaker structure, 6 transmembrane domains (TMs; indicated as numbers 1...6), pore region between TM5 and TM6, cyclic nucleotide and calmodulin-binding sites overlap (Cam/CN). **b** Ionotropic glutamate receptors: TM domains 1, 3 and 4 cross the membrane, TM2 does not cross the membrane and forms the pore region, extracellular substrate-binding site (S). **c** Two-pore channel: 12 TM domains, pore regions between TM5 and TM6 as well as between TM11 and TM12, two EF hands (they catalyse Ca^{2+}-binding), '14-3-3 protein' binding site. **d** Mechanosensitive cation channels (MCA1), EF hand, protein kinase domain

Demidchik and Maathuis 2007). A number of Ca^{2+}-permeable conductances with different electrophysiological properties have been detected in roots of monocotyledonous and dicotyledonous plant species (reviewed by Tester 1990; Schroeder and Thuleau 1991; Demidchik et al. 2002a; Demidchik 2006; Demidchik and Maathuis 2007; Jammes et al. 2011). Huge diversity of properties of these conductances has been detected that can be explained by the fact that, hypothetically, different subunits from different families of channels can combine and form functional channel. Sometimes, in same preparations, researchers observed contrasting results for Ca^{2+} currents. Moreover, the arrangement of subunits could be a dynamic process, which 'behaviour' is difficult to predict. This can explain 'instability' in electrophysiological recordings that were carried out in vivo. Some indirect experimental evidence to support this hypothesis has recently been obtained in the study of plant K^+ channels (Xicluna et al. 2006). Joint expression of AKT2 and KAT2 in *Xenopus* oocytes resulted in a formation of channels with properties that were very different from AKT2 or KAT2 expressed individually (Xicluna et al. 2006).

Although genes of putative plant Ca^{2+}-permeable channels are well known, the functional and molecular analyses of these genes are still fragmentary. The linkage between gene and specific conductance is still missing for almost all 'candidate genes'. An exceptional case is vacuolar non-selective Ca^{2+}-permeable channel

TPC1 (Peiter et al. 2005). TPC1 (see structure in Fig. 14.1c) has been demonstrated to be responsible for Ca^{2+}-permeable 'SV conductance' in tonoplast ('slow vacuolar channel'), which has been known for a long time (since 1980s; reviewed by Maathuis 2010). SV channels dominate the vacuolar Ca^{2+} release in both roots and shoots and play pivotal roles in Ca^{2+}-mediated regulation in guard cells and roots (Maathuis 2010). However, these channels that are expressed in the vacuolar membrane are unlikely to catalyse Ca^{2+} transport through the plasma membrane.

14.3 What You Have to Know Before Starting Measurement

Measurement of Ca^{2+} conductance requires some theoretical understanding of processes that underlie the plasma membrane ionic conductances. The aim of this section is to show how Ca^{2+} currents is generated and how it can be 'isolated' from the total current of the plasma membrane, which is conducted by a number of ions that pass the membrane through different transporters.

14.3.1 Cation Channels Catalyse Ca^{2+} Influx (not Efflux)

Any positively charged ions are called cations. They are critical for plants because most micro- and macronutrients are available in the soil as cations (such as Ca^{2+}, Mg^{2+}, NH_4^+, K^+, Zn^{2+}, $Fe^{2+/3+}$, $Cu^{+/2+}$, etc.) (Bergmann 1992; Tyerman and Schachtman 1992). Ion channels transport cations 'passively' that means that they do not use metabolic energy, such as ATP (Nobel 2005). Therefore, sometimes, they are called 'passive' transporters of passive ion transport systems. The driving force for passive transport is the difference in chemical potentials (a measure of concentration or activity) of specific cation outside and inside the cytoplasm. Cations diffuse through the channel from the area of their higher chemical potential (concentration) to the area of their lower chemical potential. If membrane is electrically charged, so-called 'electrochemical' potentials should be considered instead chemical potentials, which combine both difference in the distribution (gradient) of concentrations and electrical charges (see book by Nobel 2005, for details).

In biological conditions, change of electrical potential can affect the direction of movement of all ions, apart from passive transport of Ca^{2+}, which is always directed from the extracellular space to the cytosol. This is related to a huge gradient of Ca^{2+} across the plasma membrane, which is from two to four orders of magnitude. Ca^{2+} level outside the cell is 0.01–1 mM, while, in the cytoplasm, it is approximately 0.0001 mM (100 nM) (Fig. 14.1b). According to Nernst equation for divalent cations, the electric potential of 29 mV is required for maintaining one-order difference of their activity across the membrane (Nobel 2005). This means that, if extracellular Ca^{2+} concentration is 0.01 mM, the reversal potential

(potential changing direction of ion flux) for Ca^{2+} (E_{Ca}) will be +58 mV. This is much more 'positive' than those values that are detected in vivo, even in 'heavily' depolarised plant cells (e.g. depolarisation caused by salinity, pathogens or oxidative stress) (Demidchik et al. 2010). The apoplast Ca^{2+} activity is about 1–3 mM, even if [Ca^{2+}] in the soil solution is low (0.01 mM) (Demidchik unpublished). This is related to the fact that cell wall components, such as anionic groups (–COOH and others) can bind Ca^{2+}, leading to 'accumulation' of Ca^{2+} in the thin extracellular layer near the plasma membrane. This means that real E_{Ca} is likely to be above +116 mV. In electrophysiological studies, so-called 'positive shift' in reversal potential ($E_{rev.}$) is indicative to increase of Ca^{2+} permeability of the plasma membrane. An activation of Ca^{2+}-permeable channels usually causes this 'positive shift' of $E_{rev.}$ (see for example Foreman et al. 2003).

14.3.2 Isolation of Ca^{2+} Conductance from the Total Plasma Membrane Conductance

In plant plasma membranes, K^+ channel conductance dominates, if extracellular K^+ concentration is high (>5–10 mM) (Sokolik and Yurin 1981, 1986). It masks all other conductances. This domination decreases in the presence of high external [Ca^{2+}] (>1 mM) and low [K^+] (<5 mM). In these conditions, the membrane potential as well as total transmembrane current are controlled by combined ion flux through K^+ channels, NSCCs, anion channels and active transporters, such as H^+-ATPase. Moreover, Ca^{2+} causes blockage of K^+ current through K^+ channels (Shabala et al. 2006) and NSCCs (Demidchik and Tester 2002) in roots. This additionally decreases K^+ channel contribution to the total plasma membrane conductance. To investigate Ca^{2+} current, researchers can increase Ca^{2+} level in the bath (extracellular solution) to at least 10 mM (up to 100 mM in some studies) (Gelli and Blumwald 1997; White 1998; Véry and Davies 2000; Pei et al. 2000; Demidchik et al. 2002b; Foreman et al. 2003). At bath 10–100 mM Ca^{2+} and 0.1–1 mM K^+ (or in K^+-free conditions), Ca^{2+} influx current dominate over the other conductances in the root plasma membrane (Demidchik et al. 2002b, 2009).

The permeability of K^+ channels can also be inhibited by application of pharmaceuticals, such as tetraethylammonium (TEA$^+$; applied from inside and outside the plasma membrane at the concentration 1–10 mM) or Ba^{2+} (outside; >0.3 mM), which specifically block K^+ channels and probably do not affect other conductances. An application of Ba^{2+} inside the patch-clamp pipette to block K^+ channels has been reported in some studies; however, the high intracellular Ba^{2+} activity, which is probably very similar to high Ca^{2+} activity, is not 'normal' (non-physiological) and can result in affected regulation of all conductances.

Removal of K^+ from the bath and microelectrode/pipette is widely used approach to 'isolate' Ca^{2+} conductance. First, if K^+ is removed from the bath, and only Ca^{2+} is left, the total influx of cations is catalysed by Ca^{2+}-permeable channels. This is critical for analyses of the 'pure' Ca^{2+} current. Second, the presence of high [K^+] in the patch-clamp pipette often has a stimulatory effect on K^+ channels; therefore, it is better to replace pipette K^+ with 'poorly' permeable cations, such as Na^+, Cs^+ or TEA^+. Although these cations are well-known antagonists of K^+ channels (Sokolik and Yurin 1981, 1986), they do not affect the Ca^{2+} permeability (Demidchik et al. 2002a). Third, high [K^+] in either bath or pipette can decrease the quality of seal in patch-clamp assays (Demidchik and Tester 2002). Patch-clamp pipette solutions containing salts of Na^+, Cs^+ or TEA^+ in concentration of 30–100 mM provide the best quality of recordings for *Arabidopsis thaliana* and *Triticum aestivum* root Ca^{2+} currents (Demidchik et al. 2002b; Demidchik and Tester 2002; Demidchik unpublished).

Another important problem is so-called 'chloride current component', which often affects recordings of Ca^{2+} currents. This is related to the activity of anion channels, which catalyse passive transport of Cl^- and other anions from cytoplasm to extracellular space. This Cl^- transport generates negative current that is of same polarity as Ca^{2+} influx currents. Cl^- is not abundant in plant cells (cytosolic Cl^- is 1–5 mM), but it can increase due to Cl^- leakage from KCl-filled microelectrodes and patch-clamp pipettes. To avoid the problem of Cl^- conductance, a number of methods can be used. Application of anion channel blockers, which are large and poorly permeable anions, has routinely been used to distinguish between cation and anion channel mediated currents (reviewed by Tyerman 1992; Demidchik and Tester 2002; Demidchik et al. 2002a). Nevertheless, this approach is not always effective because Cl^- channel blockers affect the viability of protoplasts and change regulation of cation channels. Another solution has been routinely used in patch-clamp studies which is the decrease of Cl^- concentration in the patch-clamp pipette from 50–100 to 5–10 mM (Demidchik and Tester 2002c). This can be achieved by using gluconate salts instead chlorides. Best results have been obtained for 30–80 mM Na-gluconate in the pipette solution. However, the complete removal of Cl^- from patch-clamp pipette is impossible because Cl^- is required for the electrical coupling between solution and AgCl-coated wire (electrical contact in the pipette). Moreover, high Cl^- in the pipette contributes to lower values or the pipette resistance and higher signal-to-noise ratios that are very important for good quality voltage clamp recordings. Therefore, when researchers decrease the Cl^- in the pipette, they should take into account possible consequences. Good signal-to-noise ratios can be achieved up to 10 mM Cl^- (not lower) in the pipette solution, if it is combined with low resistance patch-clamp pipettes (usually, having diameter of pore, d_{pore}, higher than 0.5 μm). However, in the case of impalement techniques, such as one- or two-electrode voltage clamp, when the microelectrode tip is very thin ($d_{pore} = 30$–150 nm), at least 100–200 mM Cl^- is required to maintain adequate signal-to-noise ratios. This clearly shows that, in the case microelectrode impalement techniques, Cl^- leakage is 'inevitable' and quality measurements of Ca^{2+} currents can be problematic.

14.3.3 Pharmacological Analysis of Ca^{2+}-Permeable Channels

When K^+ and Cl^- currents are 'minimised', researchers may also want to ensure that they are indeed dealing with Ca^{2+} conductance. Lanthanides, such as Gd^{3+} and La^{3+} (usually applied as chlorides) have been routinely used as antagonists of plant Ca^{2+}-permeable channels (reviewed by Demidchik et al. 2002a). These trivalent cations should be applied at concentration of 0.01–0.1 mM. Higher concentrations of these blockers, which have been used in some studies, can affect the plasma membrane lipids and cause artefacts (Cheng et al. 1999). Organic 'Ca^{2+} channel antagonists' (particularly dihydropyridines and phenylalkylamines) can also be used to detect Ca^{2+} conductances in plants; however, the high concentration of these substances (>0.1 mM) can also induce non-specific harmful effects, such as membrane lesions (reviewed by Demidchik et al. 2002a). Among organic 'Ca^{2+} channel antagonists', nifedipine, verapamil and diltiazem, have been most widely used in plant electrophysiology. However, not all Ca^{2+}-permeable channels are sensitivity to these antagonists. Besides, these inhibitors can block K^+ channels in some preparations.

Although test with one blocker is not very informative, a complex pharmacological analysis, which should include tests with TEA^+, anion channel blocker (such as 'SITS', 'DIDS', 'NPPB' or flufenamic acid), Gd^{3+}, nifedipine, verapamil and diltiazem is a very useful and effective tool for investigation of Ca^{2+}-permeable channels in plants. Only complex analysis can confirm that the channel is indeed Ca^{2+}-permeable. Note that, in some root cells, an application of inhibitors of active transport systems (particularly electrogenic H^+-ATPase) may additionally be required. In patch-clamp experiments with root protoplasts, H^+-ATPase activity is very low or not present. However, in studies with microelectrode impalement techniques, when researchers investigate intact tissues, H^+-ATPase activity can still be significant, and application of H^+-ATPase inhibitors, such as 'FCCP' (1–10 μM), 'CCCP' (1–10 μM), 'DCCD' (50–300 μM) or Cu^{2+} (3–10 μM), is critically important.

14.3.4 Different Types of Root Ca^{2+}-Permeable Cation Channels and Their Current–Voltage Relationships

The functionality of electrical conductor, which is a material conducting electrical currents, can be characterised by the current (charge transferred per surface, per time; I, SI units are Amperes—A) induced in response to the change of voltage (difference in charges; V, U or E, SI units are Volts—V). The conductivity (or specific conductance; G, SI units are Siemens—S, $G = I/U$) characterises the capacity of a given material to conduct current. It is opposite to resistivity (or specific resistance; $R = 1/G$, SI units are Ohms—Ω, $R = U/I$), which demonstrates how material opposes the flow of electric current. Some of conductors

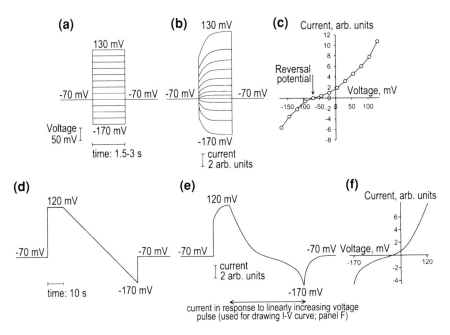

Fig. 14.2 Schemes of voltage-clamp protocols for investigation of plant cation channels. Holding potential in all cases is −70 mV. **a** Voltage step protocol (perpendicular or square voltage pulses): voltage is sequentially clamped at 13 levels from −170 to 130 mV for the time of 1.5–3 s. **b** Thirteen current curves in response to voltage steps (from 'a'). **c** I–V curve plotted using voltage (from 'a') and steady-state current values (from 'b'). **d** Voltage ramp protocol (gradual voltage change): voltage is pre-clamped at 120 mV and, then, gradually decreased from 120 to −170 mV sequentially clamped at 13 levels from −170 to 130 mV for the time of 1.5–3 s. **e** Current in response to voltage ramp pulse (from 'd'). **f** Ramp I–V curve plotted using values of current (from 'd') and voltage (from 'e')

allow more ion passage than others. For example, Cu is 'better' conductor than Al; this also means that Al has higher resistivity. 'Ideal' conductors (such as metals) have linear dependence of current from voltage (also called 'I–V dependence' or 'I–V curve'). This means than their conductance remains constant at different voltages.

In semiconductors, conductivity is regulated by specific factors/conditions, including the magnitude of voltage. Some ion channels behave similarly to voltage-dependent semiconductors and show different conductance at different voltages. This group of channel is called voltage-dependent channels. To investigate voltage dependence of cation channels as well as other biophysical characteristics, the I–V of the plasma membrane has to be measured and plotted (Fig. 14.2). There are three principal kinds of I–V curves (I–Vs): (1) step or 'steady-state' I–V (Fig. 14.2a–c), (2) ramp I–V (Fig. 14.2d–f) and (3) tail-current I–V which is, sometimes, also called instantaneous I–V (Demidchik et al. 1997).

Steady-state I–V ('step I–V') can be obtained by application of 'perpendicular' (sometimes also called 'square') pulses of hyperpolarising or depolarising voltage

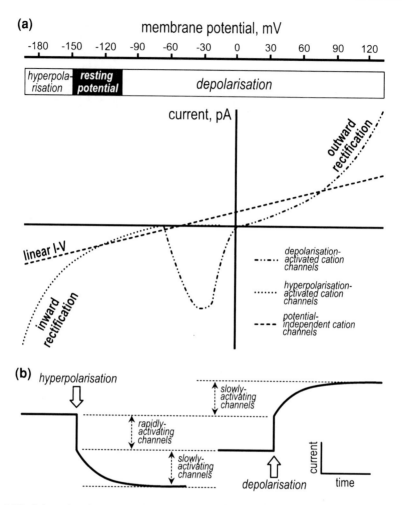

Fig. 14.3 Schematic picture of cation currents through the plasma membrane of root cells. **a** Current–voltage curves of three different Ca^{2+}-permeable channels. **b** Time courses of activation of currents through slowly- and rapidly-activating cation channels

(so-called 'voltage step protocol' in Fig. 14.2a) to the plasma membrane. In the case of 'perpendicular pulse', the voltage is rapidly changed from one level ('holding voltage/potential'—HP) to another (test voltage or pulse voltage), and then back to HP. In most patch-clamp studies, HP has been maintained at the level of -70 to -100 mV. These HP values improve the quality of 'giga seal' isolation. In one- or two-electrode voltage clamp (impalement techniques), the HP is between -100 and -150 mV. Test voltage is usually applied as series of 10–20 pulses (steps) from -300 to $+120$ mV. These pulses induce currents, which, in most cases, show time-dependent activation (such as shown in Figs. 14.2b, 14.3b). There are also time independent or instantaneous currents, activating very rapidly—within few millisecond after application of voltage pulse (Demidchik and Tester 2002). To record 'high quality' steady-state I–Vs, voltage pulse should be long enough, at least 1.5–3 s, to allow full current activation and reaching 'steady state' (Figs. 14.2b, 14.3b). For some channels, which have slow activation kinetics of currents, these pulses can be much longer (up to several minutes). Steady state values of current (I) and values of test voltage (V) can be plotted against each other and their relationship is called I–V curve (Figs. 14.2c, 14.3a).

Voltage ramp is a 'gradually changing' voltage pulse (Fig. 14.2d). The change of current over time in response to this 'voltage ramp protocol' is demonstrated in Fig. 14.2e. Plotting this current (Fig. 14.2e) against voltage (Fig. 14.2d), which is possible by using electrophysiological software, such as Clampex (developed by Molecular Devices; previous name of this company is Axon Instruments) or Patchmaster (developed by HEKA; previous name of this company is List Instruments). Voltage ramp protocol may also include a pre-pulse (holding a cell for several seconds at hyperpolarisation of depolarisation), which is required to allow currents to activate (shown in Fig. 14.2d). In patch-clamp assays, ramp protocol is usually much faster than perpendicular pulse protocol and less damaging to the membrane because long-term holding voltage at very negative or very positive levels may induce membrane breakdown or loss of giga seal. However, perpendicular pulse protocol allows much more accurate measurement of I–Vs. Moreover, only perpendicular pulse protocol makes it possible to measure kinetics of the current and conclude about whether the studied channel is slowly or rapidly activated.

Several approaches can be used to obtain the tail-current I–V (Sokolik and Yurin 1981, 1986; Demidchik et al. 1997). The voltage should be hold at the level that induces maximal channel activation, for example, -180 to -250 mV for the hyperpolarisation activated cation channels or -20 to $+100$ mV for the depolarisation activated cation channels. Then, voltage step or ramp protocol can be applied, but they should have very short (20–50 ms) pulses of test voltage. To obtain I–V curve, current, which is measured at the end of the pulse, is plotted against the voltage. Tail-current I–V are not used in studies of channels with instantaneous activation kinetics; however, they are important for assessing channel selectivity (by the shift of E_{rev}) in slowly activating cation channels (Hille 2003).

Three main types of Ca^{2+}-permeable channels with different voltage dependence have been found in plant roots (Fig. 14.3a), which are as follows: (1) hyperpolarisation activated Ca^{2+}-permeable channels (HACCs; conduct more current at voltages below -130 to -150 mV), (2) depolarisation activated Ca^{2+}-permeable channels (conduct more current at voltages above -70 to -90 mV; DACCs), (3) voltage-independent Ca^{2+}-permeable channels (these are not semi-conductors and equally conduct current along the voltage axis; VICCs) (Fig. 14.3a). In root cells, all three types of Ca^{2+} conductances can co-exist (Miedema et al. 2008).

Hyperpolarisation and depolarisation are states (or processes) that are related to increase or decrease, respectively, of resting potential (RP), which is a potential that is measured in the cell without voltage clamping or generation of action potential. VICCs dominate at resting potential because both HACCs and DACCs are not active at these voltages (Demidchik et al. 2002b). VICCs transport Ca^{2+} for 'nutritional' need and seem to be 'permanently' active. They are insensitive to all organic Ca^{2+} channel antagonists, Cl^- channel inhibitors and TEA^+, but blocked by Gd^{3+}. These channels might also be involved in transport of other physiologically important cations, such as Mg^{2+} and Zn^{2+}. They are also involved in the regulation of the basal Ca^{2+} level, which is critical for plant cell signalling (Demidchik et al. 2002b).

Nonlinear parts of I–V curve are related to the phenomenon which is called 'current rectification' (when current flows better in one of direction); therefore, channels are also named by the direction of their rectification: inwardly or outwardly rectifying. For further details read the book by Hille (2003). Inwardly rectifying Ca^{2+}-permeable channels usually, but not always, are the same as HACCs because they conduct Ca^{2+} currents better at hyperpolarisation (Gelli and Blumwald 1997; Véry and Davies 2000; Demidchik et al. 2002b, 2009; Foreman et al. 2003). Hypothetically, outwardly rectifying Ca^{2+}-permeable channels conducting Ca^{2+} efflux current may exist in plant cells ('outward rectification' in Fig. 14.3a); however, their role in cell physiology would not be significant because E_{Ca} is always above +100 mV. Romano et al. (1998) have suggested that K^+-permeable outwardly rectifying channels (in *Arabidopsis thaliana* leaves) are involved in Ca^{2+} influx; however, this interesting hypothesis has not been tested in roots.

A special case for research on Ca^{2+}-permeable channels is a 'bell-like' I–V of some DACCs (Fig. 14.3a). Discovered in 1970s in giant alga (Lunevsky et al. 1977, 1983), DACCs have been identified in roots of higher plants (Thuleau et al. 1994; Thion et al. 1998; Miedema et al. 2008). Activity of these channels is very unstable, in patch-clamp conditions, currents through DACCs can only be observed during first 10–15 min after formation of contact between pipette solution and cytoplasm (Miedema et al. 2008). This may be related to the state of the cytoskeleton, which can regulate DACC activity (Thion et al. 1998). DACCs have not been observed in all cells; therefore to investigate their conductance, a number of cells have to be tested. When studying DACCs, ATP (as Mg^{2+} salt; 1–2 mM) can be added to the pipette solution to increase their lifetime in patches (Miedema et al. 2008).

Ca^{2+}-permeable channels are not only different in the voltage dependence, but also in the time dependence of the activation. They form two major classes, which are 'slowly' and 'rapidly' activating Ca^{2+} permeable cation channels, respectively (reviewed by Demidchik and Maathuis 2007). Slowly activating channels, after the change of voltage (depolarisation or hyperpolarisation), react by activation of conductance during few minutes (1–2 min) while the reaction of rapidly activating channels as fast as few milliseconds (Demidchik and Tester 2002; reviewed by Demidchik et al. 2002a) (Fig. 14.3b).

14.4 Electrophysiological Techniques for Studying Root Ca^{2+}-Permeable Channels

The central question of any research on plant Ca^{2+}-permeable conductances is how Ca^{2+} is transported through the channel and how this transport is regulated by internal and external factors. Even if genes are known, the electrophysiological analyses cannot be omitted because only these analyses show how channel functions in living cells. To investigate this, electrical currents or net fluxes mediated by cation movement through the channel should be measured. There are two main approaches in electrophysiology of plant cells: extracellular and intracellular recordings. Electrophysiological techniques for extracellular recordings measure changes in electrical field or ion concentration outside the cell, which are related to the functional activity of the population of cation channels. Specific physical size and shape of a given tissue, as well as access only to external layers of cells may seriously limit the use of extracellular recordings. Intracellular recordings can be performed by impaling a cell with one or more fine-tipped electrodes (so-called 'impalement techniques'), or by sticking a cell to the glass micropipette (patch-clamp technique). Intracellular recordings allow measuring membrane voltage and current. Moreover, it is possible to control these parameters and investigate the behaviour of channels, for example, at different voltages (very important for understanding of voltage dependence of a given channel). Root cation channels have also been characterised electrophysiologically by incorporation of purified plasma membrane proteins into the planar lipid bilayers or artificial lysosomes (vesicles) that mimic the cell membrane. This technique is not discussed here because it is reviewed elsewhere (Tester 1990; White 1998). Although the use of lipid bilayers have confirmed that Ca^{2+}-permeable channels are protein with cationic conductances, it did not clarify properties of these channels in intact cells, as well as it did not indicate to genes encoding them.

14.4.1 Measurement of Field-Potentials

In recording with extracellular electrodes, an electrode is placed in the extracellular medium and field-potentials contributed by the action potentials of many small cells are measured (Zawadzki 1980). If electrodes contact the tissue, they are called 'surface contact electrodes'. This technique has been extensively used for examination of generalised electrical responses of roots and leaves to biotic stresses and gravity (Iwabuchi et al. 1989; Favre et al. 2001). An advantage of this technique is that it is non-destructive for cells; however, the information that is obtained is difficult to interpret. Basically, changes in electric field-potentials demonstrate that 'something has happened' but it does not show 'what exactly has happened'.

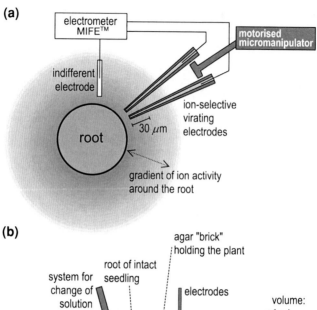

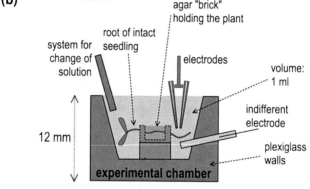

Fig. 14.4 Schematic picture of extracellular vibrating ion-selective electrodes. **a** Scanning of Ca^{2+} gradient in the 'close proximity' to the root surface. **b** Experimental chamber for studyng root Ca^{2+}-permeable channels. This design of this chamber allows to reduce the time for solution mixing from 2 min to 30 s

14.4.2 Extracellular Ion-Selective Microelectrodes

'Extracellular ion-selective electrodes' are the most widely used extracellular electrophysiological technique (Fig. 14.4). Automated 'vibrating' extracellular ion-selective microelectrodes, or so-called MIFE® (Microelectrode Ion Flux Estimation), has been particularly important for non-invasive characterisation of Ca^{2+} influx through plant Ca^{2+}-permeable channels in root cells (Demidchik et al. 2002b, 2003a, 2009, 2010, 2011). In MIFE®, extracellular ion-selective microelectrodes with sharp tips are fabricated from borosilicate glass. Then tips should be 'broken' ('blunted') to make a wider pore ($d = 2$–3 μm). Tip is filled with ionophore for Ca^{2+} and electrode is calibrated using standard Ca^{2+} solutions (Geochem software can be used to calculate Ca^{2+} activity). The quality of MIFE electrodes can vary but 'good electrode' usually shows the response which is close to Nernstian: a 10-fold change in Ca^{2+} activity should lead to 29 mV response in the MIFE electrode. Electrode should also demonstrate a rapid response (reaching steady state in at least 1–2 s) and reading without drift.

Several MIFE electrodes with different ionophores can be used simultaneously to measure several ions, if required. Microelectrodes are mounted on motorised micromanipulator (Eppendorf or Narishige) near the surface of the root and 'vibrate' (move forward/backward) with a period of few second, measuring ionic activity at two points at given distances (usually at about 50 and 100 μM from the root surface) (Fig. 14.4). Using the magnitude of this difference and taking into account the geometry of the surface, a special MIFE® analysis software calculates net ionic flux across the membrane surface. In contrast to extracellular electrodes MIFE® provides information about the ion selectivity of fluxes and numeric data on these fluxes. MIFE® can be applied to different parts of intact root system (including root hairs and root tips) as well as protoplasts (Shabala et al. 1997, 1998; Babourina et al. 2001; Demidchik et al. 2003a).

Advantages of MIFE® for study of Ca^{2+}-permeable channels are as follows: (1) it is non-invasive (cells remain intact; therefore their responses seem to be more 'natural'), (2) experiment is relatively simple and not expensive (as compared to patch clamp), (3) it shows the selectivity of cation fluxes, (4) it allows analysing fluxes of several ions simultaneously. However, apart from its many advantages, MIFE® also has a number of limitations. First, MIFE® electrodes do not allow recording ion activities higher than 0.2–1 mM. Second, some ionophores for MIFE® electrodes are not very selective between cations. For example, Na^+-selective dye also senses K^+. Therefore, corrections should be made if several cations are present in the assay solution and their activities are recorded. Third, using MIFE® is limited by surfaces that are exposed to the bathing medium. This technique does not allow measurements of fluxes in internal tissues or layers of cells. Another problem in measurements of Ca^{2+} flux with MIFE was that these recordings required a pause of about 2 min after addition of test substances to the assay chamber for mixing of solutions. This limitation is crucial when fluxes show fast kinetics of activation, such as receptor-like amino acid or purine-activated Ca^{2+} influx. Demidchik et al. (2011) have recently developed a new MIFE®-based approach to study fast changes in plant cation fluxes. These authors have minimised the volume of the assay chamber and changed its geometry, as well as designed new rapid system for addition of solutions. Overall, these modifications to MIFE allowed decreasing the time of mixing from 2 to 0.5 min (it reduced the pause after mixing by four times).

A combination of MIFE® with patch-clamp technique has been developed in Australia by Tyerman et al. (2001). Although this approach is technically difficult, it allows simultaneous measurements of cationic conductance and corresponded fluxes that makes it possible very accurate identification of ion-specificity of the conductance. Work on coupling MIFE with other electrophysiological (two-electrode voltage-clamp) and cellular (confocal imaging) techniques is now in progress in several laboratories. Combination of MIFE®, two-electrode voltage-clamp and confocal microscopy will be a powerful tool for studying Ca^{2+}-transporting systems that lack activities in patch-clamp tests, for example, plant purine- and amino acid-induced conductances (Demidchik et al. 2004, 2009, 2011).

14.4.3 Intracellular Techniques: Measurements of Membrane Potential with Single Sharp Microelectrode

The theory of selective permeability to cations of the plant plasma membrane (reviewed by Hille 2003) was developed in the beginning of the twentieth century by laboratory of Osterhout (1908). This was followed by the first tests of the plasma membrane potential in sea green alga *Valonia macrophysa* using impalements with sharp glass microelectrodes (Osterhout et al. 1927). In the next 35 years, membrane potentials were examined in *Nitella, Nitellopsis, Chara, Valonia* and *Halicystis* (reviewed by Osterhout 1958; Hope and Walker 1975; Yurin et al. 1991). In these experiments, plant cells were impaled with one electrode and the difference in electric potentials was measured between this electrode and indifferent electrode in the extracellular solution. Measurements of plasma membrane potentials in different conditions played a key role in the development of ideas of plant mineral nutrition, signalling and hormone-induced regulation. These studies predicted existence of cation channels in plants.

Intriguingly, the electrophysiological properties of green algae and root cells are very similar. This could be due to the fact that root epidermis, as well as algae cells are, 'open systems', that directly contact with extracellular solution (pond, sea water or soil solution). For example, cation channels in *Arabidopsis thaliana* (Demidchik et al. 1997, 2001) resembles cation channels in *Nitella flexilis* (Demidchik et al. 2002b, 2003a, 2011). NSCCs and K^+ outwardly rectifying channels in these objects have similar kinetics, voltage dependence and selectivity to cations.

These days, measurement of membrane potential remains a very useful tool in plant electrophysiology. For example, using this technique, Ehrhardt et al. (1992) discovered Nod-factor-induced depolarisation of the plasma membrane of plant root cells that later on was shown to be an effect of activation of inwardly rectifying K^+ channels (Kurkdjian et al. 2000; Ivashikina et al. 2001). Measuring membrane voltage response to purines, Lew and Dearnaley (2000) have found depolarising effect of these substances on the root plasma membrane. Later on, it was shown that purine activated cation channels that are similar to animal purinoceptors catalyse this depolarisation (Demidchik et al. 2003b, 2011). Dennison and Spalding (2000) demonstrated that glutamate can depolarise the plasma membrane of *Arabidopsis thaliana* root epidermal cells. This finding has led to the discovery of plant glutamate-activated cation channels (Demidchik et al. 2004). Impalement with sharp microelectrodes was an important step for development of intracellular ion-selective electrodes, which allow to measure the activities of different cations in the cytosol and other cellular compartments (Miller et al. 2001).

14.4.4 Intracellular Techniques: Two- and One-Microelectrode Voltage-Clamp

Membrane potential measurements have predicted that cation-permeable channels exist in the plant plasma membranes (addition of cations shifted membrane potential), but it has not shown properties of these channels or mechanisms of regulation. The nature of plant cationic conductances remained unclear until the first application on plants of so-called two-electrode voltage-clamp technique by Findley (1961) and Hope (1961). This technique was previously successfully used in animal physiology by Hodgkin and Huxley (1952) for creation of 'ion channel theory'. Two-electrode voltage-clamp technique 'clamps' or maintains membrane voltage at a value the experimenter specifies. Voltage control is established using feedback through an operational amplifier circuit (Halliwell et al. 1994). The main value of voltage-clamp technique is that it allows one to measure the amount of ionic current crossing a membrane at any given voltage at a given time. Using two-electrode voltage-clamp Soviet electrophysiologists characterised for the first time plant cation channels (Sokolik and Yurin 1981, 1986), including Ca^{2+}-permeable channels (Plaks et al. 1979, 1980; Lunevsky et al. 1983) and NSCCs (Yurin et al. 1991).

Two-electrode voltage-clamp technique is much less invasive than patch-clamp technique because it does not damage cell wall, membranes and the cytosol. However, this technique has not been extensively used for characterisation of cation channels in root of higher plants. Main reasons of this include: difficulty in fabrication of electrodes and impaling, inconvenience of using small cells, preparation problems in highly organised and specialised multi-cellular tissues and organs, impossibility to manipulate the intracellular solution and electrolyte leakage from the electrode to the cytosol that was significant in cells smaller than giant alga. This has led to the fact that experimental conditions allowed using two-electrode voltage-clamp for studying cation channels only in a few higher plants. In studies of root cation channels, this technique has only been used on root hairs (Lew 1991; Meharg et al. 1994). However, in experiments on these large cells (root hairs in *Arabidopsis* are at least 0.1 mm long), another significant problem that limits application of two-electrode voltage-clamp in higher plants has been found, which is current dissipation along the length of the cells (Meharg et al. 1994). This can result in distortions of the current–voltage (I–V) curve, including consistent underestimation of the membrane current, linearization of the I–V and masking of conductance changes in the presence of transported substrates.

Recording of current using two-electrode voltage-clamp technique can be coupled with the measurements of cytosolic Ca^{2+} activity (Levchenko et al. 2005). Levchenko et al. (2005) have used three-barrelled electrodes: two barrels clamped voltage and third barrel loaded fluorescent Ca^{2+}-sensitive probe (FURA) into the cytosol. Tests were carried out on intact leaves that were not excised from the plant. Although this approach has only been applied to leaf cells, it has a great potential for characterisation of root Ca^{2+} conductances. This technique has already shown that data on ABA regulation of Ca^{2+} channels in protoplasts derived from guard cells should be revised (Levchenko et al. 2005).

Although two-electrode voltage clamp has only been applied few times to intact roots, it has been extensively used for characterisation of plant cation channels (including CNGCs) heterologously expressed in *Xenopus laevis* oocytes (Dreyer et al. 1999; Miller and Zhou 2000). Probably the most successful functional expression in *Xenopus* oocytes was carried out in the case of *Arabidopsis* K^+ channels (Véry and Sentenac 2002, 2003).

Apart from two-electrode voltage clamp, a discontinuous single electrode voltage-clamp technique (Wilson and Goldner 1975) can potentially be used for study of Ca^{2+}-permeable channels. Voltage clamping with a single microelectrode. This technique has been already applied to root cation channels (Kurkdjian et al. 2000), but it is potential in investigation of Ca^{2+} currents has not been tested.

14.4.5 Intracellular Techniques: Patch Clamp

In 1980s, two-electrode voltage clamp was replaced by patch-clamp technique that brought electrophysiological measurements to a new quality by allowing the testing of protoplasts derived form small cells and different plant tissues. Patch-clamp technique was developed in the 1970s by Erwin Neher and Bert Sakmann (Neher and Sakmann 1976) who received the Nobel Prize in 1991. With this technique a glass micropipette with large diameter at the tip forms a high resistance contact (so-called gigaohm seal, $R > 1{,}000{,}000\ \Omega$) with a membrane of the protoplast (cell wall-free plant cell) derived from plant cells (Fig. 14.5).

14.4.5.1 Protoplasts

Very clean (free of cell wall components) membrane surface is required for good quality gigaseal contact. Therefore, an important difference of plant patch-clamp assays from experiments on animals is that plant tissues should be treated with cell wall-degrading enzymes to isolate protoplasts. Cellulolytic and pectolytic enzymes are applied as a mixture. These enzymes include cellulases, which breakdown 1, 4-beta-D-glycosidic linkages in cellulose, lichenin and cereal beta-D-glucans, and pectinases or pectolyases that cleave bonds in pectines, such as polygalacturonic acid. In most cases, cellulases that are used in protoplasts isolation for patch-clamp experiments derive from pathogenic fungus *Trichoderma viride* (dark green parasitic mold), for example, Cellulase RS (Onozuka) and Y-C (Kyowa Kasei) or Cellulysin (Calbiochem). Pectolyase Y-23 (Onozuka) from *Aspergillus japonicus* is the most widely used pectolytic enzyme for protoplast isolation. Cell wall-degrading enzyme mixture normally consists of 1–3% cellulase (it could also be a mixture of different cellulases) and 0.1–0.3% pectolyase. Enzymes are commercially available as powders that require activation (reconstitution) in water for at

14 Characterisation of Root Plasma Membrane Ca^{2+}-Permeable Cation Channels

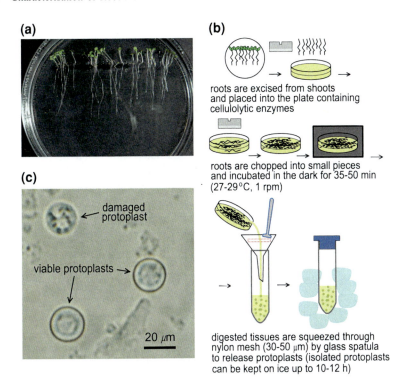

Fig. 14.5 The protocol for protoplast isolation from *Arabidopsis thaliana* roots (see also for details: Demidchik and Tester 2002; Demidchik et al. 2003). **a** Seven-day old sterile root seedlings grown vertically in 10 cm plastic dishes on the gel containing full strength Murashige and Skoog media (Duchefa), 0.35% Phytogel (gellan gum, Sigma) and 1% sucrose (pH 6 by KOH). **b** Roots are excised and placed in 5 cm plastic dish, containing cellulolytic enzyme solution. They are cut into 0.5–1 mm pieces with the razor blade and incubated in the dark. Enzymatically treated roots are squeezed by glass pestle, washed and filtered (through nylon mesh) to release protoplasts, which can be kept on ice during 10–12 h. **c** Microphotograph (Nikon Diaphot inverted microscope combined with Nikon D100 digital camera) demonstrating damaged and 'healthy' protoplasts (viable and suitable for patch-clamping). Protoplasts with a good quality gigaseal have the following properties: $d = 15$–25 μm, grey coloured, well rounded, rigid plasma membrane and free of intracellular dark bodies

least 2 h in the dark (with stirring). pH 6 (by KOH) and 28°C are found to be optimal for obtaining good quality protoplasts from *Arabidopsis* roots (Demidchik et al. 2002b). Demidchik and Tester (2002) have found that high concentration of Ca^{2+} (5–10 mM) in the enzyme solution increases the amount of isolated protoplasts. Techniques that were designed for protoplast isolation from *Arabidopsis thaliana* roots are schematically shown in Fig. 14.5. These techniques also give good results for roots of wheat and other plant species (Demidchik et al. unpublished). To test the quality of enzymatic treatment, Calcofluor White or similar cell wall binding fluorescent agents can be applied to protoplasts. Those that are suitable for patch clamping (cell wall free) will not be stained by this dye.

Intact plant cell can withstand diluted solutions but, when they do not have cell wall, they can only survive in isotonic solutions (when the intracellular osmolarity is same as osmolarity of bathing solution). Cell osmolarities vary in different plant species; therefore, isotonic solutions that are used for protoplast isolation (and then for storage of protoplasts) should be carefully tested before the experiment. For example, root protoplasts isolated from *Arabidopsis thaliana* have osmolarity of 280–320 mOsm kg^{-1} while *Triticum aestivum* root protoplasts have the osmolarity about 600 mOsm kg^{-1}. In patch-clamp studies on root plasma membranes, osmolarities have usually been found in preliminary tests and then adjusted by sugar alcohols, such as mannitol or sorbitol ($C_6H_{14}O_6$, Mol. mass 182.172). Addition of 1 g of mannitol or sorbitol to 50 ml of water gives the osmolarity of approximately 118 mOsm kg^{-1} in obtained solution. Vapour pressure osmometer is needed to measure osmolarities in all patch-clamp solutions (researchers should adjust them by addition of mannitol/sorbitol or water).

Protoplasts can also be isolated by the laser ablation technique (Kurkdjian et al. 1993). In this technique, the piece of the cell wall of preliminary plasmolysed cells (plasmolysis can be induced by high osmolarity solution) is cut off by laser. Then, after deplasmolysis step, which is induced by decrease of osmolarity, the protoplast is released that can be patch clamped. This method is described in details by Henriksen et al. (1996) and Véry and Davies (2000). The laser ablation approach has been successfully used for characterisation of Ca^{2+}-permeable cation channels in root hairs (Véry and Davies 2000; Foreman et al. 2003).

14.4.5.2 Patch-Clamp Pipettes

Good quality patch-clamp pipettes can be made by any conventional 2-stage microelectrode puller, such as Narishige, HEKA, Zeitz DMZ, WPI or Sutter Instruments (please follow to manufacturer's instructions). Two stages (Fig. 14.6a) are required to smooth edges on the pipette tip and to make it blunt (with cone-shaped tip) and having relatively wide pore ($d = 0.5–2$ μm). Two kinds of pipette glasses are used which are borosilicate glasses (such as Pyrex, Simax, Boralex, Kimax, Corning 7040 and 7052, etc.) and aluminosilicate/alumosilicate (such as WPI, Corning 1723, etc.) glasses, respectively. The general requirement for glasses is that they should provide low noise and low capacitance as well as good membrane-adhesive properties. Many of the patch-clamp studies on plant membranes have been performed with soft KIMAX-51 (Kimble Glass). Using this glass allows to avoid polishing of pipette after pulling, which is usually necessary for other glass varieties (polishing can additionally smooth and blunt the pipette, improving its adhesiveness). In classical studies, to reduce noise and capacitances pipettes, researchers covered pipettes by Sylgard (hydrophobic, translucent elastomer resin). However, wax or paraffin can give similar results (Demidchik and Tester 2002). To this end, patch-clamp pipette is 'treated' by another capillary filled with by melted paraffin/wax (Fig. 14.6b). Moreover, coating with petroleum jelly (such as Vaseline White Jelly), can also be used to this purpose.

14 Characterisation of Root Plasma Membrane Ca^{2+}-Permeable Cation Channels 359

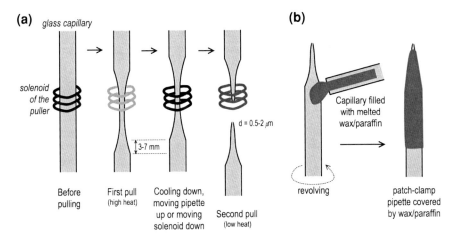

Fig. 14.6 Fabrication of patch-clamp micropipettes. **a** Pulling patch-clamp pipettes in two stages. Glass capillary (for example, KIMAX-51) is centered in the solenoid (spiral heating element of the puller). During 'first stage', controlled 'heating' and 'pulling' are applied to melt the pipette and make it 'thinner'. To recentre the heating element, it (after cooling down) can be moved down or pipette can be shifted up (depending on the puller's model). Lower heating is applied at the second stage, when the capillary is divided into two pipettes. **b** The covering of the patch-clamp pipette with wax/paraffin to reduce noise and capacitance

A number of 'recipes' for pipette solutions have been developed for patch-clamp analyses of root Ca^{2+}-permeable channels. Some recommendations are as follows: (1) pH has to be adjusted by biological buffers (1–10 mM ultrapure Tris, MES, HEPES, etc.) to the level of 7.1–7.4 (the value of cytosolic pH in most plant cells); (2) free Ca^{2+} should be calculated by Maxchelator or similar software and fixed at 100 nM by specific Ca^{2+}-binding agents, such BAPTA or EGTA; (3) K^+ should not be added to the pipette solution and Na^+ or Cs^+ are better to use instead K^+ (the reasons for this are described in the Sect. 3.2); (4) deionised water and only ultrapure chemicals from approved suppliers should be used; (5) Mg_2ATP can be added (freshly prepared) for studying HACCs, but this can be ignored in the case of NSCCs (these channels do not require intracellular ATP); (6) prepared solution should be aliquoted and kept at $-20°C$ (or $-80°C$) for not longer than 1 year.

All solutions for filling patch-clamp pipettes (pipette solutions) should be filtered through bacterial filter (0.22 μm). This facilitates the filling of pipettes with solution and improves the quality of gigaseal. After filling, the resistance of patch-clamp pipettes (usually it is 1–10 megaohms) can be tested with portable digital multimeter (any model will be fine). Some researchers use so-called 'bubble index' (a number of air bubbles passing through the tip of patch pipette in ethanol) to assess pipette resistance of (see for details Ogden and Stanfield 1994). Liquid junction potentials (they appear due to different diffusion coefficients and activities of ions in the pipette and bathing solution) have to be measured (or calculated) and corrected, where possible, before the patch-clamp experiment (see for details Ogden and Stanfield 1994). However, calculations cannot be carried out for the liquid junction potentials generated between pipette solution and the cytoplasm of the protoplast.

14.4.5.3 Patch-Clamp Set-Up and Configurations

In patch-clamp technique, only one electrode measures voltage and injects current to clamp the voltage at constant level; indifferent electrode is placed near the cell (AgCl-coated Ag wire). Patch-clamp electronic circuit is described elsewhere (see for example Ogden and Stanfield 1994). List of patch-clamp amplifier manufacturers includes the following: Molecular Devices, HEKA, Dagan, Warner Instruments, World Precision Instruments, etc. All these manufacturers produce reliable and good quality patch-clamp systems. More expensive models may have some additional features (sometimes, these features are not used), such as current clamp and two-electrode voltage clamp circuits, wide range of capacitance and series resistance corrections, precision filtering (note: electronic filtering by software gives similar results), etc.

Patch-clamp set-up usually consists of patch-clamp amplifier which is connected to computer via analogue–digital converter (interface) and controlled by professional electrophysiology software (all these components are produced by the manufacturer of amplifier), inverted microscope (most models will be fine; Nikon, Carl Zeiss, Leica, Olympus, etc.), good quality antivibration table with Faraday cage (most models will be fine; e.g. WPI, Intracel, etc.), precision micromanipulator (Narishige, Hodgkin-Huxley, Siskiyou, Sensapex, etc.), experimental chamber and perfusion system (such as WPI, AutoMate Scientific or 'homemade').

A number of experimental chambers and bathing perfusion (solution flow) systems for patch-clamp experiments have been designed by companies and individual researchers. The systems that have been adapted for studying plant ion channels can be divided into three main categories: (A) systems driven by two peristaltic pumps (first pump delivers solution while second pump simultaneously removes it), (B) systems driven by one pump (solution is delivered 'passively' via gravity down flow from the flask mounted above the experimental chamber; pump removes the solution), (C) passive/gravity system (it is similar to 'B' but solution passively leaks out through the hole in bottom of the chamber). The 'rapid perfusion systems' have also been used in studies of root Ca^{2+}-permeable channels (Demidchik et al. 2011). Similar systems are used in animal physiology for investigation of rapid electrophysiological responses, such as activation of ionotropic receptors. In this case, solutions can be changed as fast as in few milliseconds (see AutoMate Scientific website for details).

Types of patch-clamp modes are shown in Fig. 14.7. They include: cell attached, whole cell, inside-out and outside-out modes that all have some advantages and disadvantages. Inside-out and outside-out modes are also called 'excised patches'. Whole cell and outside-out modes (Fig. 14.7) are the most

14 Characterisation of Root Plasma Membrane Ca^{2+}-Permeable Cation Channels

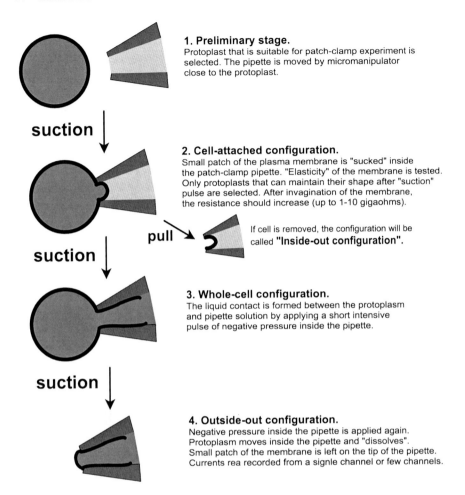

Fig. 14.7 Configurations of patch-clamp techniques that are used for study of plant cation channels. **Cell-attached configuration/mode**—*Advantages*: measurements of currents through single channels; minimal changes to native characteristics of channels because protoplast is not perfused with pipette solution. *Disadvantages*: intracellular solution and bathing media cannot be changed and controlled. **Inside-out configuration**—*Advantages*: single channel currents can be recorded; intracellular medium can be modified and controlled. *Disadvantages*: extracellular solution is not and controlled; high risk that the lack of cytosol can modify the channel properties. **Whole-cell configuration**—*Advantages*: intracellular ans extracellular solutions are controlled; behaviour of Ca^{2+} conductance (catalysed by a number of channels) can be investigated. *Disadvantages*: single channels cannot be studied; protoplast is perforated so some cell contents are diluted. **Outside-out configuration**—*Advantages*: recordings can be carried out from individual channels. Bathing solution is controlled. *Disadvantages*: intracellular solution if not controlled; high risk that channel properties are modified

frequently used patch-clamp configurations for studying plant cation channels. Patch clamp can be combined with other techniques, such as fluorescent imaging (Taylor et al. 1997; Pei et al. 2000) or MIFE (see above).

14.4.5.4 Patch Clamp and Root Ca^{2+}-Permeable Cation Channels

The first experiments with patch-clamp technique on plant cells were carried out in parallel by German and Israel laboratories in mid-1980s (Schroeder et al. 1984; Moran et al. 1984). Two types of cation channels were characterised in these early studies (one of them probably was NSCCs). After these pioneering works, many laboratories worldwide employed patch clamping for studying cation channels. Our modern knowledge about cation channels in higher plants is based on patch-clamp studies that were carried out in the last 20 years.

Tissue-specific expression of GFP in roots, application of laser ablation to root hairs, as well as development of techniques for protoplast routine isolation from the root elongation zone made it possible to patch-clamp protoplasts and compare Ca^{2+} conductances from functionally and morphologically different root cells (Kiegle et al. 2000; Véry and Davies 2000; Demidchik and Tester 2002; Demidchik et al. 2002b, 2003, 2007, 2009, 2011; Foreman et al. 2003). These studies have shown that Ca^{2+}-permeable channels are 3–5 times more active in growing cells (approximate equal current density in roots hairs and cells of elongation zone) than in mature cells (mature epidermis). This showed that Ca^{2+} influx is required for elongation growth in roots (to stimulate the vesicle movement to the growing part of the cell). Moreover, growing cells provide the first contact of plants with the new environment in which these cells can meet stresses, such as pathogens or toxicants. This can be explained by the fact that Ca^{2+} is not only a stimulator of the vesicular traffic but it is also used as a second messenger, particularly during the stress response. Another key finding is that the outer cell layer (epidermis) has a higher Ca^{2+} influx conductance than cell of cortex or pericycle. This can also be related to the role of these cells and their Ca^{2+}-permeable channels in the environmental signalling in roots. Ca^{2+}-permeable channels in roots (patch clamped in outside-out mode) have demonstrated the range of single channel conductances: (1) 5.9 pS (Demdichik et al. 2002b), (2) 15.3 pS (Demidchik et al. 2007), (3) 20 pS (Demidchik et al. 2009) and (4) 22 pS (Véry and Davies 2000). This suggests that they are encoded by different genes. In accordance with this idea, root Ca^{2+}-permeable conductances show different pharmacology, kinetics of activation and voltage dependance.

14.4.5.5 Disadvantage of Patch-Clamp Technique

Although patch-clamp provide many advantages, this technique also has some problems. One of the main problems is that treatment of plant cells by cell wall-degrading enzymes damages native channels and their regulation. This damage can be caused by two mechanisms that are as follows: (1) commercial cellulases and pectolyases contain proteases which can breakdown peptide bonds and cause protein denaturation; (2) cellulases, pectolyases and products of cell wall degradation (e.g. some short polysaccharide molecules) function as elicitors and induce oxidative burst that may cause oxidative damage of membrane proteins (Brudern and Thiel 1999; Kennedy and De Filippis 2004). Overall, these damages probably affect properties of some cation channels or cause 'instability' or disappearance of Ca^{2+} current in 'patch-clamp conditions' (see for example Demidchik et al. 2004, 2007, 2011; Miedema et al. 2008). Treatment of plants with SH-reducing agents is a possible way to avoid undesirable artefacts from the oxidative damage. However, it is not a 'panacea' because in natural conditions SH-groups are not maintained as permanently reduced, so SH-reducing agents may cause artifacts too.

Another major problem of patch-clamp technique is the replacement of the cytosolic native solution by artificial solution in whole cell, inside-out and outside-out modes. This provides the control of intracellular medium but, on the other hand, it affects the cytosolic regulation of channels. The quality of voltage clamping in patch-clamp experiments is good if protoplast membrane resistance is higher than resistance of the patch pipette. The technical problem arises when the ratio between the resistance of patch pipette and protoplast membrane increases. If the resistance of the patch pipette is higher than the resistance of the patched membrane, the voltage is manly clamped on the tip of the pipette but not on the membrane. This particularly affects I–V curve rectification and currents recorded under maximally depolarised or hyperpolarised voltages. The study of cation channels in large cells, such as *Arabidopsis* root epidermis, is particularly in danger of artefacts due to very low resistance of isolated vacuoles. Smaller patches, diluted solutions and lower resistance of the pipette help to avoid artefacts. Another obvious problem of patch-clamp technique is that not all plant cells and tissues give protoplasts suitable for patching. It is also difficult to select between protoplasts that are derived from different tissues.

14.4.6 Ca^{2+} Imaging and Aequorin Luminometry

Apart from electrophysiological techniques, a number of advanced Ca^{2+} imaging technologies, such as FRET (Monshausen et al. 2009; Rincón-Zachary et al. 2010) and dual dye ratiometric calcium imaging (Foreman et al. 2003; Miwa et al. 2006) can be used for investigation of activities of Ca^{2+}-permeable channels. These techniques will not be described here due to the text limit, but they are reviewed elsewhere (Fricker et al. 2006; Swanson et al. 2010). The main problem with all imaging techniques is that they do not show the mechanisms of channel function. They are also not designed for quantitative measurements. Even most sophisticated ratiometric analyses and cameleon constructs do not guarantee that Ca^{2+} in the cytosol (or organelles) is measured with high accuracy. First measurements using plants transformed with apoaequorin have been carried out by Knight et al. (1991). Since, then aequorin luminometry (Fig. 14.8) have been routinely 'employed' for studying Ca^{2+} entry through cation channels and various types of Ca^{2+} signalling events (Demidchik et al. 2002b, 2003, 2009). Application of Ca^{2+}-aequorin luminometry is a great addition to patch-clamp studies because it demonstrates that Ca^{2+} currents are related to phenomena at the cellular signalling

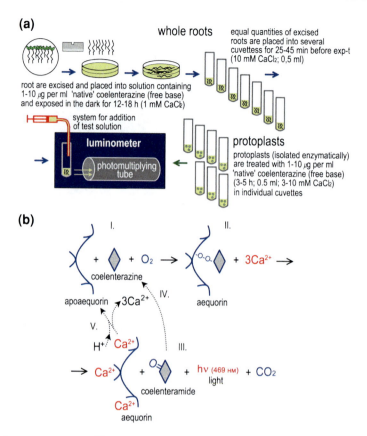

Fig. 14.8 The scheme of Ca^{2+}-aequorin luminometry. **a** Main procedures that should be carried out for measurements of cytosolic Ca^{2+} in whole roots and protoplasts. **b** Chemical reactions, which are responsible for Ca^{2+}-aequorin imaging: (I.) apoequorin (apoprotein) is expressed in the cytosol; it requires cofactor coelenterazine and O_2 for activation (formation of aequorin); (II.) one molecule of interacts with three calcium ions; (III.) after interaction with Ca^{2+}, aequorin releases coelenteramide (oxidised coelenterazine), carbon dioxide and blue light (469 hm); (IV.) coelenteramide can be reduced to coelenterazine in the cytosol; (V.) Ca^{2+} is replaced by H^+ (finish of the cycle)

level. Expression of aequorin in specific tissues allows to detect the difference in Ca^{2+} channel activities in different tissues and to investigate their physiological functions (Demidchik et al. 2004).

Aequorin emits very low light; therefore, the sensitive luminometer is required for Ca^{2+}-aequorin measurements. It can be commercial or assembled by researchers. The main part of the luminometer is high quality photo multiplying tube (PMT). Apparatuses designed and assembled by researchers using commercially available PMTs almost always require additional PMT cooling by Peltier element. Moreover, light leakage is a huge problem in 'homemade' machines. Commercial luminometers are much better. For example, Berthold luminometers do not require cooling while providing higher signal-to-noise ratio than 'homemade' luminometers. However, only few commercially available luminometers allow injecting solution to the experimental chamber (cuvette) by syringe. This means that almost all commercial machines require modification (building the injecting system). Modern commercial luminometers can be directly connected to computer and software is provided by manufacturer (e.g. WinTerm by Berthold). Recordings of photon counts can be stored as MS Excel files and additionally analysed by any scientific software. At the end of each experiment, discharge of all aequorin (by cell lysis induced by addition of 20–30% ethanol with 2 M $CaCl_2$) should be carried out for calibration. Cytosolic Ca^{2+} is calculated using the calibration equation derived empirically (Fricker et al. 1999): pCa = 0.332588 ($-\log k$) + 5.5593, where k = luminescence counts per second/total luminescence counts remaining.

14.5 Conclusions and Perspectives

Physiological characterisation of the calcium permeability and Ca^{2+}-permeable channels in root cell membranes has been progressing during almost 80 years. It adapted many methods from animal physiology but also developed plant-specific approaches, for example, MIFE, patch clamp on protoplasts, cell wall laser microsurgery and Ca^+ aequorin luminometry with intact roots. Application of voltage-clamp techniques resulted in the characterisation of three major types of root Ca^{2+}-permeable channels in a number of plant species. Electrophysiological techniques should be used with care and experimentators must take into account possible artefacts and problems. The future of plant electrophysiology is a developing new non-invasive techniques and approaches that cause less damage to the cell and Ca^{2+}-permeable channels than patch-clamping. Combining different physiological methods with electrophysiological techniques also have good perspectives.

References

Babourina O, Hawkins B, Lew RR, Newman I, Shabala S (2001) K$^+$ transport by *Arabidopsis* root hairs at low pH. Aust J Plant Physiol 28:635–641

Baluška F, Mancuso S, Volkmann D, Barlow PW (2010) Root apex transition zone: a signalling–response nexus in the root. Trends Plant Sci 15:402–408

Bergmann W (1992) Nutritional disorders of plants-visual and analytical diagnosis. G. Fisher, Jena

Brudern A, Thiel G (1999) Effect of cell-wall-digesting enzymes on physiological state and competence of maize coleoptile cells. Protoplasma 209:246–255

Carbone E (2004) Ion trafficking through T-type Ca^{2+} channels. A way to look at channel gating position. J Gen Physiol 124:619–622

Cheng Y, Liu M, Li R, Wang C, Bai C, Wang K (1999) Gadolinium induces domain and pore formation of human erythrocyte membrane: an atomic force microscopic study. Biochim Biophys Acta 1421:249–260

Davenport R (2002) Glutamate receptors in plants. Ann Bot 90:549–557

Datta S, Kim CM, Pernas M, Pires ND, Proust H, Tam T, Vijayakumar P, Dolan L (2011) Root hairs: development, growth and evolution at the plant-soil interface. Plant Soil 346:1–14

Demidchik V, Sokolik A, Yurin V (1997) The effect of Cu^{2+} on ion transport systems of the plant cell plasmalemma. Plant Physiol 114:1313–1325

Demidchik V, Sokolik A, Yurin V (2001) Characteristics of non-specific permeability and H+ ATPase inhibition induced in the plasma membrane of *Nitella flexilis* by excessive Cu2+. Planta 212:583–590

Demidchik V, Davenport RJ, Tester MA (2002a) Nonselective cation channels in plants. Annu Rev Plant Biol 53:67–107

Demidchik V, Bowen HC, Maathuis FJM, Shabala SN, Tester MA, White PJ, Davies JM (2002b) *Arabidopsis thaliana* root non-selective cation channels mediate calcium uptake and are involved in growth. Plant J 32:799–808

Demidchik V, Tester MA (2002) Sodium fluxes through nonselective cation channels in the plant plasma membrane of protoplasts from *Arabidopsis* roots. Plant Physiol 128:379–387

Demidchik V, Shabala SN, Coutts KB, Tester MA, Davies JM (2003a) Free oxygen radicals regulate plasma membrane Ca^{2+} and K^+-permeable channels in plant root cells. J Cell Sci 116:81–88

Demidchik V, Nichols C, Oliynyk M, Glover B, Davies JM (2003b) Is extracellular ATP a signalling agent in plants? Plant Physiol 133:456–461

Demidchik V, Adobea P, Tester MA (2004) Glutamate activates sodium and calcium currents in the plasma membrane of *Arabidopsis* root cells. Planta 219:167–175

Demidchik V (2006) Physiological roles of plant nonselective cation channels. In: Baluska F, Mancuso S, Volkmann D (eds) Plant neurobiology. Springer, Berlin, pp 235–248

Demidchik V, Shabala S, Davies J (2007) Spatial variation in H_2O_2 response of *Arabidopsis thaliana* root epidermal Ca^{2+} flux and plasma membrane Ca^{2+} channels. Plant J 49:377–386

Demidchik V, Shang Z, Shin R, Thompson E, Rubio L, Chivasa S, Slabas AR, Glover BJ, Schachtman DP, Shabala SN, Davies JM (2009) Plant extracellular ATP signalling by plasma membrane NADPH oxidase and Ca^{2+} channels. Plant J 58:903–913

Demidchik V, Cuin TA, Svistunenko D, Smith SJ, Miller AJ, Shabala S, Sokolik A, Yurin V (2010) *Arabidopsis* root K^+ efflux conductance activated by hydroxyl radicals: single-channel properties, genetic basis and involvement in stress-induced cell death. J Cell Sci 123:1468–1479

Demidchik V, Shang Z, Shin R, Shabala S, Davies JM (2011) Receptor-like activity evoked by extracellular ADP in *Arabidopsis thaliana* root epidermal plasma membrane. Plant Physiol 156:1375–1385

Dennison KL, Spalding EP (2000) Glutamate-gated calcium fluxes in *Arabidopsis*. Plant Physiol 124:1511–1514

Dreyer I, Horeau C, Lemaillet G, Zimmermann S, Bush DR, Rodriguez-Navarro A, Schachtman DP, Spalding EP, Sentenac H, Gaber RF (1999) Identification and characterization of plant transporters using heterologous expression systems. J Exp Bot 50:1073–1087

Ehrhardt DW, Atkinson EM, Long SR (1992) Depolarization of alfalfa root hair membrane potential by *Rhizobium meliloti* Nod factors. Science 256:998–1000

Favre P, Greppin H, Agosti RD (2001) Repetitive action potentials induced in *Arabidopsis thaliana* leaves by wounding and potassium chloride application. Plant Physiol Biochem 39:961–969

Findley GP (1961) Voltage-clamp experiments with *Nitella*. Nature 191:812–814

Fleischer A, O'Neill MA, Rudolf E (1999) The pore size of non-graminaceous plant cell walls is rapidly decreased by borate ester cross-linking of the pectic polysaccharide rhamnogalacturonan II. Plant Physiol 121:829–838

Foreman J, Demidchik V, Bothwell JHF, Mylona P, Miedema H, Torres MA, Linstead P, Costa S, Brownlee C, Jones JDG, Davies JM, Dolan L (2003) Reactive oxygen species produced by NADPH oxidase regulate plant cell growth. Nature 422:442–446

Fricker M, Plieth C, Knight H, Blancoflor E, Knight MR, White NS, Gilroy S (1999) Fluorescence and luminescence techniques to probe ion activities in plant cells. In: Mason WT (ed)

Fluorescent and luminescence probes for biological activity, 2nd edn. Academic Press, New York, pp 569–596

Fricker M, Runions J, Moore I (2006) Quantitative fluorescence microscopy: from art to science. Annu Rev Plant Biol 57:79–107

Gelli A, Blumwald E (1997) Hyperpolarization-activated Ca^{2+}-permeable channels in the plasma membrane of tomato cells. J Membr Biol 155:35–45

Halliwell JV, Plant TD, Robbins J, Standen NB (1994) Voltage clamp techniques. In: Ogden DC (ed) Microelectrode techniques, the plymouth workshop handbook, 2nd edn. Company of Biologists, Cambridge

Haswell ES, Phillips R, Rees DC (2011) Mechanosensitive channels: what can they do and how do they do it? Structure 19:1356–1369

Henriksen CH, Taylor AR, Brownlee C, Assmann SM (1996) Laser microsurgery of higher plant cell walls permits patch-clamp access. Plant Physiol 110:1063–1068

Hetherington AM, Brownlee C (2004) The generation of calcium signals in plants. Annu Rev Plant Biol 55:401–427

Hille B (2003) Ionic channels of excitable membranes. Sinauer Associates, Inc., Sunderland

Hodgkin AL, Huxley AF (1952) A quantitative description of membrane current and its application to conduction and excitation in nerve. J Physiol 117:500–544

Hope AB (1961) Ionic relation of cells of *Chara australis*. V. The action potential. Aust J Biol Sci 15:69–82

Hope AB, Walker NA (1975) The physiology of giant algal cells. Cambridge University Press, Cambridge

Ivashikina N, Becker D, Ache P, Meyerhoff O, Felle HH, Hedrich R (2001) K^+ channel profile and electrical properties of *Arabidopsis* root hairs. FEBS Lett 508:463–469

Iwabuchi A, Yano M, Shimizu H (1989) Development of extracellular electric pattern around *Lepidium* roots: its possible role in root growth and gravitropism. Protoplasma 148:98–100

Jammes F, Hu HC, Villiers F, Bouten R, Kwak JM (2011) Calcium-permeable channels in plant cells. FEBS J 278:4262–4276

Kennedy BF, De Filippis LF (2004) Tissue degradation and enzymatic activity observed during protoplast isolation in two ornamental *Grevillea* species. In Vitro Cell Dev Biol Plant 40:119–125

Kiegle E, Gilliham M, Haseloff J, Tester M (2000) Hyperpolarisation-activated calcium currents found only in cells from the elongation zone of *Arabidopsis thaliana* roots. Plant J 21:225–229

Knight MR, Campbell AK, Smith SM, Trewavas AJ (1991) Transgenic plant aequorin reports the effects of touch and cold-shock and elicitors on cytoplasmic calcium. Nature 352:524–526

Kurkdjian A, Leitz G, Manigault P, Harim A, Greulich KO (1993) Non-enzymatic access to the plasma membrane of *Medicago* root hairs by laser microsurgery. J Cell Sci 105:263–268

Kurkdjian A, Bouteau F, Pennarun AM, Convert M, Cornel D, Rona JP, Bousquet U (2000) Ion currents involved in early Nod factor response in *Medicago sativa* root hairs: a discontinuous single-electrode voltage-clamp study. Plant J 22:9–17

Levchenko V, Konrad KR, Dietrich P, Roelfsema MRG, Hedrich R (2005) Cytosolic abscisic acid activates guard cell anion channels without preceding Ca^{2+} signals. Proc Natl Acad Sci U S A 102:4203–4208

Lew RR (1991) Electrogenic transport properties of growing *Arabidopsis* root hairs—the plasma membrane proton pump and potassium channels. Plant Physiol 97:1527–1534

Lew RR, Dearnaley JDW (2000) Extracellular nucleotide effects on the electrical properties of growing *Arabidopsis thaliana* root hairs. Plant Sci 153:1–6

Miller AJ, Zhou JJ (2000) *Xenopus* oocytes as an expression system for plant transporters. Biochim Biophys Acta Biomembr 1465:343–358

Miller AJ, Cookson SJ, Smith SJ, Wells DM (2001) The use of microelectrodes to investigate compartmentation and the transport of metabolized inorganic ions in plants. J Exp Bot 52:541–549

Miwa H, Sun J, Oldroyd GE, Downie JA (2006) Analysis of Nod-factor-induced calcium signalling in root hairs of symbiotically defective mutants of *Lotus japonicus*. Mol Plant Microbe Interact 19:914–923

Lunevsky VZ, Zherelova OM, Vostrikov IY, Berestobsky GN (1983) Excitation of Characeae cell membranes as a result of activation of calcium and chloride channels. J Membr Biol 72:43–58

Lunevsky V, Aleksandrov A, Berestovsky G, Volkova S, Vostrikov I, Zherelova O (1977) Ionic mechanism of excitation of plasmalemma and tonoplast of characean algal cells. In: Thellier M, Monnier A, Demarty M, Dainty J (eds) Transmembrane ionic exchanges in plants, pp 167–172 (Colloque du C.N.R.S. No. 258)

Maathuis FJM (2010) Vaculoar cation channels: roles as signalling mechanisms and in plant nutrition. In: Demidchik V, Maathuis FJM (eds) Ion channels and plant stress responses. Springer-Verlag, Berlin, pp 191–206

Meharg AA, Maurosset L, Blatt MR (1994) Cable correction of membrane currents from root hairs of *Arabidopsis thaliana* L. J Exp Bot 45:1–6

Miedema H, Demidchik V, Véry AA, Bothwell JHF, Brownlee C, Davies JM (2008) Two voltage-dependent calcium channels co-exist in the apical plasma membrane of *Arabidopsis thaliana* root hairs. New Phytol 179:378–385

Monshausen GB, Bibikova TN, Weisenseel MH, Gilroy S (2009) Ca^{2+} regulates reactive oxygen species production and pH during mechanosensing in *Arabidopsis* roots. Plant Cell 21:2341–2356

Moran N, Ehrenshtein G, Iwasa K, Bare C, Mischke C (1984) Ion channels in plasmalemma of wheat protoplasts. Science 226:835–838

Neher E, Sakmann B (1976) Single channel currents recorded from membrane of denervated frog muscle fibres. Nature 260:799–802

Nobel PS (2005) Physicochemical and environmental plant physiology. Elsevier-Academic Press, Waltham, MA

Ogden D, Stanfield P (1994) Patch clamp techniques for single channel and whole-cell recording. In: Ogden D (ed) Microelectrode techniques: the plymouth workshop handbook. Company of Biologists Ltd, Cambridge, pp 53–78

Osterhout WJV (1908) The organization of the cell with respect to permeability. Science 38:408–409

Osterhout WJV, Damon EB, Jacques AG (1927) Dissimilarity of inner and outer protoplasmic surfaces in *Valonia*. J Gen Physiol 11:193–205

Osterhout WJV (1958) The use of aquatic plants in the study of some fundamental problems. Annu Rev Plant Physiol 8:1–11

Pei ZM, Murata Y, Benning G, Thomine S, Klusener B, Allen GJ, Grill E, Schroeder JI (2000) Calcium channels activated by hydrogen peroxide mediate abscisic acid signalling in guard cells. Nature 406:731–734

Peiter E, Maathuis FJM, Mills LN, Knight H, Pelloux M, Hetherington AM, Sanders D (2005) The vacuolar Ca2+-activated channel TPC1 regulates germination and stomatal movement. Nature 434:404–408

Plaks AV, Sokolik AI, Yurin VM (1980) Excitable calcium channels of *Nitella* cell tonoplast. Izvestiya of the Academy of Sciences of BSSR, Biology section 1:121–124

Plaks AV, Sokolik AI, Yurin VM, Goncharik MN (1979) Chloride channel activation and excitation of Nitella cell tonoplast. Rep Acad Sci BSSR 23:947–949

Rincón-Zachary M, Teaster ND, Sparks JA, Valster AH, Motes CM, Blancaflor EB (2010) Fluorescence resonance energy transfer sensitized emission of yellow cameleon 3.60 reveals root-zone-specific calcium signatures in *Arabidopsis* in response to aluminum and other trivalent cations. Plant Physiol 152:1442–1458

Romano LA, Miedema H, Assmann SM (1998) Ca^{2+}-permeable, outwardly-rectifying K^+ channels in mesophyll cells of *Arabidopsis thaliana*. Plant Cell Physiol 39:1133–1144

Schroeder JI, Hedrich R, Fernandez JM (1984) Potassium-selective single channels in guard cell protoplasts of *Vicia faba*. Nature 312:361–362

Schroeder JI, Thuleau P (1991) Ca^{2+} channels in higher plant cells. Plant Cell 3:555–559

Shabala SN, Newman IA, Morris J (1997) Oscillations in H^+ and Ca^{2+} ion fluxes around the elongation region of corn roots and effects of external pH. Plant Physiol 113:111–118

Shabala SN, Newman IA, Whittington J, Juswono UP (1998) Protoplast ion fluxes: their measurements and variation with time, position and osmoticum. Planta 204:146–152

Shabala S, Demidchik V, Shabala L, Cuin TA, Smith SJ, Miller AJ, Davies JM, Newman IA (2006) Extracellular Ca^{2+} ameliorates NaCl-induced K^+ loss from *Arabidopsis* root and leaf cells by controlling plasma membrane K^+-permeable channels. Plant Physiol 141:1653–1665

Sokolik AI, Yurin VM (1981) Transport activity of potassium channels in the plasmalemma of *Nitella* cells at rest. Soviet Plant Physiol 28:294–301

Sokolik AI, Yurin VM (1986) Potassium channels in plasmalemma of *Nitella* cells at rest. J Membr Biol 89:9–22

Swanson SJ, Choi WG, Chanoca A, Gilroy S (2010) In vivo imaging of Ca^{2+}, pH, and reactive oxygen species using fluorescent probes in plants. Annu Rev Plant Biol 62:273–297

Tester M (1990) Plant ion channels: whole-cell and single channel studies. New Phytol 114:305–340

Thion L, Mazars C, Nacry P, Bouchez D, Moreau M, Ranjeva R, Thuleau P (1998) Plasma membrane depolarization-activated calcium channels, stimulated by microtubule-depolymerizing drugs in wild-type *Arabidopsis thaliana* protoplasts, display constitutively large activities and a longer half-life in ton 2 mutant cells affected in the organization of cortical microtubules. Plant J 13:603–610

Thuleau P, Ward JM, Ranjeva R, Schroeder JI (1994) Voltage-dependent calcium-permeable channels in the plasma membrane of a higher plant cell. EMBO J 13:2970–2975

Tyerman SD (1992) Anion channels in plants. Annu Rev Plant Physiol Plant Mol Biol 43:351–373

Tyerman SD, Schachtman DP (1992) The role of ion channels in plant nutrition and prospects for their genetic manipulation. Plant Soil 146:137–144

Tyerman SD, Beilby M, Whittington J, Juswono U, Newman I, Shabala S (2001) Oscillations in proton transport revealed from simultaneous measurements of net current and net proton fluxes from isolated root protoplasts: MIFE meets patch-clamp. Austr J Plant Physiol 28:591–604

Véry AA, Davies JM (2000) Hyperpolarization-activated calcium channels at the tip of *Arabidopsis* root hairs. Proc Natl Acad Sci U S A 97:9801–9806

Véry A-A, Sentenac H (2002) Cation channels in the *Arabidopsis* plasma membrane. Trends Plant Sci 7:168–175

Véry A-A, Sentenac H (2003) Molecular mechanisms and regulation of K^+ transport in higher plants. Annu Rev Plant Biol 54:575–603

White PJ (1998) Calcium channels in the plasma membrane of root cells. Ann Bot 81:173–183

Wilson WA, Goldner MM (1975) Voltage clamping with a single microelectrode. J Neurobiol 6:411–422

Xicluna J, Lacombe B, Dreyer I, Alcon C, Jeanguenin L, Sentenac H, Thibaud JB, Cherel I (2006) Increased functional diversity of plant K^+ channels by preferential heteromerization of the shaker-like subunits AKT2 and KAT2. J Biol Chem 282:486–494

Yurin VM (1969) The effect of calcium ions and anions on bio-electrical potentials in *Nitella flexilis* cells at rest. Dissertation of candidate of biological sciences. Institute of Experimental Botany. The Academy of Sciences of BSSR, Minsk

Yurin VM, Sokolik AI, Kudryashov AP (1991) Regulation of ion transport through plant cell membranes. Science and Engineering, Minsk

Zawadzki T (1980) Action potentials in *Lupinus angustifolius* L. shoots. 5. Spread of excitation in the stem, leaves, and root. J Exp Bot 31:1371–1377

Subject Index

2,4-dinitrophenol, 14
12-oxo-phytodienoic acid, 46
3,4,5-trimethyoxybenzoic acid 8-(diethyl-amino) octyl ester (TMB-8), 100

A

Abaxial motor cells, 98, 99
Abiotic stress, 138, 144, 146, 147, 155, 176, 296, 320, 323, 326, 335
Abscisic acid (ABA), 33, 35, 59, 120, 126, 140, 156, 224, 229, 298, 367, 368
Acetylcholine, 186, 207
Actin, 38, 39, 113, 118, 123, 152, 164, 183, 186, 187, 199, 201, 203, 206, 208, 231, 236, 238, 244, 274, 275, 290, 303, 305–307, 309, 315, 317, 319–321, 337, 342
Action potential, 2, 3, 16, 19, 23, 27, 29, 30, 35, 57–61, 64, 66, 68, 69, 83, 104, 112, 113, 118, 121, 144, 170, 171, 174, 175, 177, 178, 180, 182, 183, 186–188, 196–215, 217, 219, 220, 223–225, 227–234, 236, 239–250, 277, 281, 282, 285, 286, 288, 290, 296, 298, 300, 301, 308, 325, 335, 351, 366, 367, 369
Actuators, 202, 206
Acid rain, 89, 116, 120, 167, 170, 178, 196, 197, 204, 209, 230, 245
Actin, 38, 118, 123, 186, 187, 199, 201, 203, 206, 208, 236, 244, 303, 305–307, 315, 317, 319, 321, 337
Actuating mechanisms, 55, 63, 68
Adaxial motor cells, 93, 105
Aequorin, 339, 363–365, 367

Albizzia, 88, 121, 131, 133, 140, 141, 273
Albizzia saman, 118
Aldrovanda vesiculosa, 35
Aliasing, 176
All-or-none law, 11, 41
Aloe vera, 145, 171, 191, 192, 195, 196, 205, 206
Amylase, 25
Anaesthetics, 114
Analog signal, 175
Anion channels, 2, 17, 28, 41, 127, 129, 138, 152, 180, 183, 190, 291, 333, 335, 344–346, 367, 369
Angular bending, 110
Antrhracene-9-carboxlyic acid (A-9-C), 28
Apex, 167, 169, 192, 193, 201, 292, 297, 303, 304, 309, 311, 313–315, 318–320, 360, 365
Aphid, 48, 146, 151, 157, 162, 167, 168, 171, 212–214, 219, 244
Aphid secreted salivary proteins, 151
Apoplast, 52, 58, 98, 100, 147, 186, 215, 239, 240, 241, 305, 314, 327, 329, 344
Apoplastic signal, 144, 172
Aquaporin, 3, 12, 14, 15, 28, 83, 90, 91, 94, 113, 117, 120, 121, 129, 162, 182, 188, 282, 285, 293, 295–297, 299, 300, 341
Arabidopsis, 59, 96, 107, 113, 116, 119–122, 127, 137–142, 146, 155, 156, 160, 161, 167–169, 171, 172, 199, 201, 223, 225, 227–228, 231, 238, 245, 246, 277, 285, 299–301, 307, 310–313, 316–321, 329, 330, 333, 335, 336, 341, 345, 350, 354–358, 363, 365–369

A (*cont.*)
ATPase, 58, 90, 93, 99, 113, 119–122, 128, 139, 145, 188, 190, 197, 202, 203, 211, 225, 227, 240, 245, 316, 324, 328, 344, 346, 366
Atmospheric electricity, 254
Auxin, 91
Auxin (IAA), 48
Auxin-secreting plant synapses, 309

B
BaCl$_2$, 16
Barometric pressure, 266
Bending elasticity, 5
Bilayer fusion, 6
Bimetallic couple, 6
Bioelectrochemical excitation, 175
Bioelectrochemical signals, 174
Bioelectrical impulses, 175
Biological clock, 190
Biologically closed electrical circuit, 182
Biomimetic robot, 63, 68
Biotic stress, 147, 155
Book model, 4

C
Ca^{2+}/H$^+$ antiporter, 54
Ca^{2+} -binding proteins, 147
Ca^{2+} channel, 100, 147
Ca^{2+} oscillations, 113
Ca^{2+} -permeable channel, 129, 346
Ca^{2+} signaling, 129
Calcium deficiency, 51
Calcium-dependent protein kinases (CDPKs), 155
Calcium signaling, 143
Calmodulin, 206
Calvin cycle, 43
cAMP (cyclic adenosine monophosphate), 101
cAMP-dependent protein kinase (PKA), 127
Carnivorous plants, 2
Capacitor, 19
Capture process, 376
Carbonic anhydrase, 54
Carnivorous, 33
Cell wall, 116
Cellular excitability, 208
Cellular messangers, 208
Cellular motors, 208
CCCP, 14, 16
Channel conductance, 126
Chara, 41

Charged capacitor method, 19
Chemiotropism, 173
Chemical hypothesis, 187
Chemical transmission, 234
Chewing/piercing herbivores, 143
Chloroform, 114
Chlorophyll fluorescence, 37
Chlorophyll fluorescence imaging, 40
Cilia, 23
Circadian modulation, 96
Circadian movements, 89, 95
Circadian rhythm, 127, 190, 194
Circadian oscillators, 194
Clivia miniata, 190
Closing charge, 21
Closing process, 12
CO$_2$ assimilation, 37
CO$_2$ fixation, 44
Cotariocalyx motorius, 85
Colorado potato beetle (Leptinotarsa decemlineata), 145
Collenchyma sheath, 187
Companion cells, 175
Computer modeling, 8
Concentrated movements, 187
Concentration waves, 108
Conducting bundles, 145
Conductive bundles, 175
Coupling between the pulvinus cells, 88
Crassulacean acid metabolism, 196
Cryptochrome, 129
Current clamp technique, 104
Current injection, 58
Current-voltage relationships, 346
Curvature, 5, 7, 90
Cylindrical motor cell, 91
Cytoplasm, 235
Cytoplasmic pH, 42
Cytoplasmic streaming, 235
Cytosolic Ca^{2+} influx, 147

D
Dark-adapted state, 45
DC current pulse, 104
DCMU (3-(3', 4'-dichlorophenyl)-1, 1-dimenthylurea, 39
Decision making, 68
Decision-making block, 64
Defense response, 161
Defensive function, 163
Depolarization, 51
Desmodium, 85, 86, 94, 110, 111, 117, 271
Digestion, 24, 25, 65

Subject Index

Digestion of insects, 64
Digestive secretion, 25, 30
Dilatation cycles, 249
Diurnal cycle, 125
Diurnal leaf movements, 249
Dionaea muscipula Ellis, 2
Drosera, 35
Drosera prolifera, 55
Drosera SPP, 35
Drosera tentacles, 55

E
Elastic, 12
Eastic energy, 5
Elastic properties, 184
Elasticity, 1
Electric double layer, 196
Electrical behavior, 226
Electrical charge, 19
Electrical coupling, 236
Electrical double layer, 196
Electrical memory, 19
Electrical pulse, 10, 104
Electrical signal, 20, 21, 147
Electrical signalling, 112
Electrical stimuli, 11
Electrical stiumulation, 4, 13
Electrical stimulus, 10, 55
Electrical synapses, 308
Electrical transmission, 234
Electrochemistry, 196
Electrochemical energy, 174
Electrochemical impulses, 174
Electrochemical oscillators, 196
Electrochemical signals, 174
Electrochemical transducers, 174
Electron transport chains, 54
Electronic transmission, 234
Electrostatic field, 196
Electrostimulation, 20, 25–27, 145, 183, 184, 187, 188, 191
Endogenous leaf movements, 96
Enregy, 34
Environmental stressors, 144
Epidermis, 3, 4
Equivalent electrical circuit, 145
Error-proof design, 65
Ethylene glycol-bis(B aminoethyl ether)N, N,N',N'-tetraacetic acid, 27
Excitable membranes, 174
Excitability, 14, 27, 41, 56, 174, 176, 198, 208
Excitability of the Venus flytrap, 27

External signaling, 154
Extracellular potentials, 98

F
Fast movement, 11
Fast proton transport, 3
Feedforward mechanism, 64
Fertilization, 56, 219
Fission, 6
Flame stimulation, 51
Flavin-binding aquaporin, 129
Flexor cell, 126
Floral stimulus, 282
Fluorescence, 35, 37
Fluorescence quenching, 37

G
G protein, 101
Gap junctions, 238
Gas exchange, 35
Gas switching, 36
Gaussian curvature, 6, 9
Gene expression, 143, 152
Geomagnetic field, 255
Geotropism, 173
Glycol-bis(2-aminoethylether)-N, N,N', N'-tetra-aceitic acid, 27
Graviperception, 199
Gravitropism, 113, 199
Gravity vector, 200
Guard cell, 51

H
H^+/anion antiporter, 129
H^+/K^+ antiporter, 129
H^+-ATPase, 128
H^+-cotransport, 28
H^+ extrusion, 25
H^+ pump, 28
H_2O_2, 161
Hair irritation, 40
Heat signals, 88
Helianthus annuus, 35
Heliotropism, 173
Herbicides, 175
Herbivore attack, 143, 144, 147
Herbivore damage, 143
$HgCl_2$, 14, 16
Hibiscus, 56
Hill reaction, 41
Hunting cycle, 23

H (*cont.*)
Hydraulic layers, 7
Hydraulic pressure, 73, 174
Hydraulic signal, 48
Hydro elastic mechanism, 189
Hydro mechanical movement, 188
Hydrodynamic flow, 3
Hydroelastic Curvature
    Model, 1, 4
Hydrostatic pressure, 5
Hydrotropoism, 173
Hyperpolarization, 51

I
Ion channels, 12
Ion current, 104
Ion transport oscillations, 110
Impulse transduction, 12
Initial mean curvature, 8
Inositol-1,4,5-triphosphate (IP3), 100
Insect, 23
Internal signaling, 152
Irritants, 175

J
Jasmonic acid, 35

K
K^+ channel, 16
K^+ influx, 127
K^+ transport, 126
Kaffir lily, 191
KAT1, 127
Kinetics of closing, 11

L
Lanthanum ions, 27
Lateral leaflet rhythm, 104
Lateral pulvinus, 90
Leaf curvature, 3
Leaf closing factor, 131
Leaf-feeding larvae, 145
Leaf folding, 46
Leaf movement, 33, 130
Leaf opening factor, 131
Leaflet movement, 86
Leaflet movement cycle, 90
Leaflet oscillation, 91, 104
Leaflet rhythm, 95, 101

Light-dark cycle, 190
Lima bean (P. lunatus), 163
Limit cycle, 102
Localized heat stress, 145
Long-term memory, 23
Lunisolar gravitational variations, 254
Lunisolar tide, 256
Lunisolar tidal acceleration, 271

M
Magnetic fields, 106
Magnetotropism, 173
Magnetotropic bending reactions, 107
Maize (Zea mays), 163
Maize leaf, 52
Matrix potentials, 92
Mean curvature, 6
Mechanical load, 95
Mechanical stimulation, 14, 46, 175
Mechanical stimulus, 20
Mechanical wounding, 53, 146
Mechanosensory neurons, 181
Mechanosensitive
    ion channel, 23, 182
Membrane depolarization, 126
Membrane permeability, 92
Membrane potentials, 98
Memory, 17
Memory retention, 23
Metabolomics, 143, 162
Mesophyll cells, 46, 48
Microelectrode, 212
Microfibrils, 90
Microtubules, 209
Mitrochontria, 147
Midrib, 4, 86
MIFE™, 352
Mimosa, 7
Mimosa pudica, 34
Mitogen-activated protein kinase (MAPK),
    154
Morphing, 2
Morphing structures, 7
Motor, 64
Motor cell, 51
Muscular hypothesis, 186
Myosin, 208, 303

N
Nyosin, 208, 303, 307
NaN_3, 28

Subject Index

Neomycin, 100
Nerve impulses, 174
Neural network, 208
Neurotransmitter, 108, 174, 207
Nitella, 93
NMR imaging, 7
Nonlinear dynamics, 71
Nuclease, 25
Nutrient acquisition, 64
Nyctinasty, 130
Nyctinastic leaf movement, 129, 130
Nyctinastic plant, 125, 131
Nyquist criterion, 176

O

O_2 evolution, 41
Open state, 24
Oscillations, 85
Oscillatory period, 116
Oscillatory system, 103
Oscillating system, 85
Osmoregulatory model, 128
Osmotic motor, 86
Osmotic motor hypothesis, 186
Osmotic pressure, 92
Oxygen evolving
     complex (OEC), 37

P

Parenchyma, 3
Patch-clamp, 355
Patch-clamp pipettes, 358
Pathogen, 145
Pentachlorophenol, 14
Permeability, 344
Pesticide, 197
Petiole, 188
Petiole bending, 188
Petiole movement, 188
Phase plane, 102
Phaseolus, 88
Phaseolus vulgaris, 128
Phenophytin, 197
Phloem, 46
Phloem cells, 174
Phloem unloading, 35
Phosphatidyl inositol
     signaling chain, 100
Photonastic movement, 195
Photoperiodism, 190
Photosystem II, 37

Photosynthesis, 33
Photosynthetic active radiation (PAR),
     34, 36–39, 44, 48, 49, 288
Phototropin, 128
Phototropism, 173
Physiological behavior, 226
Phytoactuators, 173
Phytochrome, 126, 178, 195, 282,
     285, 291, 292
Phytohormone, 126
Phytosensors, 173
Plant growth stimulants, 175
Plant-insect interactions, 143
Plant/lunisolar gravitational relationship, 252
Plant memory, 192
Plant synapse, 309
Plasma membrane, 147
Plasmalemma, 104, 116
Plasmalemma ATPases, 93
Plasmalemma-based receptor, 100
Plasmodesmata, 23, 116
Plastoquinone, 37
Podium cells, 24
Polar auxin transport (PAT), 303
Pollination, 219
Potassium channels, 2
Pressure, 7
Primary pulvini, 195
Proteinase, 25
Protein expressions, 159
Protein phosphorylation, 126
Proteomics, 143
Protoplasts, 126
Proton pump, 45, 128, 222
Proton transport, 3, 17, 68
Pseudomonas syringae, 146
Pulvini motor organ, 3
Pulvinus, 46, 86
Pulvinus curvature, 91
Pulvinus functions, 88
Pulvinus length, 90
Pulvinus movements, 91
Pulvinus radius, 90
Pulvinus structure, 90
Pulvinus water transport, 91
Protoplasts, 126

R

Reaction center, 46
Receptor, 327
Receptor potentials, 2
Reopening process, 68, 76

R (cont.)
Respiration, 33
Rhachis, 86
Rhythm, 103
Ricca's factor, 49
Robot, 63
Root synapse, 312
Root-to-shoot communication, 218
Root-to-shoot signaling, 219
Ruthenium red, 27

S
Salicylic acid (SA), 51
Samanea, 88
Samanea saman, 125
Sampling rate, 175
Sampling frequency, 176
Sap, 218
Schotky diode, 192
Sclerenchyma, 46
Scotonastic period, 195
Sealing, 25
Sealing process, 76
Secretory cycle, 24
Seismonsastic movements, 184
Semi-closed state, 77
Semi-closed transition state, 68
Sensitive hair, 20
Sensitive plant, 2, 33
Sensor, 64
Sensory mechanism, 173
Sensory memory, 20, 23
Sensory-motoric voltage, 303
Sieve tubes, 218
Sieve-tube elements, 175
Shaker-type K+ channel
    AKT2, 127
Short-term memory, 23
Shrunken state, 90
Short-term electrical memory, 17, 21
Signal transduction, 85, 152
Signaling hormones, 159
Single electrode voltage clamp, 356
Singularity, 103
Space plant biology, 200
SPICK1, 127
SPICK2, 127
Spruce (*Pinus glabra*), 164
Static magnetic field, 106
Stimulus, 2
Stretch-activated ion channels, 182
Stretch-inactivated ion channels, 182
Stroma, 54

Stomata, 35
Stomata guard cell, 3
Stomatal conductance, 36, 49
Stomatal opening, 127
Stromatal closure, 127
Swollen state, 90
Synapse, 241
Synaptical transmission, 241
System potentials, 144

T
TEACl, 13, 14, 16
Telegraph plant (*Codarioalyx motorius*), 85, 86, 93, 94, 110
Tertiary pulvini, 195
Tetraethylammonium chloride
    (TEACl), 12
Tightening of the trap, 24
Time sensing, 190
Tissue permeability, 92
thermotropism, 173
Three-barrelled electrodes, 355
Thigmotropism, 173
Thylakoid membrane, 41
Tomato, 333
Transcriptomics, 143
Transducer, 88
Transmembrane potential, 143
Trap closing, 23
Trap closure, 4
Trap opening, 25
Trap tightening, 24
Trapping, 11
Trap closure, 4
Transpiration, 35
Transpiration rate, 36
Transpiration rhythm, 104
Transpiration stream, 100
Trigger hair, 2, 21
Triggering process, 68, 69
Tropism, 173
Toxin, 164
Turgor, 3
Turgor pressure, 5
Two-electrode voltage-clamp, 355

U
Ultradian movements, 90
Ultradian oscillations, 96
Ultradian rhythm, 103
Uncoupler, 12, 28, 197
Undersampling, 176

V

Vacuole, 147
Variation potential, 33
Vascular bundles, 23
Venus flytrap, 1
Venus flytrap model, 71
Vicia faba, 55
Voltage-gated channel, 20, 104, 147, 180, 208, 213, 242
Voltage-gated Ca^{2+} channel, 147
Voltage-gated ion channel, 104
Volume regulation, 126

W

Water binding, 93
Water channels, 12, 90
Water conductivity, 92
Water jets, 5
Water potential, 91, 92
Water transport, 91
Water transport oscillations, 110
Water transpiration, 37
Wounding, 49, 144
Wound-induced relaxation, 209

X

Xylem, 133

Z

$ZnCl_2$, 16